Perspectives in Mathematical Logic

Ω-Group:
R. O. Gandy H. Hermes A. Levy G. H. Müller
G. E. Sacks D. S. Scott

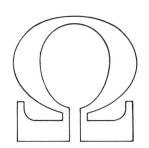

Jens E. Fenstad

General Recursion Theory

An Axiomatic Approach

Springer-Verlag
Berlin Heidelberg New York 1980

Jens E. Fenstad

Department of Mathematics
University of Oslo
Oslo—Blindern
Norway

AMS Subject Classification (1980): 03-02, 03D75

ISBN 3-540-09349-4 Springer-Verlag Berlin Heidelberg New York
ISBN 0-387-09349-4 Springer-Verlag New York Heidelberg Berlin

Library of Congress Cataloging in Publication Data.
Fenstad, Jens Erik. General recursion theory. (Perspectives in mathematical logic).
Bibliography: p. Includes index. 1. Recursion theory. I. Title.
QA9.6.F46. 511'.3. 79-13099

Typesetting: William Clowes & Sons Limited, London, Beccles and Colchester.
Printing and bookbinding: Graphischer Betrieb K. Triltsch, Würzburg
2141/3140-543210

Preface to the Series

On Perspectives. *Mathematical logic arose from a concern with the nature and the limits of rational or mathematical thought, and from a desire to systematise the modes of its expression. The pioneering investigations were diverse and largely autonomous. As time passed, and more particularly in the last two decades, interconnections between different lines of research and links with other branches of mathematics proliferated. The subject is now both rich and varied. It is the aim of the series to provide, as it were, maps or guides to this complex terrain. We shall not aim at encyclopaedic coverage; nor do we wish to prescribe, like Euclid, a definitive version of the elements of the subject. We are not committed to any particular philosophical programme. Nevertheless we have tried by critical discussion to ensure that each book represents a coherent line of thought; and that, by developing certain* themes, *it will be of greater interest than a mere assemblage of results and techniques.*

The books in the series differ in level: some are introductory some highly specialised. They also differ in scope: some offer a wide view of an area, others present a single line of thought. Each book is, at its own level, reasonably self-contained. Although no book depends on another as prerequisite, we have encouraged authors to fit their book in with other planned volumes, sometimes deliberately seeking coverage of the same material from different points of view. We have tried to attain a reasonable degree of uniformity of notation and arrangement. However, the books in the series are written by individual authors, not by the group. Plans for books are discussed and argued about at length. Later, encouragement is given and revisions suggested. But it is the authors who do the work; if, as we hope, the series proves of value, the credit will be theirs.

History of the Ω-Group. *During 1968 the idea of an integrated series of monographs on mathematical logic was first mooted. Various discussions led to a meeting at Oberwolfach in the spring of 1969. Here the founding members of the group (R. O. Gandy, A. Levy, G. H. Müller, G. E. Sacks, D. S. Scott) discussed the project in earnest and decided to go ahead with it. Professor F. K. Schmidt and Professor Hans Hermes gave us encouragement and support. Later Hans Hermes joined the group. To begin with all was fluid. How ambitious should we be? Should we write the books ourselves? How long would it take? Plans for authorless books were promoted, savaged and scrapped. Gradually there emerged a form and a method. At the end of an infinite discussion we found our name, and that of the series. We established our centre in Heidelberg. We agreed to meet twice a year together with authors, consultants and assistants, generally in Oberwolfach. We soon found the value of collaboration: on the one hand the permanence of the founding group gave coherence to the over-all plans; on the other hand the stimulus of new contributors kept the project alive and flexible. Above all, we found how intensive discussion could modify the authors' ideas and our own. Often the battle ended with a detailed plan for a better book which the author was keen to write and which would indeed contribute a perspective.*

Acknowledgements. *The confidence and support of Professor Martin Barner of the Mathematisches Forschungsinstitut at Oberwolfach and of Dr. Klaus Peters of Springer-Verlag made possible the first meeting and the preparation of a provisional plan. Encouraged by the Deutsche Forschungsgemeinschaft and the Heidelberger Akademie der Wissenschaften we submitted this plan to the Stiftung Volkswagenwerk where Dipl. Ing. Penschuck vetted our proposal; after careful investigation he became our adviser and advocate. We thank the Stiftung Volkswagenwerk for a generous grant (1970–73) which made our existence and our meetings possible.*

Since 1974 the work of the group has been supported by funds from the Heidelberg Academy; this was made possible by a special grant from the Kultusministerium von Baden-Württemberg (where Regierungsdirektor R. Goll was our counsellor). The success of the negotiations for this was largely due to the enthusiastic support of the former President of the Academy, Professor Wilhelm Doerr. We thank all those concerned.

Finally we thank the Oberwolfach Institute, which provides just the right atmosphere for our meetings, Drs. Ulrich Felgner and Klaus Glöde for all their help, and our indefatigable secretary Elfriede Ihrig.

Oberwolfach
September 1975

R. O. Gandy *H. Hermes*
A. Levy *G. H. Müller*
G. E. Sacks *D. S. Scott*

Author's Preface

This book has developed over a number of years. The aim has been to give a unified and coherent account of the many and various parts of general recursion theory.

I have not worked alone. The Recursion Theory Seminar in Oslo has for a number of years been a meeting place for an active group of younger people. Their work and enthusiasm have been an important part of the present project. I am happy to acknowledge my debts to Johan Moldestad, Dag Normann, Viggo Stoltenberg-Hansen and John Tucker. It has been a great joy for me to work together with them.

But there are other debts to acknowledge. The Oslo group learned much of their recursion theory from Peter Hinman who spent the year 1971–72 in Oslo and Peter Aczel who was here for part of the year 1973. Robin Gandy, Yiannis Moschovakis and Gerald Sacks have also in many ways helped to shape our ideas about general recursion theory.

I appreciate the invitation from the Ω-group to write this book for their series. In particular, I would like to thank Gert Müller for his friendship and helpfulness.

Finally, I would like to thank Randi Møller for her expert typing of the manuscript, David Kierstead for his valuable help with the proof-reading, and the printers and publishers for the excellent production of this volume.

Advice to the reader. The book is in principle self-contained, but we expect that any reader would have had some previous exposure to recursion theory, e.g. Rogers [136]. The books by Barwise [11], Hinman [61] and Moschovakis [115, 118] supplement our account, often from a different perspective.

The exposition is organized around a main core which is reasonably complete. This is the general theory of computations. But the main core flowers into a number of side branches intended to show how computation theories connect with and unify other parts of general recursion theory. Here we are less complete and proofs may sometimes be more in the nature of a comment on the basic ideas involved. In the main body of the text and in an Epilogue we have tried to point beyond to open problems and areas of further research.

Oslo, October 1979 — Jens Erik Fenstad

Table of Contents

Pons Asinorum

Chapter 0

On the Choice of Correct Notions for the General Theory

This is a book on general recursion theory. The approach is axiomatic, and the aim is to present *a* coherent framework for the manifold developments in ordinary and generalized recursion theory.

The starting point is an analysis of the relation

$$\{a\}(\sigma) \simeq z,$$

which is intended to assert that the "computing device" named or coded by a and acting on the input sequence $\sigma = (x_1, \ldots, x_n)$ gives z as output.

The history of this notion goes back to the very foundation of the theory of general recursion in the mid 1930's. It can be traced from the theory of Turing on idealized machine computability via Kleene's indexing and normal form theorems to present-day generalizations.

It was Kleene who in 1959 took this relation as basic in developing his theory of recursion in higher types [83]; subsequently it was adopted by Moschovakis [112] in his study of prime and search computability over more general domains.

Indexing was also behind various other abstract approaches. We mention the axiomatics of Strong [166], Wagner [169], and H. Friedman [33], and the computation theories of Y. Moschovakis [113]. The latter theory, in particular, has been very influential on our own thinking.

Historical development is one thing, conceptual analysis another: it is the purpose of this introduction to present some theoretical grounds for basing our approach to general recursion theory on the axiomatic notion of a "computation theory". The discussion here is not part of the systematic development of the theory of computation theories and so proofs will not be given, but we urge the reader to go to the various sources cited.

0.1 Finite Algorithmic Procedures

We shall start out by analyzing how to compute in the context of an arbitrary algebraic system

$$\mathfrak{A} = \langle A, \sigma_1, \ldots, \sigma_l, S_1, \ldots, S_k \rangle,$$

where the operations σ and the relations S are finitary but not necessarily total. Whenever necessary we assume that equality on A is among the basic relations S.

The notion of a *finite algorithmic procedure* (fap) is one of a number of abstract algorithms introduced by H. Friedman [34] and further studied by J. C. Shepherdson [148]. A comparative study of fap's with inductive definability and computation theories is carried through in Moldestad, Stoltenberg-Hansen and Tucker [108, 109].

To be precise a fap P is an ordered list of instructions $I_1, \ldots, I_k$ which are of two kinds:

Operational Instructions

(i) $\quad r_\mu := \sigma(r_{\lambda_1}, \ldots, r_{\lambda_m}),$

i.e. apply the m-ary operation σ to the contents of registers $r_{\lambda_1}, \ldots, r_{\lambda_m}$ and replace the contents of register r_μ by this value.

(ii) $\quad r_\mu := r_\lambda,$

i.e. replace the content of register r_μ with that of r_λ.

(iii) $\quad H,$

i.e. stop.

Conditional Instructions

(iv) $\quad$ if $S(r_{\lambda_1}, \ldots, r_{\lambda_m})$ then i else j,

i.e. if the m-ary relation S is true of the contents of $r_{\lambda_1}, \ldots, r_{\lambda_m}$ then the next instruction is I_i, otherwise it is I_j.

(v) $\quad$ if $r_\mu = r_\lambda$ then i else j,

i.e. if registers r_μ and r_λ contain the same element then the next instruction is I_i, otherwise it is I_j.

By convention, a fap P involves a finite list of registers $r_0, r_1, \ldots, r_{n-1}$ where the first few registers $r_1, \ldots, r_m$ are reserved as *input registers* and r_0 as *output register*; the remaining registers $r_{m+1}, \ldots, r_{n-1}$ are called *working registers*.

A partial function $f\colon A^m \to A$ is called fap-*computable* if there exists a fap P over the system $\mathfrak{A}$ such that for all $(a_1, \ldots, a_m)$ if $a_1, \ldots, a_m$ are loaded into registers $r_1, \ldots, r_m$, respectively, and P applied, then $f(a_1, \ldots, a_m) = a$ iff P halts and the content of the output register r_0 is a.

The class of fap-computable functions over $\mathfrak{A}$ is denoted by FAP($\mathfrak{A}$).

The general notion of a fap may be too poor to support a reasonable theory of computing over an algebraic system. There are two natural extensions of the general notion: *a finite algorithmic procedure with stacking* (fap S) first defined in [108], and *a finite algorithmic procedure with counting* (fap C) which first appeared in Friedman's paper.

In a fap S we have two new *operational instructions*

(vi) $\quad s \coloneqq (i; r_0, \ldots, r_{n-1}),$

i.e. place a copy of the contents of the registers $r_0, \ldots, r_{n-1}$ as an n-tuple in the stack register s together with the marker i; i is a natural number.

(vii) $\quad$ restore $(r_0, r_1, \ldots, r_{j-1}, r_{j+1}, \ldots, r_{n-1}),$

i.e. replace the contents of the registers $r_0, r_1, \ldots, r_{j-1}, r_{j+1}, \ldots, r_{n-1}$ by those of the last (i.e. *topmost*) n-tuple placed in the stack.

There is one new *conditional instruction*

(viii) $\quad$ if $s = \varnothing$ then i else j,

with the obvious meaning.

In a fap S program stacking is only introduced through a *stacking block* of instructions

$s \coloneqq (i; r_0, \ldots, r_{n-1})$
I_{i_1}
$\vdots$
I_{i_l}
goto k
$*: r_j \coloneqq r_0$
restore $(r_0, r_1, \ldots, r_{j-1}, r_{j+1}, \ldots, r_{n-1})$.

The meaning of this block is as follows. Let the fap S contain the registers $r_0, \ldots, r_{n-1}, s$. At the start of the block we *store* the information in $r_0, \ldots, r_{n-1}$ on top of the stack s together with its marker i. Then we use a sequence $I_{i_1}, \ldots, I_{i_l}$ to *reload* some but not necessarily all of the registers $r_0, \ldots, r_{n-1}$. This done, we are ready for a subcomputation within the total program. For this we use the *return instruction* goto k (which is an abbreviation of "if $r_\mu = r_\mu$ then k else k"). Note that I_k must be either an ordinary fap instruction outside all the blocks in the program or the first instruction of any stacking block in the program. "$*: r_j \coloneqq r_0$" is called the *exit instruction* of the block. This means that if the subcomputation is successful, i.e. that it stops and leaves r_0 non-empty, then place the content of r_0 into register r_j. We then restore the contents of $r_0, \ldots, r_{j-1}, r_{j+1}, \ldots, r_{n-1}$ and proceed with the main computation using the new content of r_j supplied by the subcomputation.

Just as above we have a notion of fap S-*computable* and a class of functions FAPS($\mathfrak{A}$).

In a fap C we add to the *algebra registers* $r_0, r_1, \ldots$ certain *counting registers* $c_0, c_1, \ldots$ which are to contain natural numbers. There are three new *operational instructions*

(ix) $\quad c_\mu := c_\lambda + 1,$

i.e. add one to the contents of c_λ and place that value in c_μ.

(x) $\quad c_\mu := c_\lambda \dot{-} 1,$

i.e. if c_λ contains 0 place 0 in c_μ, else subtract one from the contents of c_λ and place that value in c_μ.

(xi) $\quad c_\mu := 0,$

i.e. make the contents of c_μ zero.

We may add one new *conditional instruction*

(xii) $\quad$ if $c_\mu = c_\lambda$ then i else j,

again with the obvious meaning.

With a fap C program we can have computable functions of two kinds, viz. f can be a partial function with values in A,

$$f\colon \omega^n \times A^m \to A$$

or f can have values in ω,

$$f\colon \omega^n \times A^m \to \omega.$$

In both cases we use $c_1, \ldots, c_n, r_1, \ldots, r_m$ as input registers. In the first case r_0 is the output register, in the second c_0. We thus get a notion of fap C-*computable* and a class of functions FAPC($\mathfrak{A}$).

Stacking and counting can be combined to produce a notion of fap CS-*computable* and a class FAPCS($\mathfrak{A}$) over $\mathfrak{A}$.

These are the basic classes and we have the following immediate diagram of inclusions

$$\begin{array}{ccccc} & & \mathrm{FAPC}(\mathfrak{A}) & & \\ & \nearrow & & \searrow & \\ \mathrm{FAP}(\mathfrak{A}) & & & & \mathrm{FAPCS}(\mathfrak{A}). \\ & \searrow & & \nearrow & \\ & & \mathrm{FAPS}(\mathfrak{A}) & & \end{array}$$

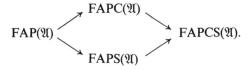

It should be clear that when $\mathfrak{A}$ includes enough arithmetic then the classes coincide. In a more general algebraic context there are interesting differences. We shall relate some basic results from [108] and [109], and, in particular, we shall discuss how inductive definability over $\mathfrak{A}$ and computation theories over $\mathfrak{A}$ fit into the above diagram.

0.2 FAP *and Inductive Definability*

The class Ind($\mathfrak{A}$) of inductively defined functions over the structure $\mathfrak{A}$ can be introduced in various but related ways. We follow the syntactic description in [105] which itself is patterned upon Platek's equational calculus [133]. It is convenient in working with partial functions on A to extend A by a new element u for undefined. We introduce a class of terms by the following clauses:

(i) Variables $x_1, x_2, \ldots$ are terms of type 0.

(ii) For each m, the m-ary partial function variables $p_1^m, p_2^m, \ldots$ are terms of type $1 \cdot m$.

(iii) For each m-ary operation σ of $\mathfrak{A}$ the function symbol $\boldsymbol{\sigma}$ is a term of type $1 \cdot m$.

(iv) For each m-ary relation S of $\mathfrak{A}$ the function symbol $\mathbf{DC}_S$ is a term of type $1 \cdot m + 2$.

(v) $\mathbf{u}$ is a term of type 0.

(vi) If t is a term of type $1 \cdot m$ and $t_1, \ldots, t_m$ are terms of type 0, then $t(t_1, \ldots, t_m)$ is a term of type 0.

(vii) If t is a term of type 0 then $\mathbf{FP}[\lambda p_i^m, y_1, \ldots, y_m \cdot t]$ is a term of type $1 \cdot m$.

If S is an m-ary relation in $\mathfrak{A}$ then

$$DC_S(a_1, \ldots, a_m, x, y) = \begin{cases} x & \text{if } S(a_1, \ldots, a_m) \\ y & \text{if} \neg\, S(a_1, \ldots, a_m). \end{cases}$$

And *FP* is the fixed-point operator. It should be clear how to interpret terms in the algebra $\mathfrak{A}$. A partial function $f\colon A^m \to A$ will be *inductively definable* if there is an algebra term (i.e. a term of type 0) with $y_1, \ldots, y_m$ as its only free variables such that for all $a_1, \ldots, a_m \in A$, $f(a_1, \ldots, a_m) = t(a_1, \ldots, a_m)$. Ind($\mathfrak{A}$) is the class of functions inductively definable over $\mathfrak{A}$.

As an example let us consider the term

$$\mathbf{FP}[\lambda p', y\mathbf{DC}_S(y, y, \mathbf{DC}_S(\boldsymbol{\sigma}_1(y), y, \boldsymbol{\sigma}_2(p'(\boldsymbol{\sigma}_1(y)), \boldsymbol{\sigma}_1(y))))](x).$$

The reader may want to verify that the following is a fap S program which computes the function defined by the above term. The program has one input register r_1 and two working registers r_2, r_3.

1. if $S(r_1)$ then 2 else 4
2. $r_0 := r_1$
3. if $s = \varnothing$ then H else *
4. $r_2 := \sigma_1(r_1)$
5. if $S(r_2)$ then 6 else 8
6. $r_0 := r_1$
7. if $s = \varnothing$ then H else *
8. $s := (1; r_0, r_1, r_2, r_3)$
9. $r_1 := \sigma_1(r_1)$
10. goto 1
11. $*: r_2 := r_0$
12. restore (r_0, r_1, r_3)
13. $r_3 := \sigma_1(r_1)$
14. $r_0 := \sigma_2(r_2, r_3)$
15. if $s = \varnothing$ then H else *

We have a single block I_8–I_{12}, and we see how we have to store previous information and use this block in the iterative computation of the value of the fixed-point function on the input value in r_1.

0.2.1 Theorem. $\mathrm{Ind}(\mathfrak{A}) = \mathrm{FAPS}(\mathfrak{A})$.

The example is no proof; see [108] for details which are far from trivial. The example should also suggest that $\mathrm{FAP}(\mathfrak{A})$ is in general a proper subclass of $\mathrm{Ind}(\mathfrak{A})$. In fact, in [108] a subclass of "direct terms" of the class of algebra terms is distinguished such that the corresponding class of *directly inductively definable* functions, $\mathrm{DInd}(\mathfrak{A})$, exactly corresponds to $\mathrm{FAP}(\mathfrak{A})$.

0.3 FAP *and Computation Theories*

Let there be given a notion $\{a\}(\sigma) \simeq z$ over some domain A, and from this let us abstract the set of all computation tuples $\Theta = \{(a, \sigma, z); \{a\}(\sigma) \simeq z\}$. A function f on A is computable under the given notion if there is an index or code a such that for all σ

$$f(\sigma) \simeq z \quad \text{iff} \quad (a, \sigma, z) \in \Theta.$$

The axiomatic approach reverses this procedure. Let there be given a set Θ of tuples (a, σ, z) over A. A function f on A is called Θ-computable if there is a code a such that the equivalence above holds.

Not every set Θ is a reasonable computation theory. We must put in some basic functions and require closure of Θ under some reasonable properties; the details are given in the first few paragraphs of Chapter 1.

As the reader will see from the general development in Chapter 1, in considering computation theories over a structure we are almost forced to also include recursive (sub-) computations on the natural numbers; having given a code set C we can inside this code set reconstruct a copy of the integers and thus have access to the recursive functions over this "successor set" (see Section 1.4).

It is therefore natural given a structure $\mathfrak{A} = \langle A; \boldsymbol{\sigma}, \mathrm{S}\rangle$ to expand it to a structure $\mathfrak{A}_\omega = \langle A \cup \omega; \boldsymbol{\sigma}, \mathrm{S}, s, p, 0\rangle$ where $s, p, 0$ are the successor, predecessor, and constant zero function on ω, respectively. We have the following relationship [109].

0.3.1 Proposition. *Let* $f\colon \omega^n \times A^m \to A$ *or* $f\colon \omega^n \times A^m \to \omega$. *Then*

(i) $f \in \mathrm{FAP}(\mathfrak{A}_\omega)$ iff f *is* fap C-*computable over* $\mathfrak{A}$.
(ii) $f \in \mathrm{FAPS}(\mathfrak{A}_\omega)$ iff f *is* fap CS-*computable over* $\mathfrak{A}$.

The question is now whether counting alone or both counting and stacking are necessary to give a computation theory for an arbitrary $\mathfrak{A}$?

To give the answer we need to introduce the notion of a *term evaluation function*. The class $T[X_1, \ldots, X_n]$ is inductively defined by the clauses:

(i) $X_1, \ldots, X_n$ belong to $T[X_1, \ldots, X_n]$;
(ii) if $t_1, \ldots, t_m$ belong to $T[X_1, \ldots, X_n]$ and σ is an m-ary operation of $\mathfrak{A}$, then $\sigma(t_1, \ldots, t_m) \in T[X_1, \ldots, X_n]$.

Each term $t(X_1, \ldots, X_n)$ defines a function $A^n \to A$ by substitution of algebra elements for indeterminates. Terms in $T[X_1, \ldots, X_n]$ can be numerically coded uniformly in n, in the sense that there is a recursive subset $\Omega \subseteq \omega$ such that for each $i \in \Omega$ there is a unique term $[i]$ in $T[X_1, \ldots, X_n]$, and the correspondence $i \to [i]$ is a surjection. And there are recursive functions which given $i \in \Omega$ allow us to effectively write down the corresponding term $[i]$. Define $E_n\colon \Omega \times A^n \to A$ by $E_n(i, \mathbf{a}) = [i](\mathbf{a})$.

0.3.2 Proposition. $\mathrm{FAP}(\mathfrak{A}_\omega)$ *is a computation theory iff* E_n *is uniformly* fap C-*computable.*

This is proved in [109]. One way is rather straightforward. The difficult part comes in verifying that $\mathrm{FAP}(\mathfrak{A}_\omega)$ has a universal function. The problem is that in the absence of a computable pairing scheme a machine with a fixed number of registers may not be able to simulate a machine with an arbitrarily large number of registers. One way of getting around this problem is by letting the simulating machine manipulate codes for terms instead of actually executing the simulated operations. For codes for terms are natural numbers for which we do have the required pairing function. At some points there is a need to evaluate terms and it is exactly here the computability of the E_n-functions is needed.

But E_n is uniformly fap CS-computable. This is, in fact, by 0.2.1 and 0.3.1 reducible to showing that E_n belongs to $\mathrm{Ind}(\mathfrak{A}_\omega)$. E_n has the following inductive definition

$$E_n(i, \mathbf{a}) = \begin{cases} a_j & \text{if } i \text{ codes } X_j \\ \sigma_j(E_n(i_1, \mathbf{a}), \ldots) & \text{if } [i] = \sigma_j([i_1], \ldots) \\ u & \text{if } i \text{ does not code a term or codes the empty term.} \end{cases}$$

Pulling the results together we arrive at the following characterization.

0.3.3 Theorem. FAPS($\mathfrak{A}_\omega$) *is the minimal computation theory over* $\mathfrak{A}_\omega$ *with code set* ω.

The mathematical part of the theory is clear, we have located the exact position of Ind($\mathfrak{A}$) and the minimal computation theory in the diagram of Section 0.1. We showed in 0.2.1 that Ind($\mathfrak{A}$) = FAPS($\mathfrak{A}$), and it follows from 0.3.3 that FAPSC($\mathfrak{A}$) is the class of functions computable in the minimal computation theory over $\mathfrak{A}$ with code set ω.

Going beyond is a matter of personal taste and preference. Our opinion is that computations in general should be allowed to use both elements of the structure and natural numbers (e.g. the *order* of an element in a group). Term evaluation should also be computable, hence we seem to be led to FAPCS($\mathfrak{A}$), i.e. to a computation theory over $\mathfrak{A}$.

0.3.4 Example. We conclude with an example due to J. V. Tucker (see his [168] for a more general result). He first shows that if $\mathfrak{A}$ is locally finite, then the halting problem for FAPS($\mathfrak{A}$) is fap CS-decidable.

The argument is based upon the simple fact that the number of state descriptions in a fap S computation is effectively bounded by a fap CS-computable function (because the order of an element in $\mathfrak{A}$ is fap CS-computable).

Let $\mathfrak{A}_0$ be the group of all roots of unity. This is surely a locally finite structure. We have the following relationships:

$$\text{Ind}(\mathfrak{A}_0) = \text{FAPS}(\mathfrak{A}_0) \subsetneqq \text{FAPC}(\mathfrak{A}_0) = \text{FAPCS}(\mathfrak{A}_0).$$

The last equality is typical in algebraic contexts, term evaluation E_n will be computable. It remains to prove that FAPS($\mathfrak{A}_0$) $\neq$ FAPCS($\mathfrak{A}_0$). This follows from the fact that $\mathfrak{A}_0$ is a computable group in the sense of Rabin-Mal'cev [99], i.e. it has a recursive coordination $\alpha\colon \Omega_\alpha \to \mathfrak{A}_0$, where we can choose α to be a bijection on a recursive subset $\Omega_\alpha \subseteq \omega$, and the "pull-back" of the group operation in $\mathfrak{A}_0$ is recursive on Ω_α.

We now observe that any fap CS-function pulls back to a recursive function. Since there are r.e. sets which are not recursive, it follows from the initial observation that the halting problem for FAPS($\mathfrak{A}_0$) is fap CS-decidable, that FAPS($\mathfrak{A}_0$) $\neq$ FAPCS($\mathfrak{A}_0$).

We shall in the main part of the book be interested in recursion theories over domains which include the natural numbers. From this point of view the preceding discussion gives "sufficient" theoretical grounds for basing the general theory on the axiomatic notion of a computation theory.

There have, however, been recent discussions of an inductive definability approach to recursion in higher types. It may be useful to compare these approaches to the computation-theoretic point of view.

0.4 Platek's Thesis

The aim of this approach is to study definability/computability over an arbitrary domain Ob using the fixed-point operator. But fixed-points at one level may be obtained from fixed-points of higher levels. This leads to the hierarchy HC of hereditarily consistent functionals over Ob as the natural domain of the general theory.

Remark. Platek's thesis was never published, we follow the discussion in J. Moldestad *Computations in Higher Types* [105].

The hierarchy HC is defined inductively as follows. Any type symbol $\tau \neq 0$ can be written in the form $\tau_1 \to (\tau_2 \to \ldots (\tau_k \to 0) \ldots)$. The level of τ is defined by $l(\tau) = \max\{l(\tau_i) + 1\}$. Monotonicity for partial functions is defined as usual. Then $HC(\tau)$ is defined to be the set of all partial monotone functions defined on a subset of $HC(\tau_1) \times \ldots \times HC(\tau_k)$ and with values in Ob.

The fixed-point operator at type τ is an element $FP \in HC((\tau \to \tau) \to \tau)$, such that when FP is applied to an element $f \in HC(\tau \to \tau)$ it produces the least fixed-point $FP(f)$ of f, which will be an element of $HC(\tau)$.

Platek's index-free approach to recursion theory can now be introduced. Let $\mathscr{B} \subseteq HC$. Then the recursion theory generated by $\mathscr{B}$, which we will denote by $\mathscr{R}_\omega(\mathscr{B})$, is the least set extending $\mathscr{B}$, closed under composition, and containing the function DC (definition by cases), the combinators $I(f) = f$, $K(f, g) = f$, and $S(f, g, h) = f(h)(g(h))$, and the fixed-point operator FP.

To set out the relationship with computation theories we quote two results from Moldestad's study. The first is Platek's reduction theorem.

0.4.1 Reduction Theorem. *Let $\mathscr{B}$ contain some basic functions (a coding scheme, the characteristic function of the natural numbers, the successor and predecessor functions). If $\mathscr{B} \subseteq HC^{l+2}$, then*

$$\mathscr{R}_\omega(\mathscr{B})^{l+3} = \mathscr{R}_{l+1}(\mathscr{B})^{l+3}.$$

If we are interested in objects of level at most $l + 3$, then we need only apply the fixed-point operator up to type $l + 1$.

0.4.2 Equivalence Theorem. *There exists a computation theory Θ (derived from Kleene's schemata* S1–S9 *in* [83]) *such that*

$$\mathscr{R}_\omega(\{h_1, \ldots, h_k\}) = \Theta[h_1, \ldots, h_k]$$

for all finite lists of HC objects $h_1, \ldots, h_k$.

We shall in a moment return to the reduction theorem. But first we spell out the content of the equivalence theorem. It seems that we can draw the following conclusions. Let $\mathscr{B} \subseteq HC$.

1 $\mathscr{R}_\omega(\mathscr{B}) = \bigcup \{\mathscr{R}_\omega(\mathscr{B}_0) : \mathscr{B}_0 \text{ finite subset of } \mathscr{B}\}$.
2 For finite $\mathscr{B}_0$, $\mathscr{R}_\omega(\mathscr{B}_0) = \Theta[\mathscr{B}_0]$.
3 $\mathscr{R}_\omega(\mathscr{B}) \subseteq \Theta[\mathscr{B}]$, but in general $\subsetneq$.

Only the third assertion requires a comment. The notation $\Theta[\mathscr{B}]$ is somewhat ambiguous. We must assume that $\mathscr{B}$ is given as a *list*, i.e. with a specific enumeration. This means that in any precise version of $\Theta[\mathscr{B}]$ we have the enumeration function of the list $\mathscr{B}$. But this enumeration function is not necessarily in $\mathscr{R}_\omega(\mathscr{B})$.

Back to the reduction theorem. This result shows that the framework of computation theories is adequate if the aim is to study computability/definability over some given domain. We need not climb up through the hierarchy HC. A computation theory Θ can be considered as a set of functions $\Theta \subseteq HC^1$. Then, by the reduction theorem

$$\mathscr{R}_\omega(\Theta)^1 = \mathscr{R}_1(\Theta)^1 = \Theta,$$

the last equality being true since Θ satisfies the first recursion theorem; see Theorems 1.7.8 and 1.7.9 of Chapter 1.

Remark. Platek obtains Kleene's theory of recursion in higher types as a "pull-back" from his theory on HC. We shall return to this matter in Chapter 4.

0.5 Recent Developments in Inductive Definability

The index-free approach of Platek is conceptually of great importance in the development of generalized recursion. The theory has, however, some weak points. Recently improved and largely equivalent versions have been published independently by Y. Moschovakis and S. Feferman.

The relevant papers of Moschovakis are the joint contribution with Kechris, *Recursion in higher types* [77], and the paper *On the basic notions in the theory of induction* [117]. Feferman's version is presented in *Inductive schemata and recursively continuous functionals* [25].

Feferman summarizes his criticism of Platek in the following points.

a The structure of natural numbers is included as part—there could be more general situations, e.g. applications in algebra.

b Inductive definability of relations is not accounted for.

c Platek's pull-back of Kleene's theory of recursion in higher types from *HC* is complicated and *ad hoc.*

We believe that point *a* is adequately dealt with in Section 0.3 above. We shall eventually return to point *b*. In connection with *c* we just state our complete agreement and that we will return to it in Chapter 4.

As a basis for a comparison with computation theories we shall discuss a result from *Recursion in higher types*.

A partial monotone functional $\Phi(\mathbf{x}, f, \mathbf{g})$ defines in the usual way a fixed-point $\Phi^\infty(\mathbf{x}, \mathbf{g})$. Let $\mathscr{F}$ be a class of functionals. If Φ belongs to $\mathscr{F}$, we call Φ^∞ an $\mathscr{F}$-fixed-point. The class $\mathrm{Ind}(\mathscr{F})$ will consist of all functionals $\Psi(\mathbf{x}, \mathbf{g})$ for which there exists an $\mathscr{F}$-fixed-point $\Phi^\infty(\mathbf{u}, \mathbf{x}, \mathbf{g})$ and constants $\mathbf{n}$ from ω such that

$$\Psi(\mathbf{x}, \mathbf{g}) = \Phi^\infty(\mathbf{n}, \mathbf{x}, \mathbf{g}).$$

The induction completeness theorem tells us that $\mathrm{Ind}(\mathscr{F})$ is closed under inductive definability.

Not every collection $\mathscr{F}$ gives a reasonable recursion theory. A class $\mathscr{F}$ is called *suitable* if it contains the following initial objects: characteristic function of ω, the identity on ω, the successor function on ω, the characteristic function of equality on ω, and the evaluation functional. In addition $\mathscr{F}$ is required to be closed under addition of variables, composition, definition by cases, substitution of projections, and functional substitution.

$\mathrm{Ind}(\mathscr{F})$ is said to have the *enumeration property* (is *ω-parametrized*) if for each $n \geqslant 1$ there is some $\varphi(e, x_1, \ldots, x_n) \in \mathrm{Ind}(\mathscr{F})$ such that a function f belongs to $\mathrm{Ind}(\mathscr{F})$ iff there exists some $e \in \omega$ such that

$$f(\mathbf{x}) = \varphi(e, \mathbf{x}).$$

0.5.1 Enumeration Theorem [77]. *Let $\mathscr{F}$ be a suitable class of functionals on a domain A including ω. If $\mathscr{F}$ is finitely generated and admits a coding scheme, then* $\mathrm{Ind}(\mathscr{F})$ *has the enumeration property.*

Remark. One notices that the machinery provided by the requirement of suitability of $\mathscr{F}$ corresponds to a large extent to what we have put into our general notion of a computation theory. There is one difference, in the computation-theoretic approach we have adopted the enumeration property and proved the first recursion theorem, whereas in the inductive approach the first recursion theorem is an axiom and enumeration a theorem. One may argue what is "philosophically" the most basic or natural, mathematically they serve the same purposes provided, we should add, there is enough coding machinery available.

The enumeration theorem for $\mathrm{Ind}(\mathscr{F})$ leads to the same situation as pointed out in connection with Platek's $\mathscr{R}_\omega(\mathscr{B})$. The enumeration theorem leads to a computation theory Θ, such that if $\mathscr{F}_0$ is a finite basis for the finitely generated class $\mathscr{F}$ (and $\mathscr{F}$ is suitable and admits a coding scheme), then

$$\mathrm{Ind}(\mathscr{F}) = \Theta[\mathscr{F}_0].$$

Both Moschovakis [117] and Feferman [25] include inductive definability in relations, see point b above. Moschovakis' approach is based on the notion of *induction algebra*, which is a structure

$$\mathbb{F} = \langle \{X^\alpha\}_{\alpha\in I}, \{\leqslant^\alpha\}_{\alpha\in I}, \{\vee^\alpha\}_{\alpha\in I}, \mathscr{F}\rangle$$

where each $\leqslant^\alpha$ is a partial ordering on X^α in which every chain has a least upper bound, and $\vee^\alpha$ is a supremum operator on X^α, i.e. $x \leqslant x \vee y$ and if $x \leqslant y$, then $x \vee y = y$, for all $x, y \in X^\alpha$. $\mathscr{F}$ is a class of operations, i.e. maps of the form $f\colon X^{\alpha_1} \times \ldots \times X^{\alpha_n} \to X^\alpha$.

The two main examples are the induction algebras of *relations* and induction algebras of *partial functions*. The case of partial functions was discussed above. In the case of relations we start with a domain A, let X^0 equal the set of truth values $\{T, F\}$, and for $n \geqslant 1$, set X^n = all n-ary relations on A. In this case $\leqslant$ is set inclusion and $\vee$ is set union.

Let Φ be a class of second-order relations on A. To each $\varphi(\mathbf{R}, \mathbf{s})$ in Φ we associate an operation f by

$$f(\mathbf{R}) = \{\mathbf{s} : \varphi(\mathbf{R}, \mathbf{s})\}.$$

(Conversely, an operation f determines a relation by the equivalence $\varphi(\mathbf{R}, \mathbf{s})$ iff $\mathbf{s} \in f(\mathbf{R})$.)

In this way—provided suitable conditions are imposed on Φ—we get an induction algebra of relations on A.

Furthermore, provided the classes we start with are "rich enough" in structure, there will be an enumeration theorem for the finitely generated algebras, and, hence, a computation theoretic equivalent. But there could, in principle, be more general situations. This, however, was discussed at length in Sections 0.2–0.3. See also Moldestad–Tucker [110] for a discussion of other general approaches and how they are related to the present development.

We should, perhaps, add one more comment. Codes, indices are usually claimed to be *ad hoc* and, hence, conceptually unsatisfactory. And a comparison with an intrinsic versus coordinate based treatment in geometry is often made. But is the analogy really to the point! In our discussion of Platek's $\mathscr{R}_\omega(\mathscr{B})$—and a similar result holds for Ind($\mathscr{F}$)—we concluded that

1. $\mathscr{R}_\omega(\mathscr{B}) = \bigcup \mathscr{R}_\omega(\mathscr{B}_0),$

where $\mathscr{B}_0$ is a finite subset of $\mathscr{B}$, and for finite $\mathscr{B}_0$

2. $\mathscr{R}_\omega(\mathscr{B}_0) = \Theta[\mathscr{R}_0].$

In 2 the codes are introduced as a systematic, even canonical way of referring to the objects in the finite list $\mathscr{B}_0$ and to the operations generating $\mathscr{R}_\omega(\mathscr{B}_0)$ out of $\mathscr{B}_0$. 1 and 2 together say that the *global theory* $\mathscr{R}_\omega(\mathscr{B})$ admits *natural local co-*

ordinates $\Theta[h_1, \ldots, h_k]$ suitable for more "delicate", i.e. computation-theoretic investigations of the theory, degree structure, computation in higher types, etc.

This seems to be a reasonable analogy with geometry. But is the analogy complete? Are there any properties *in the large* of generalized recursion theory? Or is everything local, i.e. computation-theoretic?

With these remarks we let our case for computation theories rest. Some brief comments on the plan for our exposition.

Part A sets out the *general theory*. Chapter 1 gives the combinatorial part leading up to the simple representation theorem and a general version of the first recursion theorem. In Chapter 2 we add the notions of subcomputation and length of a computation and give a second representation theorem, viz. a representation in terms of partial type-2 functionals over the domain which preserves not only the computable objects, but also the full structure of subcomputations.

Part B discusses *finite theories* on one (Chapter 3) and two (Chapter 4) types. Finite theories on one type are the general version of hyperarithmic theory or, equivalently, the theory of recursion in ${}^2E$. The finite theories on two types are a general version of recursion in higher types, but are also a suitable framework for various developments in second order definability. The relationship to Spector 1- and 2-classes is explained.

Part C is devoted to *infinite theories*, i.e. general versions of ORT (ordinary recursion theory) and admissibility theory. Chapter 5 gives the basic facts including the imbedding theorem of finite theories on one type into infinite theories, and also various results connected with the "abstract 1-section" theorem. In Chapter 6 a general account of degree theory is given including some recent excursions into inadmissibility theory.

Part D treats *set recursion* and *computations in higher types*. Chapter 7 discusses reflection phenomena and proves the general plus-1 and plus-2 theorems. Set recursion and its connection to recursion in higher types is the topic of Chapter 8. Some discussion of degree theory in higher types is also included.

A collector is never satisfied until all specimens have been collected, neatly arranged, and labelled. This was never our intention. The book is introductory. The aim is to provide a reasonably unified view. Not the only possible one, but one broad and detailed enough to serve as a basis and general framework. Beyond this the reader must proceed by himself.

Remark. Our approach to the general theory has been developed over a number of years; see the reports [26], [27], [28], and [29].

Part A

General Theory

Chapter 1

General Theory: Combinatorial Part

This chapter contains the basic definitions, examples, and constructions of the general theory of computations—and two theorems: In Section 1.6 we prove the simple representation theorem (1.6.3) which is used in Section 1.7 to prove a general version of the first recursion theorem (1.7.8, 1.7.9).

1.1 Basic Definitions

In a typical computation situation there are three separate parts. There is a *computing device* or *machine* M which acts upon an *input* σ from some fixed input alphabet or domain A_1. After "computing" for some time M may give an *output* z belonging to some fixed output alphabet or domain A_2.

The input σ is usually a finite sequence $\sigma = (x_1, \ldots, x_n)$ of elements from A_1, the output a single element from A_2. In many cases there may be practical advantages in having distinct input and output alphabets. In a theoretical analysis this is superfluous, hence we shall from now on assume that $A_1 = A_2$ is a fixed set of objects, the *computation domain*.

The computing device M is usually one of a class of similar computing "machines". It was a basic insight of the early theories of computation that the machines could be *coded* by elements of the computation domain. This idea will be our starting point: The basic axioms shall provide an analysis of the relation

$$\{a\}(\sigma) \simeq z,$$

which is *intended* to assert that the computing device named or coded by a and acting on the input sequence $\sigma = (x_1, \ldots, x_n)$ gives z as output.

1.1.1 Definition. A *computation domain* is a structure

$$\mathfrak{A} = \langle A, C; 0, 1 \rangle$$

where A is a non-empty set, C is a subset of A, and 0, 1 are two designated elements of C.

C is called the set of *codes*. It may or may not be equal to A. In ordinary Turing machine theory on ω the natural numbers, the set of codes C is equal to ω, i.e. every element of the computation domain codes a Turing machine. In higher types the set of codes may still be ω, which is now a proper subset of the computation domain, which consists of numbers, functions, and functionals of various types.

In the general case we have given the two sets A and C such that $\varnothing \neq C \subseteq A$. Usually C contains (an isomorphic copy of) ω. At the present level of generality we shall not make this a part of the definition of computation domain. We also remark that in many examples the output set is (a subset of) the code set.

To facilitate the presentation we introduce some notational conventions. We use

$x, y, z, \ldots$	for elements in A.
$a, b, c, \ldots$	for elements in C.
$\sigma, \tau, \ldots$	for finite sequences from A.

In particular, we let σ, τ or (σ, τ) denote the concatenation of sequences. By (a, σ, z), where $a \in C$, $z \in A$ and $\sigma = (x_1, \ldots, x_n)$ is a finite sequence from A, we understand the sequence

$$(a, x_1, \ldots, x_n, z).$$

As usual $\text{lh}(\sigma) =$ the length of the sequence σ.

1.1.2 Definition. Θ is a *computation set* over $\mathfrak{A}$ if Θ is a set of tuples (a, σ, z), where $a \in C$, $\sigma = (x_1, \ldots, x_n)$, each $x_i \in A$, $z \in A$, and $\text{lh}(a, \sigma, z) \geqslant 2$.

Requiring $\text{lh}(a, \sigma, z) \geqslant 2$ means that we have a code a and an output z present, but not necessarily an input sequence.

1.1.3 Remark. At this stage we need not make any requirement of single-valuedness, hence given a and σ there may be more than one z such that $(a, \sigma, z) \in \Theta$. However, in most cases we will require that Θ is single-valued, e.g. in analyzing the subcomputations of computations in functionals.

Let Θ be a computation set over the domain $\mathfrak{A}$. To every $a \in C$ and every natural number $n \geqslant 0$ we can associate a *partial multiple-valued* function (pmv function) $\{a\}_\Theta^n$ as follows

$$\{a\}_\Theta^n(\sigma) \simeq z \quad \text{iff} \quad \text{lh}(\sigma) = n \quad \text{and} \quad (a, \sigma, z) \in \Theta.$$

A pmv function f is really a map from A, or a suitable cartesian product over A, to subsets of A (including the empty subset). The functional notation then has the following meaning

$$
\begin{array}{rll}
f(\sigma) \simeq z & \text{means} & z \in f(\sigma) \\
f(\sigma) = g(\sigma) & \text{means} & \forall z[f(\sigma) \simeq z \text{ iff } g(\sigma) \simeq z] \\
f(\sigma) = z & \text{means} & f(\sigma) = \{z\} \\
f \subseteq g & \text{means} & \forall\sigma\forall z[f(\sigma) \simeq z \Rightarrow g(\sigma) \simeq z].
\end{array}
$$

By a *mapping* we understand a total, single-valued function.

1.1.4 Definition. Let Θ be a computation set over $\mathfrak{A}$. A pmv function f is Θ-*computable* if for some $\hat{f} \in C$ we have

$$f(\sigma) \simeq z \quad \text{iff} \quad (\hat{f}, \sigma, z) \in \Theta.$$

We call $\hat{f}$ a Θ-*code* for f and write $f = \{\hat{f}\}_\Theta^n$, where $n = \mathrm{lh}(\sigma)$ is the number of arguments of f.

A partial multiple-valued (pmv) functional on A

$$\varphi(\mathbf{f}, \sigma) = \varphi(f_1, \ldots, f_l, x_1, \ldots, x_m),$$

maps pmv functions on A and elements of A to subsets of A. φ is called *consistent* if

$$f_1 \subseteq g_1, \ldots, f_l \subseteq g_l, \varphi(\mathbf{f}, \sigma) \simeq z \quad \Rightarrow \quad \varphi(\mathbf{g}, \sigma) \simeq z.$$

1.1.5 Definition. Let Θ be a computation set over $\mathfrak{A}$. A pmv consistent functional φ is called *weakly Θ-computable* if there exists a code $\hat{\varphi} \in C$ such that for all $e_1, \ldots, e_l \in C$ and all sequences $\sigma = (x_1, \ldots, x_n)$ from A we have

$$\varphi(\{e_1\}_\Theta^{n_1}, \ldots, \{e_l\}_\Theta^{n_l}, \sigma) \simeq z \quad \text{iff} \quad \{\hat{\varphi}\}_\Theta^{l+n}(e_1, \ldots, e_l, \sigma) \simeq z.$$

We see that φ is weakly Θ-computable if we can calculate φ on Θ-computable functions by calculating on the codes of the functions.

The notion of *weak* Θ-computability presupposes a notion of *strong* Θ-computability. Such a notion will be introduced in Section 1.7 below. The distinction between weak and strong computability will also be of importance in analyzing computations in higher types.

We will now consider some specific functions and functionals.

1.1.6 Definition by Cases (on the code set C).

$$DC(x, a, b, c) = \begin{cases} 1, & \text{if not all } a, b, c \in C \\ a, & \text{if } x = c \text{ and all } a, b, c \in C \\ b, & \text{if } x \neq c \text{ and all } a, b, c \in C. \end{cases}$$

Outright definition by cases makes equality on A Θ-computable. This we may not always want, e.g. in higher type theories. We note that DC is a mapping, i.e. total and single-valued.

1.1.7 Composition.

$$\mathbf{C}^n(f, g, \sigma) = f(g(\sigma), \sigma),$$

where $n = \mathrm{lh}(\sigma)$.

1.1.8 Permutation.

$$\mathbf{P}^m_{n,j}(f, \sigma, \tau) = f(\sigma^j).$$

Here $n, m \geqslant 0$, $\mathrm{lh}(\sigma) = n$, $\mathrm{lh}(\tau) = m$, $0 \leqslant j < n$, and $(x_1, \ldots, x_n)^j = (x_{j+1}, x_1, \ldots, x_j, x_{j+2}, \ldots, x_n)$. This functional performs many tasks, e.g. $\mathbf{P}^0_{n,0}$ is the evaluation functional $\mathbf{P}^0_{n,0}(f, \sigma) \simeq f(\sigma)$, as $\sigma^0 = \sigma$.

Next we consider a property which a computation set Θ on $\mathfrak{A}$ may or may not have.

1.1.9 Iteration Property. For each $m, n \geqslant 0$ there exists a mapping S^n_m such that for all $a \in C$ and sequences σ from C and all τ from A,

$$\{a\}^{n+m}_\Theta(\sigma, \tau) = \{S^n_m(a, \sigma)\}^m_\Theta(\tau),$$

where $n = \mathrm{lh}(\sigma)$ and $m = \mathrm{lh}(\tau)$.

1.1.10 Definition. Let Θ be a computation set on the domain $\mathfrak{A}$. Θ is called a *precomputation theory* on $\mathfrak{A}$ if

(i) for each n, j $(0 \leqslant j < n)$ and m, DC, $\mathbf{C}^n$, and $\mathbf{P}^m_{n,j}$ are Θ-computable with Θ-codes d, c_n, and $p_{n,j,m}$, respectively;

(ii) Θ satisfies the iteration property, i.e. for each n, m there is a Θ-code $s_{n,m}$ for a mapping S^n_m with property 1.1.9 above.

We make some remarks: **(1)** d, c_n, $p_{n,j,m}$ are all different members of C, hence C is infinite. **(2)** The use of functionals could be eliminated, we might directly refer to the codes. **(3)** If $\langle N, s\rangle$ (i.e. the natural numbers (or a copy) with the successor function is in the structure $\mathfrak{A}$, $N \subseteq C \subseteq A$, we may require that the codes c_n, $p_{n,j,m}$, and $s_{n,m}$ are Θ-computable mappings of the parameters n, j, m. This is a uniformity requirement which will be important in later sections. **(4)** $P^0_{n,0}$ is the evaluation functional. The Θ-computability of it gives us the following enumeration property: There is for each natural number n a code $p_{n,0,0} \in C$ such that for all $a \in C$ and all σ from A

$$\{p_{n,0,0}\}^{n+1}_\Theta(a, \sigma) \simeq \{a\}^n_\Theta(\sigma).$$

1.2 Some Computable Functions

We shall give several examples of functions which are Θ-computable for any precomputation theory Θ.

1.2.1 The Characteristic Function of the Code Set C. This function is defined as follows:

$$\chi_C^{(n)}(y, \sigma) = \begin{cases} 0, & \text{if } y \in C \\ 1, & \text{if } y \notin C \end{cases}$$

where $n = \mathrm{lh}(\sigma)$. A code k_n for $\chi_C^{(n)}$ is given by

$$k_n = S_{n+1}^1(p_{1,0,n}, S_1^3(d, 0, 0, 0)),$$

as the following calculation shows:

$$\begin{aligned} \{k_n\}_\Theta^{n+1}(y, \sigma) &= \{S_{n+1}^1(p_{1,0,n}, S_1^3(d, 0, 0, 0))\}_\Theta^{n+1}(y, \sigma) \\ &= \{p_{1,0,n}\}_\Theta^{n+1}(S_1^3(d, 0, 0, 0), y, \sigma) \\ &= \{S_1^3(d, 0, 0, 0)\}_\Theta^1(y) \\ &= DC(0, 0, 0, y) = \begin{cases} 0, & \text{if } y \in C \\ 1, & \text{if } y \notin C. \end{cases} \end{aligned}$$

Most such calculations will be omitted in the sequel.

1.2.2 Equality on C. For each $a \in C$ we can define a function E_a such that

$$E_a(x) = \begin{cases} 0, & \text{if } x = a \\ 1, & \text{if } x \neq a. \end{cases}$$

We see that $DC(x, 0, 1, a)$ almost does what we want. It has to be "straightened out" by using $\mathbf{P}_{n,j}^m$ a number of times in conjunction with the iteration property.

1.2.3 Identity Function on C. For each n, let

$$I_n(y, \sigma) = \begin{cases} y, & \text{if } y \in C \\ 1, & \text{if } y \notin C, \end{cases}$$

where $n = \mathrm{lh}(\sigma)$. The following is code for I_n

$$i_n = S_{n+1}^3(p_{n+3,2,0}, S_{n+3}^2(c_{n+3}, d_n, k_{n+2}), 0, 0),$$

as this calculation shows (note that d_n is the code for $DC_n(x, a, b, c, \sigma)$, $n = \mathrm{lh}(\sigma)$, which for $n > 0$ can be obtained from DC by a suitable application of $\mathbf{P}_{n,j}^m$)

$$\begin{aligned} \{i_n\}_\Theta^{n+1}(y, \sigma) &= \{p_{n+3,2,0}\}_\Theta^{n+4}(S_{n+3}^2(c_{n+3}, d_n, k_{n+2}), 0, 0, y, \sigma) \\ &= \{c_{n+3}\}_\Theta^{n+5}(d_n, k_{n+2}, y, 0, 0, \sigma) \\ &= \{d_n\}_\Theta^{n+4}(\{k_{n+2}\}_\Theta^{n+3}(y, 0, 0, \sigma), y, 0, 0, \sigma) \\ &= \begin{cases} y, & \text{if } y \in C \\ 1, & \text{if } y \notin C. \end{cases} \end{aligned}$$

1.2.4 Constant Functions on C. The constant functions can be introduced in different ways. For $y \in C$ we see that

$$DC(x, y, y, 0) = y,$$

for all x. We also see that for $y \in C$

$$\{S_n^1(i_n, y)\}_\Theta^n(\sigma) = \{i_n\}_\Theta^{n+1}(y, \sigma) = y,$$

for all sequences σ.

1.2.5 Ordered Pair on C. The existence for arbitrary precomputation theories of an ordered pair function on C with inverses, follows from the axioms. First choose a code $e \in C$ such that

$$\{e\}(a, b, c, x) = DC(x, a, b, c).$$

We omit the super- and subscript on $\{a\}_\Theta^n$ when the meaning is clear from the context. Define for $a, b \in C$:

(i) $\qquad M(a, b) = S_2^2(e, a, b).$

Next we define for $z \in C$ the "inverses"

(ii) $\qquad K(z) = \{z\}(e, e)$
(iii) $\qquad L(z) = \{z\}(e, e'),\quad$ where $\quad e' \neq e, e' \in C.$

More accurately we should define K and L by combining DC and composition, e.g.

$$K(z) = \begin{cases} \{z\}(e, e) & \text{if} \quad z \in C \\ 1 & \text{if} \quad z \notin C, \end{cases}$$

where $\{z\}(e, e)$ as a function of z can be obtained using $\mathbf{P}_{n,j}^m$ and S_n^m in a suitable way (the enumeration property of $\mathbf{P}$).

M as a map from $C \times C$ to C is 1-1, since $M(a, b) = M(a', b')$ implies that

$$\{S_2^2(e, a, b)\}(x, y) = \{S_2^2(e, a', b')\}(x, y),$$

for all x, y. Hence $a = a'$ and $b = b'$. Further we see that

$$K(M(a, b)) = \{S_2^2(e, a, b)\}(e, e) = \{e\}(a, b, e, e) = a,$$
$$L(M(a, b)) = \{S_2^2(e, a, b)\}(e, e') = \{e\}(a, b, e, e') = b.$$

Thus we have ordered pair and inverses, although they may not be the most natural choices in a specifically given precomputation theory. In particular, if

$A = C$, the pairing function M will be defined for all $a, b \in A$ with the correct properties.

1.2.6 The Fixed-point Theorem. We conclude our list of elementary examples by adapting the usual proof of the fixed-point theorem to arbitrary computation theories.

Theorem. *Let* Θ *be a precomputation theory on* $\mathfrak{A}$. *For every* $n + 1$*-ary* Θ*-computable* pmv *function* f *there exists an* $a \in C$ *such that for all* σ *from* A

$$\{a\}^n_\Theta(\sigma) = f(a, \sigma).$$

Proof. Define f_1 and S with Θ-codes $\hat{f_1}$, $\hat{S}$, respectively such that

$$f_1(x, y, \sigma) = f(x, \sigma)$$
$$S(y, \sigma) = S^1_1(y, y), \quad y \in C.$$

Let

$$a = S^1_1(S^2_{n+1}(c_{n+1}, \hat{f_1}, \hat{S}), S^2_{n+1}(c_{n+1}, \hat{f_1}, \hat{S})),$$

a simple calculation proves the theorem.

1.3 Semicomputable Relations

For the sake of completeness we include a short discussion of computable and semicomputable relations within the context of a general precomputation theory. This is a topic to which we will return at greater length in later parts of this work.

1.3.1 Definition. Let Θ be a precomputation theory over a domain $\mathfrak{A} = \langle A, C; 0, 1\rangle$.

(i) A relation $R(\sigma)$ is Θ-*semicomputable* if there is a Θ-computable function f such that

$$R(\sigma) \text{ iff } f(\sigma) \simeq 0.$$

(ii) A relation $R(\sigma)$ is Θ-*computable* if there is a Θ-computable mapping f such that

$$R(\sigma) \text{ iff } f(\sigma) = 0.$$

Not many results about Θ-semicomputable sets can be proved in this generality.

It is, however, possible to show that if $A = C$, then the relation $(a, \sigma, z) \in \Theta$ is Θ-semicomputable.

1.3.2 Example. Let Θ be a precomputation theory on a domain $\mathfrak{A}$ where $A = C$. We show that the relation

$$(a, \sigma, z) \in \Theta,$$

is Θ-semicomputable.

First define a function with code e such that

$$\{e\}(u, v, \sigma) = \begin{cases} 0 & \text{if } u = v \\ 1 & \text{if } u \neq v. \end{cases}$$

Next, let a^* be a Θ-code, computable from a, such that

$$\{a^*\}(z, \sigma) \simeq t \quad \text{iff} \quad \{a\}(\sigma) \simeq t.$$

We now observe that

$$\begin{aligned} C(\{e\}, \{a^*\}, z, \sigma) \simeq 0 \quad &\text{iff} \quad \{e\}(\{a^*\}(z, \sigma), z, \sigma) \simeq 0 \\ &\text{iff} \quad \{a\}(\sigma) \simeq z \\ &\text{iff} \quad (a, \sigma, z) \in \Theta. \end{aligned}$$

One way of obtaining a well-behaved theory is to add selection operators. Examples show that it is too restrictive to add a single-valued selection operator. Hence, we have here a case where multiple-valuedness could serve a real purpose. However, we shall in later parts of this study see greater advantages in retaining single-valuedness of Θ and use other means to obtain a good theory for the computable and semicomputable relations.

1.3.3 Definition. Let Θ be a precomputation theory over a domain $\mathfrak{A} = \langle A, C; 0, 1\rangle$. An *n-ary selection operator* for Θ is an $n + 1$-ary Θ-computable pmv function $q(a, \sigma)$, with Θ-code $\hat{q}$, such that:

> If there is an x such that $\{a\}_\Theta^{n+1}(x, \sigma) \simeq 0$, then $q(a, \sigma)\downarrow$, and for all x such that $q(a, \sigma) \simeq x$ we have $\{a\}_\Theta^{n+1}(x, \sigma) \simeq 0$.

In general $q(a, \sigma)$ may be a subset of the set of *all* x such that $\{a\}_\Theta^{n+1}(x, \sigma) \simeq 0$.

1.3.4 Theorem. *Let Θ be a precomputation theory over $\mathfrak{A}$ having selection operators.*

(i) *If $R(x, \sigma)$ is Θ-semicomputable, then so is $\exists x R(x, \sigma)$.*
(ii) *If $R(\sigma)$ and $S(\sigma)$ are Θ-semicomputable, then so is $R(\sigma) \vee S(\sigma)$.*
(iii) *R is Θ-computable iff R and $\neg R$ are Θ-semicomputable.*

We indicate the proofs: (i) Let r be a code such that $R(x, \sigma)$ iff $\{r\}(x, \sigma) \simeq 0$. Then $\exists x R(x, \sigma)$ iff $\{r\}(q(r, \sigma), \sigma) \simeq 0$.

(ii) Let r and s be Θ-codes for R and S, respectively. Let $f(r, s)$ be a code, Θ-computable in r and s, for a Θ-computable mapping such that $\{f(r, s)\}(0) = r$ and $\{f(r, s)\}(x) = s$, $x \neq 0$. Then $R(\sigma) \vee S(\sigma)$ iff $\exists v[\{\{f(r, s)\}(v)\}(\sigma) \simeq 0]$.

(iii) Let r and s be codes for R and $\neg R$, respectively. Let $m(r, s)$ be a code (Θ-computable in r,s) for the following Θ-semicomputable relation

$$(\{r\}(\sigma) \simeq 0 \wedge t = 0) \vee (\{s\}(\sigma) \simeq 0 \wedge t = 1).$$

Using the selection operator we get a Θ-computable mapping $f(\sigma) = q(m(r, s), \sigma)$ such that

$$R(\sigma) \quad \text{iff} \quad f(\sigma) = 0.$$

So much for the general theory. We turn our attention to an investigation of the structure of an arbitrary precomputation theory. But first we look at theories over ω.

1.4 Computing Over the Integers

In this section we shall briefly consider precomputation theories over the integers, i.e. we assume that $A = C = \omega$, and that the designated elements "0" and "1" really are 0 and 1. We shall also restrict ourselves to precomputation theories in which the successor function $s(x) = x + 1$ is computable. We will show that such theories are closed under the μ-operator, the predecessor function, and primitive recursion.

1.4.1 The μ-Operator. Let $f(\sigma, y)$ be any Θ-computable function. Define via the Fixed-point Theorem 1.2.6 a function $h(\sigma, y)$ by the condition

$$h(\sigma, y) = \begin{cases} 0 & \text{if } f(\sigma, y) = 0 \\ h(\sigma, y + 1) + 1 & \text{if } f(\sigma, y) \neq 0. \end{cases}$$

We see that $\mu y[f(\sigma, y) = 0] = h(\sigma, 0)$.

1.4.2 The Predecessor Function. We would like to define $p(x)$ using the μ-operator as follows

$$p(x) = \begin{cases} 0 & \text{if } x = 0 \\ \mu y[y + 1 = x] & \text{if } x \neq 0. \end{cases}$$

But this does not exactly fit into the format of 1.4.1. We would have to replace the condition $y + 1 = x$ by e.g. $x \dot{-} (y + 1) = 0$. The trouble is that $a \dot{-} b$ is usually defined using the predecessor function. However, we get around this difficulty using the idea of the construction in 1.4.1. Define (by substitution into DC):

$$h(x, y) = \begin{cases} 0 & \text{if } s(y) = x \\ h(x, s(y)) + 1 & \text{if } s(y) \neq x \end{cases}$$

then $p(x) = h(x, 0) = \mu y[y + 1 = x]$.

1.4.3 Primitive Recursion. With the predecessor function at hand, primitive recursion follows by a simple application of the Fixed-point Theorem 1.2.6. Let $g(\sigma)$ and $h(x, y, \sigma)$ be two given Θ-computable functions. We define

$$f(y, \sigma) = \begin{cases} g(\sigma) & \text{if } y = 0 \\ h(f(p(y), \sigma), p(y), \sigma) & \text{if } y \neq 0. \end{cases}$$

Remark. We seem to be using DC in all the examples 1.4.1 to 1.4.3. But the careful reader will observe that DC is not quite enough. We have to use WDC, weak definition by cases, which by 2.7.4 is available.

It is now possible to state the following *minimality* result.

1.4.4 Theorem. *Let Θ be a precomputation theory over ω, and let f be a (Kleene) partial recursive function. Then f is Θ-computable.*

By the normal form theorem for partial recursive functions any such function can be represented in the form

$$f(\sigma) \simeq U(\mu y[g(\sigma, y) = 0]),$$

where g and U are primitive recursive functions. The proof now follows from the closure properties in 1.4.1 and 1.4.3.

Note that the theorem states an *extensional* result. We have not yet introduced a notion of "equivalence" or "extension" between theories. But granted a notion of extension $\Theta \leqslant H$ and granted that the set PR of partial recursive functions over ω is organized into a computation theory in some reasonable way, we would expect (and it is, indeed, true) that $PR \leqslant \Theta$, for all precomputation theories over ω.

1.4.5 Remark. We have so far assumed that the domain has the form $\mathfrak{A} = \langle A, C; 0, 1\rangle$. If we want to include the integers and the successor function, it is more natural to consider domains of the form $\mathfrak{A} = \langle A, C, N; s\rangle$, where $N \subseteq C \subseteq A$ and $\langle N, s \upharpoonright N\rangle$ is (an isomorphic copy of) the integers, and s is defined as

(i) $$s(x) = \begin{cases} x + 1 & \text{if } x \in N \\ 0 & \text{if } x \notin N. \end{cases}$$

Then the Θ-computability of $s(x)$ implies that the characteristic function of N is Θ-computable.

If Θ is a theory over the more general type of domain $\mathfrak{A} = \langle A, C; 0, 1\rangle$, it is possible to construct a *successor* set, but this set will in general only be Θ-semicomputable and not Θ-computable.

We use the fact 1.2.5 that we have pairing on the code set C. First define a successor function

$$x' = M(0, x).$$

This defines x' for $x \in C$—extend to all of A by some suitable convention.

Next define

$$\begin{aligned}
&\mathbf{0} = M(1, 0)\\
&\mathbf{1} = \mathbf{0}' = M(0, M(1, 0))\\
&\mathbf{2} = \mathbf{1}' = M(0, M(0, M(1, 0)))\\
&\ldots
\end{aligned}$$

It is easy to verify that $\mathbf{0} \neq \mathbf{1}$, $\mathbf{1} \neq \mathbf{2}$, etc. Externally we have constructed a successor set $\mathbf{N} = \{\mathbf{0}, \mathbf{1}, \mathbf{2}, \ldots\}$, and $\langle \mathbf{N}, '\rangle$ is isomorphic to the integers and the successor function. But we cannot, in general, restrict $s(x)$ as in (i) above and still preserve its Θ-computability, $\mathbf{N}$ need not be Θ-computable.

Expanding on this possibility we may make a contact with the theory of non-standard models and non-transitive admissible sets and see how they, in fact, fall under the scope of our axioms.

However, our main interest is towards the "hard core" of recursion theory, and we shall freely assume enough coding apparatus to make life easy—even if it is not always strictly necessary (see e.g. Definition 1.5.1 and the following Remark 1.5.2). But if the reader should insist on the extreme generality of Section 1.2 we recommend that he looks at Moldestad [106] to appreciate what pathologies then may obtain.

1.5 Inductively Defined Theories

We shall now retreat a bit from the, perhaps, too great generality of Sections 1.1 to 1.3. From now on we shall assume that our precomputation theories are *singlevalued*, i.e.

$$\text{if} \quad (a, \sigma, z) \in \Theta \quad \text{and} \quad (a, \sigma, w) \in \Theta, \quad \text{then} \quad z = w.$$

We shall further assume that a copy $\langle N, s \restriction N\rangle$ of the integers is included in the computation domain, and that the domain is provided with a coding scheme

$\langle M, K, L\rangle$. Definition 1.1.1 will thus be revised as follows:

1.5.1 Definition. A *computation domain* is a structure

$$\mathfrak{A} = \langle A, C, N; s, M, K, L\rangle,$$

where $N \subseteq C \subseteq A$ and $\langle N, s \restriction N\rangle$ is (an isomorphic copy of) the integers with s as *successor function* (defined as in 1.4.5).

$\langle M, K, L\rangle$ is a *pairing structure* on $\mathfrak{A}$, i.e.

(1) M is a pairing function on A, i.e. M is total and $M(a, b) = M(a_1, b_1)$ implies $a = a_1$ and $b = b_1$.

(2) K and L are inverses to M, i.e. $K(M(a, b)) = a$ and $L(M(a, b)) = b$.

We extend the notational conventions following Definition 1.1.1 using $m, n, k, l, i, j, \ldots$ for elements in N.

1.5.2 Remarks. We shall see that we need a pairing structure on all of A in order to construct a universal function for a given precomputation theory (see Section 1.6.2). For many purposes we need only assume a C-restricted pairing structure, i.e. M is an ordered pair on C with K and L as inverses on C. For the general theory we could get away with a domain of the form $\mathfrak{A} = \langle A, C, N; s\rangle$, since a pairing structure on C exists by 1.2.5. However, as we shall note below, we would have to introduce some *external* pairing in constructing the theory $PR[\mathbf{f}]$, the precomputation theory "generated" by the functions $\mathbf{f}$. Hence, since we almost always have some coding scheme in mind over a given domain, we might as well include a pairing structure $\langle M, K, L\rangle$ as part of the data specifying the computation domain $\mathfrak{A}$.

The new definition of a computation domain necessitates certain changes in the definition of a precomputation theory. We must insist that the successor function s and the pairing structure $\langle M, K, L\rangle$ are Θ-computable. We shall also assume that the codes c_n, $p_{n,j,m}$, and $s_{n,m}$ (see Definition 1.1.10) are Θ-computable mappings of the parameters. This allows for more uniformity in defining computable functions and relations, and is quite essential in the subsequent theory.

The "new" Definition 1.1.10 is as follows:

1.5.3 Definition. Let Θ be a (single-valued) computation set on the domain $\mathfrak{A} = \langle A, C, N; s, M, K, L\rangle$. Θ is called a *precomputation theory* on $\mathfrak{A}$ if there exist Θ-computable mappings p_1, p_2, and p_3 such that

(i) for each n, j $(0 \leqslant j < n)$ and m the functions and functionals $s, M, K, L, DC, \mathbf{C}^n$, and $\mathbf{P}^m_{n,j}$ are Θ-computable with Θ-codes $\hat{s}, m, k, l, d, c_n = p_1(n)$, and $p_{n,j,m} = p_2(j, n, m)$ respectively;

(ii) Θ satisfies the iteration property (1.1.9), i.e. for each $n, m \geqslant 0$ $s_{n,m} = p_3(n, m)$ is a Θ-code for a mapping S_m^n such that for all $a \in C$, all sequences σ from C, and all τ from A,

$$\{a\}_\Theta^{n+m}(\sigma, \tau) = \{S_m^n(a, \sigma)\}_\Theta^m(\tau),$$

where $n = \mathrm{lh}(\sigma)$ and $m = \mathrm{lh}(\tau)$.

The pairing structure $\langle M, K, L\rangle$ will now be extended to a coding and decoding of arbitrary n-tuples. There are many ways of doing this. We do it by means of the following two auxiliary functions:

M^* is defined by iteration of M:

$$M^*(\ ) = 1,$$
$$M^*(x_1, \ldots, x_{n+1}) = M(x_1, M^*(x_2, \ldots, x_{n+1})).$$

By primitive recursion we define functions L_i, $i \in N$:

$$L_1(x) = L(x),$$
$$L_{n+1}(x) = L(L_n(x)).$$

1.5.4 Definition. Ordered n-tuples and inverses are defined as follows:

A $\langle x_1, \ldots, x_n\rangle = M(n, M^*(x_1, \ldots, x_n))$.
B $\mathrm{lh}(x) = K(x)$.
C $(x)_i = K(L_i(x))$, if $i \neq \mathrm{lh}(x)$ or $\mathrm{lh}(x) \notin N$.
$(x)_i = L(L_i(x))$, if $i = \mathrm{lh}(x) \in N$.

1.5.5 Remark. By adapting the arguments of 1.4.2 we see that any precomputation theory Θ has a predecessor function on N. Hence, by 1.4.3, Θ will be closed under primitive recursion on N. Therefore the auxiliary functions M^* and L_i, $i \in N$, and the extended pairing structure of 1.5.4, are Θ-computable for every precomputation theory Θ.

With these preliminaries out of the way we shall, given any sequence

$$\mathbf{f} = f_1, \ldots, f_l,$$

of partial functions, construct a precomputation theory on $\mathfrak{A}$ *generated* by the given list **f**.

1.5.6 Construction of $\Gamma_\mathbf{f}(\Theta)$. Let $\mathbf{f} = f_1, \ldots, f_l$ be a list of functions on A. $\Gamma_\mathbf{f}$ shall be a monotone operator acting on arbitrary sets of tuples (a, σ, z), where $\mathrm{lh}(a, \sigma, z) \geqslant 2$. The definition is by the following set of clauses:

1 $(\langle 1, 0\rangle, x, s(x)) \in \Gamma_\mathbf{f}(\Theta)$
2 $(\langle 2, 0\rangle, x, y, M(x, y)) \in \Gamma_\mathbf{f}(\Theta)$

3 $(\langle 3, 0\rangle, x, K(x)) \in \Gamma_{\mathbf{f}}(\Theta)$
4 $(\langle 4, 0\rangle, x, L(x)) \in \Gamma_{\mathbf{f}}(\Theta)$
5 $(\langle 5, 0\rangle, x, a, b, c, DC(x, a, b, c)) \in \Gamma_{\mathbf{f}}(\Theta)$
6 If $\exists u[(\hat{g}, \sigma, u) \in \Theta$ and $(\hat{f}, u, \sigma, z) \in \Theta]$, then $(\langle 6, 0\rangle, \hat{f}, \hat{g}, \sigma, z) \in \Gamma_{\mathbf{f}}(\Theta)$
7 Let $0 \leqslant j < n$ and τ any m-tuple from A, if $(\hat{f}, x_{j+1}, x_1, \ldots, x_j, x_{j+2}, \ldots, x_n, z) \in \Theta$, then $(\langle 7, j\rangle, \hat{f}, x_1, \ldots, x_n, \tau, z) \in \Gamma_{\mathbf{f}}(\Theta)$
8 If $a, x_1, \ldots, x_n \in C$ and $(a, x_1, \ldots, x_n, y_1, \ldots, y_m, z) \in \Theta$, then $(\langle 8, a, x_1, \ldots, x_n\rangle, y_1, \ldots, y_m, z) \in \Gamma_{\mathbf{f}}(\Theta)$
9 Let $\mathbf{f} = f_1, \ldots, f_l$ be the given list of functions. If $f_i(t_1, \ldots, t_{n_i}) \simeq z$, $1 \leqslant i \leqslant l$, then $(\langle 9, i\rangle, t_1, \ldots, t_{n_i}, z) \in \Gamma_{\mathbf{f}}(\Theta)$.

We note that clauses **1** to **5** introduces the basic functions; the successor function s, the pairing structure $\langle M, K, L\rangle$, and definition by cases DC. s and $\langle M, K, L\rangle$ are part of the data provided by the structure $\mathfrak{A}$. Clauses **6** and **7** introduces the functionals $\mathbf{C}^n$ and $\mathbf{P}^m_{n,j}$, respectively. Clause **8** introduces the iteration property, i.e. the functions S^n_m. Finally, clause **9** introduces the list **f**.

The operator $\Gamma_{\mathbf{f}}$ is monotone and has a least fixed-point $\Theta^*_{\mathbf{f}} = \bigcup \Theta^{\xi}_{\mathbf{f}}$, where $\Theta^{\xi}_{\mathbf{f}} = \Gamma_{\mathbf{f}}(\bigcup_{\eta<\xi} \Theta^{\eta}_{\mathbf{f}})$. In the simple setting of 1.5.6, $\Theta^*_{\mathbf{f}} = \Theta^{\omega}_{\mathbf{f}}$.

1.5.7 Definition. Let $\mathfrak{A}$ be a computation domain and $\mathbf{f} = f_1, \ldots, f_l$ a list of functions of A. The computation set generated by **f** over $\mathfrak{A}$, which will be called the *prime computation set* in **f**, is defined as the least fixed-point of $\Gamma_{\mathbf{f}}$ and denoted by PR[**f**], i.e.

$$\mathrm{PR}[\mathbf{f}] = \Theta^*_{\mathbf{f}} = \text{least fixed-point of } \Gamma_{\mathbf{f}},$$

Associated with the theory PR[**f**] are two notions which will play an important role in our subsequent investigations, *subcomputation* and *length of computation*.

1.5.8 Definition. Let PR[**f**] be a computation set as in Definition 1.5.7. For each $(a, \sigma, z) \in \mathrm{PR}[\mathbf{f}]$ we set

$$|a, \sigma, z|_{\mathrm{PR}[\mathbf{f}]} = \text{least } \xi \text{ such that } (a, \sigma, z) \in \Theta^{\xi}_{\mathbf{f}}.$$

$|a, \sigma, z|_{\mathrm{PR}[\mathbf{f}]}$ is called the *length of the computation* (a, σ, z). For a $\Gamma_{\mathbf{f}}$ as in 1.5.6 the length is a natural number.

1.5.9 Definition. For each $(a, \sigma, z) \in \mathrm{PR}[\mathbf{f}]$ we define the set of *immediate subcomputations.*

(i) If $a = \langle 1, 0\rangle, \langle 2, 0\rangle, \langle 3, 0\rangle, \langle 4, 0\rangle, \langle 5, 0\rangle$, or $\langle 9, 0\rangle$, then (a, σ, z) has no immediate subcomputations.

(ii) $(\langle 6, 0\rangle, \hat{f}, \hat{g}, \sigma, z)$ has $(\hat{g}, \sigma, u)$ and $(\hat{f}, u, \sigma, z)$ as immediate subcomputations. (Note u is uniquely determined.)

(iii) $(\langle 7, j\rangle, \hat{f}, x_1, \ldots, x_n, \tau, z)$ has $(\hat{f}, x_{j+1}, x_1, \ldots, x_j, x_{j+2}, \ldots, x_n, z)$ as immediate subcomputation.

(iv) $(\langle 8, a, x_1, \ldots, x_n\rangle, y_1, \ldots, y_m, z)$ has $(a, x_1, \ldots, x_n, y_1, \ldots, y_m, z)$ as immediate subcomputation.

The *subcomputation* relation is defined as the transitive closure of the immediate subcomputation relation. If (a, σ, z) is a subcomputation of (b, τ, w), we write $(a, \sigma, z) <_{\mathrm{PR}[\mathbf{f}]} (b, \tau, w)$. The *subcomputation tree* of a $(a, \sigma, z) \in \mathrm{PR}[\mathbf{f}]$ is the set of subcomputations of (a, σ, z) with the relation $<_{\mathrm{PR}[\mathbf{f}]}$.

We note that if $(a, \sigma, z) <_{\mathrm{PR}[\mathbf{f}]} (b, \tau, w)$, then $|a, \sigma, z|_{\mathrm{PR}[\mathbf{f}]} < |b, \tau, w|_{\mathrm{PR}[\mathbf{f}]}$, but the converse is not necessarily true.

1.5.10 Proposition. *The prime computation set* PR[**f**] *is a precomputation theory on* $\mathfrak{A}$ *in which each function of the list* **f** *is computable.*

Proofs of facts such as the above proposition tend to be long, tedious, and entirely a matter of routine. They will therefore mostly be omitted. Since this is the first case and since there are a few details to be observed, we shall, however, this time indicate a few of the steps. To show that each f_i in **f** is PR[**f**]-computable, we must find a code $\hat{f}_i$ and verify that

(i) $$f_i(t_1, \ldots, t_{n_i}) \simeq z \quad \text{iff} \quad (\hat{f}_i, t_1, \ldots, t_{n_i}, z) \in \mathrm{PR}[\mathbf{f}].$$

The obvious choice for the code is $\hat{f}_i = \langle 9, i\rangle$. Since $\mathfrak{A}$ has a pairing structure, $\langle 9, i\rangle = M(2, M(9, M(i, 1))) \in C$. The implication from left to right in (i) is an immediate consequence of clause **9** in 1.5.6, the reverse implication must formally be proved by induction on the length of the computation $(\hat{f}, t_1, \ldots, t_{n_i}, z) \in \mathrm{PR}[\mathbf{f}]$.

In a similar way we verify that s, $\langle M, K, L\rangle$, DC, $\mathbf{C}^n$, and $\mathbf{P}^m_{n,j}$ are PR[**f**]-computable and that the codes $c_n = \langle 6, 0\rangle$ and $p_{n,j,m} = \langle 7, j\rangle$ are PR[**f**]-computable mappings of the parameters, the latter is, of course (!), routine to do. The reader is invited to verify that

$$\hat{p}_1 = \langle 8, \langle 7, 3\rangle, \langle 5, 0\rangle, \langle 6, 0\rangle, \langle 6, 0\rangle, 0\rangle,$$

is an appropriate choice of code for the mapping $p_1(n) = \langle 6, 0\rangle$.

This leaves the iteration property: We have to construct a code $\hat{p}_3$ for a mapping $p_3(n, m)$ which itself shall be a code in the theory PR[**f**] for the mapping

$$S^n_m(a, x_1, \ldots, x_n) = \langle 8, a, x_1, \ldots, x_n\rangle.$$

For this we need the extended coding structure 1.5.4. So our task will be to construct codes for various functions there needed. We do this by first verifying that the *set* PR[**f**] has the fixed-point property 1.2.6. Next we construct in PR[**f**] a code for the predecessor function on N, 1.4.2. Finally we adapt the argument of 1.4.3 to show that the *set* PR[**f**] is closed under primitive recursion on N. With this at hand we can, granted the necessary strength of will, construct codes in PR[**f**] for the functions in 1.5.4, from which we easily obtain the code $\hat{p}_3$.

1.5.11 Remarks. (1) The reader should observe that we have repeatedly *used* the

S^n_m-function (*via* clause **8** of 1.5.6) in order to prove that it has an index computable in the theory. (2) We refer back to Remark 1.5.2: It would have been possible to introduce the clauses **1–7** and **9** of 1.5.6 by choosing appropriate elements of the code set C but, without having some sort of pairing mechanism at disposal, it is difficult to see how to introduce the S^n_m-function such that: (i) the code for S^n_m is a computable function of n, m, and (ii) such that the iteration property

$$\{S^n_m(a, x_1, \ldots, x_n)\}(y_1, \ldots, y_m) = \{a\}(x_1, \ldots, x_n, y_1, \ldots, y_m)$$

will obtain. This is the reason for including the pairing structure $\langle M, K, L\rangle$ in Definition 1.5.1.

1.6 A Simple Representation Theorem

We now turn to a converse proposition 1.5.8: Every precomputation theory on $\mathfrak{A}$ is of the form PR[**f**] for a suitable list **f**. In order to state the result precisely we need a notion of *equivalence* for precomputation theories.

1.6.1 Definition. Let Θ and H be two precomputation theories on the same domain $\mathfrak{A}$. We say that H *extends* Θ,

$$\Theta \leqslant H,$$

if there is an H-computable mapping $p(a, n)$ such that

$$(a, \sigma, z) \in \Theta \quad \text{iff} \quad (p(a, n), \sigma, z) \in H,$$

where $n = \mathrm{lh}(\sigma)$.

If $\Theta \leqslant H$ and $H \leqslant \Theta$, we say that Θ and H are *equivalent*, in symbols

$$\Theta \sim H.$$

Equivalent theories have the same computable functions. What 1.6.1 adds is that we have a computable transformation on codes, i.e. if $\Theta \sim H$ and f is Θ-computable with code $\hat{f}_\Theta$, we can inside H compute a code $\hat{f}_H = p(\hat{f}_\Theta, n)$ for f as an H-computable function. It is usually this stronger property we establish when we prove the "equivalence" between given "computation" theories, e.g. the equivalence between Turing computability and μ-recursion.

1.6.2 Universal Function. Implicit in the Θ-computability of the functional $\mathbf{P}^m_{n,j}$ lies the fact that each precomputation theory has an enumerating function, i.e. for each n there is a Θ-computable function f_n such that

$$(a, \sigma, z) \in \Theta \quad \text{iff} \quad f_n(a, \sigma) \simeq z.$$

A code for f_n can be computed in Θ from n. It is therefore possible to code up the functions f_n, $n \in N$, inside Θ to get a Θ-computable function f such that

$$(a, \sigma, z) \in \Theta \quad \text{iff} \quad f_n(a, \sigma) \simeq z$$
$$\text{iff} \quad f(n, \langle a, \sigma\rangle) \simeq z.$$

Construct the theory PR[f]; it is not difficult inside this theory to define a function $p(a, n)$ such that $f(n, \langle a, \sigma\rangle) \simeq z$ iff $\{p(a, n)\}(\sigma) \simeq z$.

We conclude *that every precomputation theory* Θ *is reducible to some theory* PR[f], *where f is a Θ-computable partial function.*

A converse seems at first obvious. If f is Θ-computable, surely PR[f] $\leqslant \Theta$. It is indeed true, but needs a careful proof. We have to analyze the computation procedure given by PR[f] and see that it can be carried out *inside* Θ.

1.6.3 Theorem: Simple Representation. *Let Θ be a precomputation theory on the domain $\mathfrak{A}$. There exists a Θ-computable function f such that $\Theta \sim$ PR[f].*

"Simple" in Theorem 1.6.3 will stand in contrast to "faithful" in Theorem 2.7.3. See 2.7.5 for further comments.

To prove 1.6.3 it is sufficient to show that if f is any Θ-computable function, then PR[f] $\leqslant \Theta$. The proof will be divided into four steps.

A. Program for the Proof. Recall from 1.5.6 and 1.5.7 that PR[f] $= \bigcup_{n<\omega} \Gamma_n$, where $\Gamma_0 = \varnothing$ and $\Gamma_{n+1} = \Gamma_f(\Gamma_n)$. We will construct a Θ-computable (partial) function $r(n, \langle a, \sigma\rangle)$ such that if $r(n, \langle a, \sigma\rangle) \simeq 0$, then $(a, \sigma, z) \in \Gamma_n$, where z is the unique value of the computation $\{a\}^n_{\mathrm{PR}[f]}(\sigma)$. We shall further see to that if $r(n, \langle a, \sigma\rangle)\downarrow$, then $r(m, \langle a, \sigma\rangle)\downarrow$ for all $m < n$. Define

$$q(\langle a, \sigma\rangle) \simeq \mu n[r(n, \langle a, \sigma\rangle) \simeq 0].$$

Since the μ-operator on N is Θ-computable, q is also Θ-computable. We note that if $(a, \sigma, z) \in$ PR[f], then $q(\langle a, \sigma\rangle)$ gives the "first" stage of the inductive generation of PR[f] at which this can be decided.

Our next task will be to define a Θ-computable function $p(n, \langle a, \sigma\rangle)$ which, whenever $(a, \sigma, z) \in \Gamma_n$, calculates z from n and the pair $\langle a, \sigma\rangle$.

Combining p and q shows that

$$(a, \sigma, z) \in \mathrm{PR}[f] \quad \text{iff} \quad p(q(\langle a, \sigma\rangle), \langle a, \sigma\rangle) \simeq z$$
$$\text{iff} \quad (t(a, n), \sigma, z) \in \Theta,$$

where $t(a, n)$ is easily constructed from the line above. We see that $t(a, n)$ is Θ-computable and PR[f] $\leqslant \Theta$ *via* this t.

B. How to Compute Inside PR[f]. To prepare for the construction of r, p and q we illustrate how to decide in Γ_6 if there is some z such that $(\langle 6, 0\rangle, \sigma, z) \in \Gamma_6$. First of all σ must be of the form $\sigma = \hat{f}, \hat{g}, \sigma_1$, where $\hat{f}, \hat{g} \in C$ (this can be decided in Θ). What we so far have been given is an *incomplete* tuple $(\langle 6, 0\rangle, \hat{f}, \hat{g}, \sigma_1, \Box_1)$, where $\Box_1$ is a "blank" to be filled in by some z, if there is indeed some computation $(\langle 6, 0\rangle, \sigma, z)$ which has been put into PR[f] at stage 6, i.e. into Γ_6. Our search may look like the following diagram; i.e. we are trying to construct the subcomputation tree of $(\langle 6, 0\rangle, \hat{f}, \hat{g}, \sigma_1, \Box_1)$:

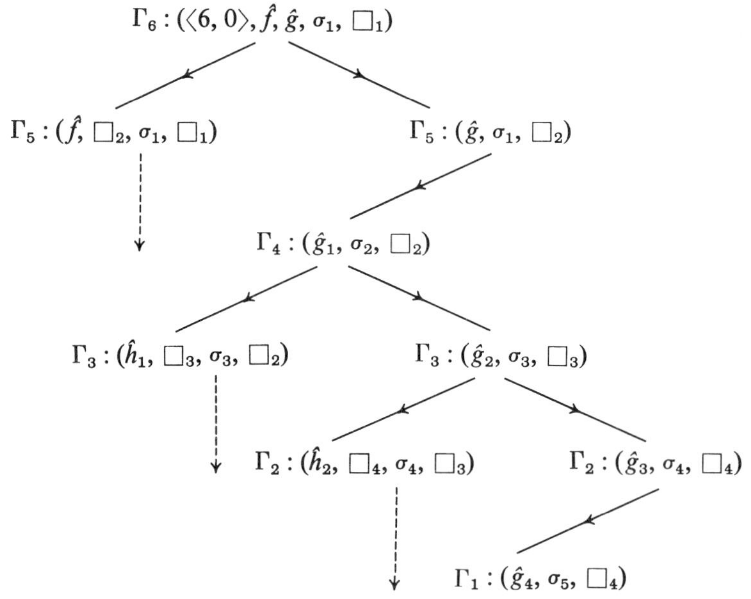

At stage Γ_6 we have the incomplete tuple $(\langle 6, 0\rangle, \hat{f}, \hat{g}, \sigma_1, \Box_1)$. There must be two *immediate subcomputations* $(\hat{g}, \sigma_1, \Box_2)$, where the blank $\Box_1$ has disappeared, but a new blank $\Box_2$ is introduced, and the "doubly indeterminate" tuple $(\hat{f}, \Box_2, \sigma_1, \Box_1)$. We first try to decide $(\hat{g}, \sigma_1, \Box_2)$. Suppose that it has one immediate predecessor, i.e. $\hat{g}$ must be of the form $\langle 7, -\rangle$ or $\langle 8, -\rangle$. Then we can uniquely read off what we have to decide at stage Γ_4, say $(\hat{g}_1, \sigma_2, \Box_2)$. Note that the blank $\Box_2$ is simply carried on. The code $\hat{g}_1$ may again indicate a substitution, which means that at stage Γ_3 we must investigate tuples $(\hat{h}_1, \Box_3, \sigma_3, \Box_2)$ and $(\hat{g}_2, \sigma_3, \Box_3)$. In the latter tuple $\Box_2$ has disappeared, but a new blank $\Box_3$ is introduced. This may again be a substitution, which leads us to investigate tuples $(\hat{h}_2, \Box_4, \sigma_4, \Box_3)$ and $(\hat{g}_3, \sigma_4, \Box_4)$ at stage Γ_2. Suppose that $\hat{g}_3$ is of the form $\langle 7, -\rangle$ or $\langle 8, -\rangle$. Then we must decide some $(\hat{g}_4, \sigma_5, \Box_4)$ at stage Γ_1. Note that everything in the diagram except the blanks is uniquely determined by the given data $\langle 6, 0\rangle$, σ.

Now we can start working backwards: $\hat{g}_4$ must be $\langle 1, 0\rangle$, $\langle 2, 0\rangle$, $\langle 3, 0\rangle$, $\langle 4, 0\rangle$, $\langle 5, 0\rangle$, or $\langle 9, 0\rangle$. Otherwise we can give the answer false to the original question. If $\hat{g}_4$ is either $\langle 1, 0\rangle$, $\langle 2, 0\rangle$, $\langle 3, 0\rangle$, $\langle 4, 0\rangle$, or $\langle 5, 0\rangle$, we can immediately fill in the blank $\Box_4$, as being the value of either s, M, K, L, or DC. If $\hat{g}_4 = \langle 9, 0\rangle$, we must ask Θ to supply the value for $\Box_4$, if such exists, i.e. if σ_5 has the appropriate length and $f(\sigma_5)\downarrow$.

Granted success at stage Γ_1, we move back to Γ_2 and fill in the blank $\square_4$. Then we must try to decide $(\hat{h}_2, \square_4, \sigma_4, \square_3)$. If we succeed in this, we move back to Γ_3 and fill in $\square_3$. Then we must attempt $(\hat{h}_1, \square_3, \sigma_3, \square_2)$. Again the process either succeeds, or we get an answer no, or the process is undefined by asking an inappropriate question about f. But note that the possible undefinabilities turn up in a way uniquely determined by the given data $\langle 6, 0\rangle, \sigma$. So if an undefinability turns up in deciding $(\langle 6, 0\rangle, \sigma, \square_1)$ at stage Γ_n, then it turns up in every later attempt to decide $(\langle 6, 0\rangle, \sigma, \square_1)$, i.e. at every stage Γ_m, $m \geqslant n$.

We now see how to work our way backwards filling in the blanks. And the following conclusion emerges: Either we get an answer YES, i.e. we are able to complete the computation $\{\langle 6, 0\rangle\}(\sigma)$ at stage Γ_6, or we get NO, in which case we may be able to complete $\{\langle 6, 0\rangle\}(\sigma)$ at some later stage, or we are blocked by UNDEFINED, and then the process will not be defined at any later stage.

If we get the answer YES, the process will provide us with the unique z such that $(\langle 6, 0\rangle, \sigma, z) \in \Gamma_6$. And we will then have succeeded in constructing the full subcomputation tree.

This completes our description of how to compute inside PR[f]. It remains to give the formal definitions of p, q, and r.

C. Definition of the Function $p(n, \langle a, \sigma\rangle)$. The function will be defined by induction on n. Since $\Gamma_0 = \varnothing$, we start with $n = 1$.

$n = 1$. In this case we set

$$
\begin{aligned}
&p(0, \langle\langle 1, 0\rangle, x\rangle = s(x)\\
&p(0, \langle\langle 2, 0\rangle, x, y\rangle) = M(x, y)\\
&p(0, \langle\langle 3, 0\rangle, x\rangle) = K(x)\\
&p(0, \langle\langle 4, 0\rangle, x\rangle) = L(x)\\
&p(0, \langle\langle 5, 0\rangle, x, a, b, c\rangle) = DC(x, a, b, c)\\
&p(0, \langle\langle 9, 0\rangle, \sigma\rangle \simeq f(\sigma).
\end{aligned}
$$

In all other cases p is undefined.

$n + 1, n \geqslant 1$. In all cases other than the ones treated below p is undefined.

$a = \langle 1, 0\rangle, \langle 2, 0\rangle, \langle 3, 0\rangle, \langle 4, 0\rangle, \langle 5, 0\rangle$, or $\langle 9, 0\rangle$: Define $p(n + 1, \langle a, \sigma\rangle)$ as in case $n = 1$.

$a = \langle 7, j\rangle$ or $\langle 8, a', x_1, \ldots, x_n\rangle$: Here we have one immediate subcomputation and the value of $p(n + 1, -)$ can be referred back to the value of $p(n, -)$:

$p(n + 1, \langle\langle 7, j\rangle, \hat{f}, \sigma, \tau\rangle) \simeq p(n, \langle \hat{f}, \sigma^j\rangle)$

$p(n + 1, \langle\langle 8, a', x_1, \ldots, x_n\rangle, \tau\rangle) \simeq p(n, \langle a', x_1, \ldots, x_n, \tau\rangle)$.

$a = \langle 6, 0\rangle$: Let the given data be $\langle 6, 0\rangle, \hat{f}, \hat{g}, \sigma$. We set

$p(n + 1, \langle\langle 6, 0\rangle, \hat{f}, \hat{g}, \sigma\rangle) \simeq p(n, \langle \hat{f}, p(n, \langle \hat{g}, \sigma\rangle), \sigma\rangle)$.

We see that we have a well-defined computation procedure in Θ. And we prove, by induction on n, that $(a, \sigma, z) \in \Gamma_n$ iff $p(n, \langle a, \sigma\rangle)\downarrow$ and $p(n, \langle a, \sigma\rangle) \simeq z$.

D. Definition of $r(n, \langle a, \sigma\rangle)$. The definition of $r(n, \langle a, \sigma\rangle)$ is almost identical to the definition of $p(n, \langle a, \sigma\rangle)$. There is, however, one crucial difference. In *B* we were led to three possibilities: Given $\langle a, \sigma\rangle$ and a stage Γ_n we either got the answer YES, the answer NO, or the answer UNDEFINED. In *C* we left $p(n, \langle a, \sigma\rangle)$ undefined if the answer was NO. In the case of $r(n, \langle a, \sigma\rangle)$ we give the value 1 in this case. And we set $r(n, \langle a, \sigma\rangle) = 0$ if the answer is YES. In this way we not only get that $(a, \sigma, z) \in \Gamma_n$ for some (and hence unique) z iff $r(n, \langle a, \sigma\rangle) \simeq 0$; but we also ensure that if $r(n, \langle a, \sigma\rangle) \simeq 0$, then $r(m, \langle a, \sigma\rangle)\downarrow$ for all $m < n$.

This completes the proof of Theorem 1.6.3.

1.7 The First Recursion Theorem

The aim of this section is to prove a version of the first fixed-point theorem for arbitrary precomputation theories. Our strategy will be to use the simple representation Theorem 1.6.3, viz. that any theory Θ is equivalent to a theory of the form PR[g], where g is some partial function on the domain.

First we have some invariance questions to settle:

1.7.1 Proposition. (i) *If* PR[g] $\leqslant$ PR[h] *and* f *is* PR[h]*-computable, then* PR[g,f] $\leqslant$ PR[h].

(ii) PR[f] $\leqslant$ PR[f, h].
(iii) *If* PR[g] $\leqslant$ PR[h], *then* PR[g,f] $\leqslant$ PR[h,f].

(i) and (ii) follows immediately from what we proved in 1.6.3, A–D, i.e. if a function f is Θ-computable, then PR[f] $\leqslant \Theta$. (i) now follows since PR[g] $\leqslant$ PR[h] implies that g is PR[h]-computable; (ii) is equally obvious. (iii) is a corollary of (i) and (ii): g and f will both be PR[h,f]-computable, hence PR[g,f] $\leqslant$ PR[h,f].

1.7.2 Remark. In proving Theorems 1.7.8 and 1.7.9 we need more detailed information about "subcomputations", which easily follows from an analysis of the arguments in 1.6.3, A–D. If f is PR[g]-computable with code $\hat{f}$, 1.7.1 tells us that PR[g,f] $\leqslant$ PR[g] *via* some PR[g]-computable mapping $p(a, n)$. From 1.6.3 A–D it follows that we can safely assume that $p(\langle g, 0\rangle, n) = \hat{f}$, $\langle g, 0\rangle$ being the PR[g,f]-code for f.

We further note that the reduction can be arranged so that *subcomputations* are preserved. This needs some explanation: If $(a, \sigma, z) \in$ PR[**h**], for some functions **h**, we construct the *tree of subcomputations* of (a, σ, z), which is uniquely determined from the tuple (a, σ, z). A tuple (b, τ, w) is a subcomputation of (a, σ, z) if it occurs as a node in this tree (see Definitions 1.5.8 and 1.5.9).

That subcomputations are preserved now means that if (b, τ, w) is a subcomputation of (a, σ, z) in PR[g,f], then $(p(b, n'), \tau, w)$ is a subcomputation of $(p(a, n), \sigma, z)$ in PR[g]. In particular, $|p(b, n'), \tau, w|_{\mathrm{PR}[g]} < |p(a, n), \sigma, z|_{\mathrm{PR}[g]}$.

This is a common feature of reductions. One theory is reduced to or imbedded into another theory by imitating step-by-step the computations of the first inside the second. We shall in a more systematic way return to this in the next section on computation theories.

1.7.3 Definition. Let Θ be a precomputation theory on $\mathfrak{A}$, and assume by 1.6.3 that $\Theta \sim \mathrm{PR}[g]$, for some suitable partial function g. Let f be any function, we set

$$\Theta[f] = \mathrm{PR}[g, f].$$

By 1.7.1 $\Theta[f]$ is unique up to equivalence, i.e. if $\Theta \sim \mathrm{PR}[g]$ and $\Theta \sim \mathrm{PR}[h]$, then $\mathrm{PR}[g, f] \sim \mathrm{PR}[h, f]$.

1.7.4 Definition. Let Θ be a precomputation theory on $\mathfrak{A}$. A functional φ is called *strongly Θ-computable* if there exists a code $\hat{\varphi}$ such that the function $\lambda x \cdot \hat{\varphi}(f, x)$ is $\Theta[f]$-computable with code $\hat{\varphi}$ for all f.

To remove any ambiguity let us agree on the following convention: $\Theta[f]$ is some theory $\mathrm{PR}[g, f]$, where g is such that $\Theta \sim \mathrm{PR}[g]$. We assume that in the construction of $\mathrm{PR}[g, f]$ f has code $\langle 9, 0\rangle$ and g has code $\langle 9, 1\rangle$, see 1.5.6.

1.7.5 Remarks. We observe the following invariance properties:

(i) Let $\Theta \sim H : f$ is Θ-computable iff f is H-computable.
(ii) Let $\Theta \sim H : \varphi$ is strongly Θ-computable iff φ is strongly H-computable.

Here (i) is immediate from the definition. (ii) follows from the following fact: Assume that $\Theta \sim \mathrm{PR}[g]$ and $H \sim \mathrm{PR}[h]$ and that $\Theta \leqslant H$, i.e. $\mathrm{PR}[g] \leqslant \mathrm{PR}[h]$. Hence $\mathrm{PR}[g, f] \leqslant \mathrm{PR}[h, f]$ by some mapping p which is independent of the particular function f, since f has the same "label" in each theory $\mathrm{PR}[g, f]$. But then any strongly Θ-computable φ is also strongly H-computable.

From 1.7.5 we can always assume that Θ has the form $\mathrm{PR}[g]$. Before turning to the fixed-point theorem we shall make a remark on strong versus weak computability.

1.7.6 Proposition. *If φ is strongly* $\mathrm{PR}[g]$*-computable, then φ is weakly* $\mathrm{PR}[g]$*-computable.*

Let $\hat{\varphi}$ be a code for φ as a strongly $\mathrm{PR}[g]$-computable functional, and let f be $\mathrm{PR}[g]$-computable with code $\hat{f}$. From 1.7.1 it follows that $\mathrm{PR}[g, f] \leqslant \mathrm{PR}[g]$ *via* a procedure which is a $\mathrm{PR}[g]$-computable function of $\hat{f}$. So from $\hat{\varphi}$ and $\hat{f}$ we can in $\mathrm{PR}[g]$ compute a code $p(\hat{\varphi}, \hat{f})$ for the function $\lambda x \cdot \varphi(f, x)$. By usual code manipulations we can find a code $\hat{\varphi}_1$ (as a function of $\hat{\varphi}$) such that $\{\hat{\varphi}_1\}(\hat{f}, x) \simeq \{p(\hat{\varphi}, \hat{f})\}(x)$. $\hat{\varphi}_1$ will then serve as a $\mathrm{PR}[g]$-code for φ as a weakly computable functional.

1.7.7 Example. Weak does not imply strong. This follows from simple cardinality considerations. Let Θ be ORT ($=$ ordinary recursion theory over ω). There are countably many strongly computable ORT-functionals, since the code set is countable. But there are uncountably many weakly Θ-computable functionals: Weak computability is a requirement concerning the computable functions, outside this set the functional can behave exactly as it pleases.

We return to the main topic of this section.

1.7.8 Theorem. *Every consistent strongly* PR[g]*-computable functional* φ *has a least fixed-point, i.e. there exists a* f^* *such that*

(i) $\varphi(f^*, x) = f^*(x)$,
(ii) $\varphi(f, x) = f(x)$ *implies that* $f^* \subseteq f$.

The proof is rather standard: Let $f_0 = \varnothing$, $f_{n+1}(x) = \varphi(f_n, x)$, and set

$$f^*(x) = \lim f_n(x).$$

Since φ is strongly PR[g]-computable, φ is consistent, i.e. if $f \subseteq h$ and $\varphi(f, x)\downarrow$, then $\varphi(h, x)\downarrow$ and $\varphi(f, x) = \varphi(h, x)$. From this we see that $f_n \subseteq f_{n+1}$ for all n, hence f^* is well defined.

Proof of (i). First suppose that $f^*(x)\downarrow$, then $f^*(x) = f_{n+1}(x) = \varphi(f_n, x) = \varphi(f^*, x)$, by the consistency of φ. Conversely, suppose that $\varphi(f^*, x)\downarrow$ with value z. This means that $(\hat{\varphi}, x, z) \in \mathrm{PR}[g, f^*]$. f^* has by our convention code $\langle 9, 0\rangle$ in PR[g, f^*]. Let U be the set of pairs $\langle u, v\rangle$ such that $(\langle 9, 0\rangle, u, v)$ is a subcomputation of $(\hat{\varphi}, x, z)$. U is a finite set and there exists some f_n in the approximation to f^* such that $U \subseteq f_n$. It follows that $(\hat{\varphi}, x, z) \in \mathrm{PR}[g, f_n]$, i.e. $\varphi(f_n, x) = z$. We get $\varphi(f^*, x) = \varphi(f_n, x) = f_{n+1}(x) = f^*(x)$.

Proof of (ii). Let f be any fixed-point for φ, we prove by induction that $f_n \subseteq f$, for all n. Hence $f^* = \lim f_n \subseteq f$. For the induction step we assume that $f_{n+1}(x)\downarrow$. Then $f_{n+1}(x) = \varphi(f_n, x) = \varphi(f, x) = f(x)$, where the middle equality follows from the induction hypothesis.

1.7.9 Theorem. *Let* φ *be strongly* PR[g]*-computable. The least fixed-point* f^* *for* φ *is* PR[g]*-computable.*

Proof. In the proof of Proposition 1.7.6 we noted that if f is any PR[g]-computable function with code $\hat{f}$, there is a reduction procedure $\mathrm{PR}[g, f] \leqslant \mathrm{PR}[g]$ which is a PR[g]-computable function of $\hat{f}$. This means, in particular, that there is a PR[g]-computable function p such that $p(\hat{\varphi}, \hat{f})$ is a PR[g]-code for the function $\lambda x \cdot \varphi(f, x)$, where $\hat{\varphi}$ is a code for φ as strongly PR[g]-computable.

We know that a least fixed-point f^* exists. It remains to construct a code for it as a PR[g]-computable function. The idea is simple, we want to construct a code $\widehat{f^*}$ such that $\{\widehat{f^*}\}_{\mathrm{PR}[g]} = \{p(\hat{\varphi}, \widehat{f^*}\}_{\mathrm{PR}[g]}(x)$. The right-hand side is here a PR[g]-

computable function $t(\widehat{f^*}, x)$. We can therefore use the Fixed-point Theorem 1.2.6 to find a code $\widehat{f^*}$ satisfying the equation and with the further property

(i) $$|p(\hat{\varphi}, \widehat{f^*}), x, z|_{\mathrm{PR}[g]} < |\widehat{f^*}, x, z|_{\mathrm{PR}[g]}.$$

The code $\widehat{f^*}$ defines a PR[g]-computable function $\{\widehat{f^*}\}_{\mathrm{PR}[g]}$. We will show that it equals the least fixed-point $\widehat{f^*}$ by verifying: (a) $\{\widehat{f^*}\}$ is a fixed-point for φ; (b) $\{\widehat{f^*}\} \subseteq \widehat{f^*}$.

(a) A simple calculation shows that $\{\widehat{f^*}\}$ is a fixed-point

$$\begin{aligned} \{\widehat{f^*}\}_{\mathrm{PR}[g]}(x) = z \quad &\text{iff} \quad \{p(\hat{\varphi}, \widehat{f^*})\}_{\mathrm{PR}[g]}(x) = z \\ &\text{iff} \quad \{\hat{\varphi}\}_{\mathrm{PR}[g,\{\widehat{f^*}\}]}(x) = z \\ &\text{iff} \quad \varphi(\{\widehat{f^*}\}, x) = z. \end{aligned}$$

(b) Suppose that $\{\widehat{f^*}\}(x) = z$, we must show that $f^*(x) = z$. This we do by induction assuming as induction hypothesis that this holds for all tuples $(\widehat{f^*}, u, v)$ such that

$$|\widehat{f^*}, u, v|_{\mathrm{PR}[g]} < |\widehat{f^*}, x, z|_{\mathrm{PR}[g]}.$$

Let $h \subseteq \{\widehat{f^*}\}$ be the subfunction needed in the given computation $\varphi(\{\widehat{f^*}\}, x) = z$ in PR[$g, \{\widehat{f^*}\}$]. We note that $\varphi(h, x) = z$. Consider now the computation $(p(\hat{\varphi}, \widehat{f^*}), x, z)$ in PR[g], we may conclude from Remark 1.7.2 that

(ii) $$|\widehat{f^*}, u, v|_{\mathrm{PR}[g]} < |p(\hat{\varphi}, \widehat{f^*}), x, z|_{\mathrm{PR}[g]},$$

for all pairs $\langle u, v\rangle \in h$.

This implies that $h \subseteq f^*$: If $h(u) = v$, then $\{\widehat{f^*}\}(u) = v$, and since $|\widehat{f^*}, u, v|_{\mathrm{PR}[g]} < |\widehat{f^*}, x, z|_{\mathrm{PR}[g]}$ (because of (i) and (ii)), the induction hypothesis tells us that $f^*(u) = v$.

From $\varphi(h, x) = z$ and $h \subseteq f^*$, the consistency of φ entails that $\varphi(f^*, x) = z$, i.e. $f^*(x) = z$. We conclude that $\{\widehat{f^*}\} \subseteq f^*$.

1.7.10 Remark. The reader will note that Theorems 1.7.8 and 1.7.9 with their proofs are in essence an adaptation of Kleene's presentation in *Introduction to Metamathematics* [78], §66. Our notion of strong Θ-computability corresponds to his notion of uniformity (see the discussion in §47 of *IMM*). In the next section on computation theories we will discuss a version of the first recursion theorem for weakly Θ-computable functionals.

1.7.11 Remark. Theorem 1.7.8 constructs the least fixed-point. The construction of Theorem 1.2.6 does not always lead to a "least" or "unique" solution. Consider the function

$$f(a, \sigma) = a,$$

in ORT over ω. Let a_0 be the fixed-point calculated following the proof of 1.2.6. Let a_1 be the fixed-point of $f'(a, \sigma) = 1 \cdot a$. We note that $a_0 \neq a_1$, that $\{a_0\}$ and $\{a_1\}$ are both total functions and fixed-points of f. But for no input σ is $\{a_0\}(\sigma) = \{a_1\}(\sigma)$.

If, however, a occurs in $f(a, \sigma)$ only through a part $\{a\}$, then the procedure of 1.2.6 is known to lead to the least fixed-point.

1.7.12 Remark. The material in this chapter is the basic core of any exposition of general recursion theory. The sources for the various combinatorial tricks are lost in the ancient history of recursion theory and the λ-calculus.

Our exposition is very much influenced by Kleene [83] and Moschovakis. [112, 113]. We should also mention the work of Strong [166], Wagner [169], and H. Friedman [33]. There is also an unpublished study by Aczel [2] on "enumeration systems" which contains the representation theorem 1.6.3 in the special case of domain ω. The thesis of L. Sasso [145] contains a great deal of material relevant for this part of the theory. In particular, he has the normal form theorem for a function recursive in a partial function.

Chapter 2

General Theory: Subcomputations

This chapter adds a new notion to the general theory, viz. the notion of *subcomputation.* We develop the elementary theory including a general version of the first recursion theorem, and ending up with a representation theorem which is "faithful" in the sense that it preserves the full structure of subcomputations.

2.1 Subcomputations

In Chapter 1 we took as our basic relation

$$\{a\}(\sigma) \simeq z,$$

asserting that the computing device a acting on the input sequence σ gives z as output. We wrote down for the set Θ of all computation tuples (a, σ, z) a set of axioms and were able to derive within this framework a number of results of elementary recursion theory, leading up to a *simple representation theorem* for any such Θ.

However, many arguments from the more advanced parts of recursion theory seem to require an analysis not only of the computation tuple, but of the whole structure of "subcomputations" of a given computation tuple. In fact, such an analysis was involved in the proof of the first recursion Theorem 1.7.9 *via* the representation Theorem 1.6.3 (see Definition 1.5.9).

In his paper *Axioms for computation theories—first draft* [113] Moschovakis emphasized the fact that whatever computations may be, they have assigned a well-defined *length*, which is always an ordinal, finite or infinite. Thus he proposed to add as a further primitive a map from the set Θ of computation tuples to the ordinals, denoting by $|a, \sigma, z|_\Theta$ the ordinal associated with the tuple $(a, \sigma, z) \in \Theta$.

We shall, in addition, abstract another but related aspect of the notion of computation and add as a further primitive a relation between computation tuples

$$(a', \sigma', z') < (a, \sigma, z),$$

which is intended to express that (a', σ', z') is a "subcomputation" of (a, σ, z), i.e. the computation (a, σ, z) depends upon the previous computation (a', σ', z').

2.1.1 Definition. Let $\mathfrak{A} = \langle A, C, N; s, M, K, L\rangle$ be a computation domain. A *computation structure* $\langle \Theta, <_\Theta \rangle$ over $\mathfrak{A}$ is a pair where Θ is a computation set and $<_\Theta$ is a transitive and well-founded relation on Θ.

Remark. The notation $\langle \Theta, <_\Theta \rangle$ is clumsy. Whenever the context permits we shall write $<$ rather than $<_\Theta$ and use the shorter Θ for $\langle \Theta, <_\Theta \rangle$ or $\langle \Theta, < \rangle$.

Note that if $(a, \sigma, z) \in \Theta$, the set

$$S_{(a,\sigma,z)} = \{(a', \sigma', z') : (a', \sigma', z') < (a, \sigma, z)\},$$

is a well-founded transitive set, the set of "subcomputations" of (a, σ, z). $S_{(a,\sigma,z)}$ has an associated ordinal $|a, \sigma, z|_\Theta$, which may be called the "length" of the computation (a, σ, z).

The notion of computable function carries over unchanged to the present setting. In the definition of Θ-computable functional we make an addition.

2.1.2 Definition. Let $\langle \Theta, < \rangle$ be a computation structure on a domain $\mathfrak{A}$. A consistent functional φ is called *weakly Θ-computable* if there exists a $\hat{\varphi} \in C$ such that for all $e_1, \ldots, e_l \in C$ and all sequences $\sigma = (x_1, \ldots, x_n)$ from A we have

(a) $\varphi(\{e_1\}_\Theta^{n_1}, \ldots, \{e_l\}_\Theta^{n_l}, \sigma) \simeq z$ iff $\{\hat{\varphi}\}_\Theta^{l+n}(e_1, \ldots, e_l, \sigma) \simeq z$.

(b) If $\varphi(\{e_1\}_\Theta^{n_1}, \ldots, \{e_l\}_\Theta^{n_l}, \sigma) \simeq z$, then there exist functions $g_1, \ldots, g_l$ such that

(i) $g_1 \subseteq \{e_1\}_\Theta^{n_1}, \ldots, g_l \subseteq \{e_l\}_\Theta^{n_l}$ and $\varphi(\mathbf{g}, \sigma) \simeq z$;

(ii) for all $i = 1, \ldots, l$, if $g_i(t_1, \ldots, t_{n_i}) \simeq u_i$, then $(e_i, t_1, \ldots, t_{n_i}, u_i) < (\hat{\varphi}, e_1, \ldots, e_l, \sigma, z)$.

2.1.3 Remark. We may motivate clause (b) of the above definition by reflecting on how we compute in a theory PR[**f**], where $\mathbf{f} = f_1, \ldots, f_l$ is a list of partial functions, see Definitions 1.5.6–1.5.9. Let $(a, \sigma, z) \in$ PR[**f**] and consider its subcomputation tree. Among the subcomputations of (a, σ, z) are various tuples $(\hat{f}_i, t_1, \ldots, t_{n_i}, u_i)$, where $\hat{f}_i$ is the code in PR[**f**] of f_i and $f_i(t_1, \ldots, t_{n_i}) \simeq u_i$. For each $i = 1, \ldots, l$ let $g_i = \{(t_1, \ldots, t_{n_i}, u_i) : (\hat{f}_i, t_1, \ldots, t_{n_i}, u_i) <_{\mathrm{PR}[\mathbf{f}]} (a, \sigma, z)\}$. Then $g_1, \ldots, g_l$ are subfunctions of $f_1, \ldots, f_l$ encoding all the information necessary to compute $\{a\}(\sigma) \simeq z$, and, moreover, $(a, \sigma, z) \in$ PR[**g**].

We are now ready for the definition of a computation theory.

2.1.4 Definition. A computation structure $\langle \Theta, <_\Theta \rangle$ on the domain $\mathfrak{A}$ is called a *computation theory* on $\mathfrak{A}$ if there exist Θ-computable mappings p_1, p_2, and p_3 such that

(a) the functions s, M, K, L, DC are Θ-computable with Θ-codes $\hat{s}, m, k, l, d$, respectively;

(b) the functionals $\mathbf{C}^n$ and $\mathbf{P}^m_{n,j}$ are weakly Θ-computable with Θ-codes $c_n = p_1(n)$ and $p_{n,j,m} = p_2(n, j, m)$;

(c) Θ satisfies the following iteration property: For all n, m $p_3(n, m)$ is a Θ-code for a mapping $S^n_m(a, x_1, \ldots, x_n)$ such that for all a, σ in C and τ in A, where $\text{lh}(\sigma) = n$ and $\text{lh}(\tau) = m$,

(i) $\{a\}^{n+m}_\Theta(\sigma, \tau) \simeq \{S^n_m(a, \sigma)\}^m_\Theta(\tau)$;
(ii) if $\{a\}^{n+m}_\Theta(\sigma, \tau) \simeq z$, then $(a, \sigma, \tau, z) < (S^n_m(a, \sigma), \tau, z)$.

Note that (c), (ii) is the only point where the definition reads differently from the precomputation case (Definition 1.5.3). And further that the condition enforced axiomatically in (c), (ii) is one which is naturally satisfied for constructed recursion theories, in particular, theories of the type PR[**f**] where $\{a\}(\sigma, \tau) \simeq z$ occurs as a "subcomputation" of $\{S^n_m(a, \sigma)\}(\tau) \simeq z$.

2.1.5 Remark. Elementary results about precomputation theories extend in an obvious fashion to computation theories, in particular, the results of Section 1.2 do so.

2.2 Inductively Defined Theories

In Section 1.5 we constructed from a given list of functions $\mathbf{f} = f_1, \ldots, f_l$ a theory PR[**f**], the prime computation theory generated by the list **f**. Using the Definition 1.5.9 of subcomputation for PR[**f**] it is obvious that the theory there constructed is a computation theory in the sense of Definition 2.1.4.

We will now extend the construction to cover also the case of functionals. *In this section we restrict ourselves to functionals defined only on total functions.* The general case leads into several knotty problems concerning subcomputations in partial objects of higher types.

2.2.1 Construction of $\Gamma_{f,\varphi}(\Theta)$. Let $\mathfrak{A}$ be a computation domain, f a partial function on A, and φ a functional on A defined only on total functions. A monotone operator $\Gamma_{f,\varphi}$ is introduced by the following set of clauses:

1. $(\langle 1, 0\rangle, x, s(x)) \in \Gamma_{f,\varphi}(\Theta)$.
2. $(\langle 2, 0\rangle, x, y, M(x, y)) \in \Gamma_{f,\varphi}(\Theta)$.
3. $(\langle 3, 0\rangle, x, K(x)) \in \Gamma_{f,\varphi}(\Theta)$.
4. $(\langle 4, 0\rangle, x, L(x)) \in \Gamma_{f,\varphi}(\Theta)$.
5. $(\langle 5, 0\rangle\ x, a, b, c, DC(x, a, b, c)) \in \Gamma_{f,\varphi}(\Theta)$.
6. If $\exists u[(\hat{g}, \sigma, u) \in \Theta$ and $(\hat{f}, u, \sigma, z) \in \Theta]$, then $(\langle 6, 0\rangle, \hat{f}, \hat{g}, \sigma, z) \in \Gamma_{f,\varphi}(\Theta)$.
7. Let $0 \leqslant j < n$ and τ any m-tuple from A, if $(\hat{f}, x_{j+1}, x_1, \ldots, x_j, x_{j+2}, \ldots, x_n, z) \in \Theta$, then $(\langle 7, j\rangle, \hat{f}, x_1, \ldots, x_n, \tau, z) \in \Gamma_{f,\varphi}(\Theta)$.
8. If $a, x_1, \ldots, x_n \in C$ and $(a, x_1, \ldots, x_n, y_1, \ldots, y_m, z) \in \Theta$, then $(\langle 8, a, x_1, \ldots, x_n\rangle, y_1, \ldots, y_m, z) \in \Gamma_{f,\varphi}(\Theta)$.
9. If $f(t_1, \ldots, t_n) \simeq z$, then $(\langle 9, 0\rangle, t_1, \ldots, t_n, z) \in \Gamma_{f,\varphi}(\Theta)$.

10. If for all $u \in A$ there is a $v \in A$ such that $(\hat{g}, u, v) \in \Theta$ and $\varphi(\{\hat{g}\}) \simeq z$, then $(\langle 10, 0\rangle, \hat{g}, z) \in \Gamma_{f,\varphi}(\Theta)$.

We note that clauses 1–9 are exactly the same as those in Definition 1.5.6. Clause 10 introduces the functional φ. For partial φ we would instead have a clause

10′. If there exists a partial function g on A such that $\forall u \forall v[g(u) \simeq v \Rightarrow (\hat{g}, u, v) \in \Theta]$ and $\varphi(g) \simeq z$, then $(\langle 10, 0\rangle, \hat{g}, z) \in \Gamma_{f,\varphi}(\Theta)$.

This is actually the format of clauses 6 and 7. For 6 the subfunctions asserted to exist can simply be taken as $\{\langle \sigma, u\rangle\}$ and $\{\langle u, \sigma, z\rangle\}$, respectively.

2.2.2 Remark. As Moschovakis has pointed out in [113], in many connections it may be more natural to use a stronger notion of computable functional. Let φ be a functional acting on one unary function and which is weakly Θ-computable for some theory Θ. Let $g(x, y)$ be a Θ-computable function. Consider the function

$$f(x) = \varphi(\lambda y \cdot g(x, y)).$$

If $A = C$, this is Θ-computable; we have $f(x) = \{\hat{\varphi}\}(S_1^1(\hat{g}, x))$. If $A \neq C$ we may need the following stronger notion: Let $\varphi(f)$ be a consistent unary functional on unary functions. For each n we define φ^n by

$$\varphi^n(f, \sigma) = \varphi(\lambda y \cdot f(\sigma, y)).$$

φ is called *uniformly (weakly) Θ-computable* if there is a Θ-computable mapping $p(n)$ such that for each n, φ^n is weakly Θ-computable with Θ-code $p(n)$. It is immediate how to change clause 10 to accommodate the stronger notion.

2.2.3 Definition. The computation set generated by f and φ over $\mathfrak{A}$, which will be called the *prime computation set* in f and φ and be denoted by PR$[f, \varphi]$, is defined as the least fixed-point of the operator $\Gamma_{f,\varphi}$.

For each $(a, \sigma, z) \in \text{PR}[f, \varphi]$ we set

$$|a, \sigma, z|_{\text{PR}[f,\varphi]} = \text{least } \xi \text{ such that } (a, \sigma, z) \in \Theta_{f,\varphi}^{\xi}.$$

The ordinal number $|a, \sigma, z|_{\text{PR}[f,\varphi]}$ is called *the length of the computation* (a, σ, z).

Note that Definition 2.2.3 makes sense whether we restrict ourselves to φ acting on total functions, i.e. clause 10 of 2.2.1 or allow partial φ as in clause 10′. In either case the notion of length is well defined, even if there is no unique choice of subfunction (as in clauses 6 and 7).

In the next definition we restrict ourselves to functionals acting only on total functions.

2.2.4 Definition. For each $(a, \sigma, z) \in \mathrm{PR}[f, \varphi]$ we introduce the set of *immediate subcomputations.*

(i) If $a = \langle 1, 0\rangle, \langle 2, 0\rangle, \langle 3, 0\rangle, \langle 4, 0\rangle, \langle 5, 0\rangle$, or $\langle 9, 0\rangle$, then (a, σ, z) has no immediate subcomputations.

(ii) $(\langle 6, 0\rangle, \hat{f}, \hat{g}, \sigma, z)$ has $(\hat{g}, \sigma, u)$ and $(\hat{f}, u, \sigma, z)$ as immediate subcomputations. (Note that u is uniquely determined.)

(iii) $(\langle 7, j\rangle, \hat{f}, x_1, \ldots, x_n, \tau, z)$ has $(\hat{f}, x_{j+1}, x_1, \ldots, x_j, x_{j+2}, \ldots, x_n, z)$ as immediate subcomputation.

(iv) $(\langle 8, a, x_1, \ldots, x_n\rangle, y_1, \ldots, y_m, z)$ has $(a, x_1, \ldots, x_n, y_1, \ldots, y_m, z)$ as immediate subcomputation.

(v) $(\langle 10, 0\rangle, \hat{g}, z)$ has as immediate subcomputations the set of all tuples $(\hat{g}, u, v)$, where u runs over the whole domain A. (Note that by clause 10 in 2.2.1 $|\hat{g}, u, v| < |\langle 10, 0\rangle, \hat{g}, z|$ for all u, v.)

The *subcomputation* relation is the transitive closure of the immediate subcomputation relation, and is denoted by $<_{\mathrm{PR}[f,\varphi]}$.

We note that if $(a, \sigma, z) <_{\mathrm{PR}[f,\varphi]} (b, \tau, w)$, then $|a, \sigma, z|_{\mathrm{PR}[f,\varphi]} < |b, \tau, w|_{\mathrm{PR}[f,\varphi]}$ but not conversely.

2.2.5 Proposition. *The prime computation set* PR[f, φ] *is a computation theory on* $\mathfrak{A}$ *in which f is computable and φ is weakly computable.*

We verify that φ is weakly PR[f, φ]-computable with code $\langle 10, 0\rangle$. First we must show that

$$\varphi(\{\hat{g}\}) \simeq z \quad \text{iff} \quad (\langle 10, 0\rangle, \hat{g}, z) \in \mathrm{PR}[f, \varphi].$$

Assume $\varphi(\{\hat{g}\}) \simeq z$. Then $\{\hat{g}\}$ satisfies the premiss of clause 10 with $\Theta = \mathrm{PR}[f, \varphi]$. Thus $(\langle 10, 0\rangle, \hat{g}, z) \in \Gamma_{f,\varphi}(\mathrm{PR}[f, \varphi]) = \mathrm{PR}[f, \varphi]$, as PR[$f$, φ] is a fixed-point for $\Gamma_{f,\varphi}$. For the converse assume that $(\langle 10, 0\rangle, \hat{g}, z) \in \mathrm{PR}[f, \varphi]$. But this is the case only if $\varphi(\{\hat{g}\}) \simeq z$.

The condition on subcomputations follows immediately from clause (v) in Definition 2.2.4.

We make one final construction.

2.2.6 Definition. Let H be computation theory on a domain $\mathfrak{A}$ and let $\mathbf{f}$ and $\boldsymbol{\varphi}$ be lists of functions and functionals over A. We define a theory $H[\mathbf{f}, \boldsymbol{\varphi}]$ on $\mathfrak{A}$ in the following way. Let $\Gamma_{\mathbf{f},\boldsymbol{\varphi}}$ be the operator of Definition 2.2.1. Set

$$\Theta^{\xi} = \Gamma_{\mathbf{f},\boldsymbol{\varphi}}\Bigl(\bigcup_{\eta<\xi} \Theta^{\eta}\Bigr) \cup \{(\langle 11, a\rangle, \sigma, z) : (a, \sigma, z) \in H \text{ and } |a, \sigma, z|_H \leqslant \xi\}$$

We now define

$$H[\mathbf{f}, \boldsymbol{\varphi}] = \bigcup_{\xi} \Theta^{\xi}.$$

The length function for $H[\mathbf{f}, \boldsymbol{\varphi}]$ is defined as in 2.2.3. In defining the subcomputation relation we add the following clause to Definition 2.2.4.

$$(b, \tau, w) <_{H[\mathbf{f},\boldsymbol{\varphi}]} (\langle 11, a\rangle, \sigma, z) \quad \text{iff} \quad b \text{ is of the form } \langle 11, b'\rangle \quad \text{and} \quad (b', \tau, w) <_H (a, \sigma, z).$$

We should expect that $H[f, \varphi]$ is in some suitable sense the "least extension" of H in which f and φ are computable. A first step is to introduce the appropriate modification of Definition 1.6.1.

2.2.7 Definition. Let $\langle \Theta, <_\Theta\rangle$ and $\langle H, <_H\rangle$ be two computation theories on a common domain $\mathfrak{A}$. We say that $\langle H, <_H\rangle$ *extends* $\langle \Theta, <_\Theta\rangle$, in symbols,

$$\langle \Theta, <_\Theta\rangle \leqslant \langle H, <_H\rangle,$$

if there is an H-computable mapping $p(a, n)$ such that

1. $(a, \sigma, z) \in \Theta \quad \text{iff} \quad (p(a, n), \sigma, z) \in H,$

where $n = \mathrm{lh}(\sigma)$, and

2. if $(a, \sigma, z) \in \Theta$ and $(b, \tau, w) <_\Theta (a, \sigma, z)$, then $(p(b, m), \tau, w) <_H (p(a, n), \sigma, z)$.

If $\langle \Theta, <_\Theta\rangle \leqslant \langle H, <_H\rangle$ and $\langle H, <_H\rangle \leqslant \langle \Theta, <_\Theta\rangle$, we say that the theories are equivalent, $\langle \Theta, <_H\rangle \sim \langle H, <_\Theta\rangle$.

Definitions 2.2.6 and 2.2.7 combine to give the following elementary proposition.

2.2.8 Proposition. *Let $\langle \Theta, <_\Theta\rangle$ be a computation theory on a domain $\mathfrak{A}$, and let the lists* $\mathbf{f}$ *and* $\boldsymbol{\varphi}$ *be given. Then $\langle \Theta[\mathbf{f}, \boldsymbol{\varphi}], <_{\Theta[\mathbf{f},\boldsymbol{\varphi}]}\rangle$ is an extension of $\langle \Theta, <_\Theta\rangle$.*

The imbedding map is simply $p(a, n) = \langle 11, a\rangle$. That the extension is the *least* one in which $\mathbf{f}$ and $\boldsymbol{\varphi}$ are computable will be proved in Section 2.6.

2.3 The First Recursion Theorem

The following recursion theorem will play a central role in our theory. In its present form it is due to Moschovakis [113]. The reader may find it instructive to compare the present version with the corresponding result 1.7.9 for precomputation theories.

2.3.1 First Recursion Theorem. *Let $\langle \Theta, <\rangle$ be a computation theory on $\mathfrak{A}$. Let*

$\varphi(f, x)$ be a consistent weakly Θ-computable functional. Let f^* be the least solution of the equation

$$\varphi(f, x) = f(x), \quad \textit{all} \quad x \in A.$$

Then f^* is Θ-computable.

For the proof we first remark that a least solution exists and can be defined inductively as $f^* = \bigcup f^\xi$, where

$$f^\xi(x) = \varphi\Big(\bigcup_{\eta<\xi} f^\eta, x\Big),$$

and $(\bigcup_\eta f^\eta)(x) \simeq z$ iff $\exists\eta[f^\eta(x) \simeq z]$. We will show that $f^*(x)$ is Θ-computable.

First construct a Θ-computable function $g(a, b, x)$ with Θ-code $\hat{g}$ such that for all $a, b \in C$ and all $x \in A$

(i) $g(a, b, x) = \{a\}_\Theta(\{b\}_\Theta(0), x)$,
(ii) if $g(a, b, x) \simeq z$, then $(a, \{b\}_\Theta(0), x, z) <_\Theta (\hat{g}, a, b, x, z)$.

Let $\hat{\varphi}$ be a Θ-code for φ and choose by the second recursion theorem a code $\hat{p}$ such that for all t

$$\{\hat{p}\}_\Theta(t) = S_1^2(\hat{g}, \hat{\varphi}, \hat{p}).$$

Finally, set

$$c = \{\hat{p}\}_\Theta(0) = S_1^2(\hat{g}, \hat{\varphi}, \hat{p}).$$

We will show that

$$f^*(x) \simeq z \quad \text{iff} \quad \{c\}_\Theta(x) \simeq z.$$

We do this by showing: (a) $\{c\}_\Theta$ is a fixed-point for φ, and (b) $\{c\}_\Theta \subseteq f^*$.

(a) Let $\{\hat{\varphi}\}_\Theta(c, x) \simeq z$; by (i) above, noting that $c = \{\hat{p}\}_\Theta(0)$, we get $g(\hat{\varphi}, \hat{p}, x) \simeq z$. Thus $\{S_1^2(\hat{g}, \hat{\varphi}, \hat{p})\}_\Theta(x) \simeq z$, i.e. $\{c\}_\Theta(x) \simeq z$. Running the argument in reverse, we see that $\{c\}_\Theta$ is a fixed-point for φ.

(b) The proof that $\{c\}_\Theta \subseteq f^*$ proceeds by induction on subcomputations. Let $\{c\}_\Theta(x) \simeq z$. This means that $g(\hat{\varphi}, \hat{p}, x) \simeq z$, and hence $\{\hat{\varphi}\}_\Theta(c, x) \simeq z$. By (ii) above

$$(\hat{\varphi}, c, x, z) <_\Theta (\hat{g}, \hat{\varphi}, \hat{p}, x, z).$$

Since $(\hat{g}, \hat{\varphi}, \hat{p}, x, z) <_\Theta (S_1^2(\hat{g}, \hat{\varphi}, \hat{p}), x, z)$ and $c = S_1^2(\hat{g}, \hat{\varphi}, \hat{p})$, we get

$$(\hat{\varphi}, c, x, z) <_\Theta (c, x, z).$$

φ is Θ-computable. Hence there is a partial function $h \subseteq \{c\}_\Theta$ such that $\varphi(h, x) \simeq z$ and $(c, u, v) <_\Theta (\hat{\varphi}, c, x, z)$ for all pairs u, v such that $h(u) \simeq v$. This means that

$$\text{if } h(u) \simeq v, \quad \text{then} \quad \{c\}_\Theta(u) \simeq v \quad \text{and} \quad (c, u, v) <_\Theta (c, x, z).$$

By the induction hypothesis we thus get:

$$\text{if } h(u) \simeq v, \quad \text{then} \quad f^*(u) \simeq v,$$

i.e. $h \subseteq f^*$. The consistency of φ then gives $\varphi(f^*, x) \simeq z$, which by the fixed-point property of f^* further entails that $f^*(x) \simeq z$. This proves that $\{c\}_\Theta \subseteq f^*$.

2.4 Semicomputable Relations

At this point it may be appropriate to include a brief discussion of Θ-computable and Θ-semicomputable relations. For comparison the reader should refer back to the discussion in Section 1.3. We will in later parts deal with the topic in greater depth. Here are a few rather superficial preliminary remarks.

2.4.1 Definition. Let $\langle \Theta, < \rangle$ be a computation theory on a domain $\mathfrak{A}$. A relation $R(\sigma)$ is called *Θ-semicomputable* if there exists a Θ-computable (partial) function f such that

$$R(\sigma) \quad \text{iff} \quad f(\sigma) \simeq 0.$$

A relation $R(\sigma)$ is called *Θ-computable* if there is a Θ-computable mapping f such that

$$R(\sigma) \quad \text{iff} \quad f(\sigma) = 0.$$

We note that the condition $f(\sigma) \simeq 0$ in the definition of semicomputable could have been replaced by the condition $f(\sigma)\downarrow$, i.e. "is defined".

There are (at least) three basic properties a decent notion of semicomputability and computability for relations should satisfy.

A. If $R(x, \sigma)$ is Θ-semicomputable, then so is $\exists x R(x, \sigma)$.
B. If $R(x, \sigma)$ and $S(x, \sigma)$ are Θ-semicomputable, then so is $R(x, \sigma) \vee S(x, \sigma)$.
C. A relation R is Θ-computable iff R and $\neg R$ are both Θ-semicomputable.

In Section 1.3 we showed in Theorem 1.3.4 that these properties hold if the theory in question has *selection operators*. The discussion there carries over without any significant change to the present context.

As we remarked in 1.3 it would be extremely restrictive to require a single-valued selection operator. And we are not inclined to save the situation by making our theories multiple-valued as in 1.3. This would lead to unwanted complications in analyzing the subcomputation relation.

Short of introducing property A axiomatically, there seems to be no way of saving it without a selection operator. The $\exists$-quantifier ranges over the whole domain A.

B can be saved if the Θ-semicomputable relations are closed under $\exists$-quantification over the natural numbers N. The proof is essentially the same as in 1.3.4.

2.4.2 Proposition. *Let Θ be a computation theory on $\mathfrak{A}$ and assume that the Θ-semicomputable relations are closed under existential quantification over N. If both $R(\sigma)$ and $S(\sigma)$ are Θ-semicomputable, then so is $R(\sigma) \vee S(\sigma)$.*

For the proof, let r and s be Θ-codes for R and S, respectively. Let $f(r, s)$ be a Θ-code, Θ-computable in r, s, for a Θ-computable mapping such that $\{f(r, s)\}_\Theta(0) = r$ and $\{f(r, s)\}_\Theta(x) = s$, $x \neq 0$. Then $R(\sigma) \vee S(\sigma)$ iff $\exists n \in N[\{\{f(r, s)\}_\Theta(n)\}(\sigma) \simeq 0]$.

In a similar way we see that C will be saved if Θ admits a selection operator over N.

2.4.3 Definition. Let Θ be a computation theory on $\mathfrak{A}$. An n-ary *selection operator for Θ over N* is an $n + 1$-ary Θ-computable function $q(a, \sigma)$ with Θ-code $\hat{q}$ such that

(i) if $q(a, \sigma)\downarrow$, then $q(a, \sigma) \in N$;
(ii) if $\exists n \in N\{a\}_\Theta(n, \sigma) \simeq 0$, then $q(a, \sigma)\downarrow$ and $\{a\}_\Theta(q(a, \sigma), \sigma) \simeq 0$.
(iii) if $\{a\}_\Theta(n, \sigma) \simeq 0$ and $q(a, \sigma) \simeq n$, then $(a, n, \sigma, 0) <_\Theta (\hat{q}, a, \sigma, n)$.

2.4.4 Proposition. *Let Θ be a computation theory on $\mathfrak{A}$ admitting a selection operator over N. A relation R is Θ-computable iff R and $\neg R$ are Θ-semicomputable.*

The proof is exactly as in 1.3.4. Let r, s be codes for $R, \neg R$, respectively. Let $m(r, s)$ be a code (Θ-computable in r, s) for the Θ-semicomputable relation

$$(\{r\}_\Theta(\sigma) \simeq 0 \wedge t = 0) \vee (\{s\}_\Theta(\sigma) \simeq 0 \wedge t = 1).$$

Using the selection operator we get a Θ-computable *mapping* $f(\sigma) = q(m(r, s), \sigma)$ such that

$$R(\sigma) \quad \text{iff} \quad f(\sigma) = 0.$$

2.4.5 Remark. The hypothesis of Proposition 2.4.4 implies the hypothesis of Proposition 2.4.2 (see 2.4.3 (ii)). In the next chapter we shall study an important class of theories which admit selection operators over N.

2.5 Finiteness

The importance of the notion of *finiteness* in general recursion theory was strongly emphasized by G. Kreisel (see e.g. [89, 91]), and many of our further developments will testify to his insight. In this section we shall, following Moschovakis [113], introduce a notion of finiteness for general computation theories and prove a few basic properties.

In order that our notion of finiteness shall be well behaved we must restrict the class of computation theories somewhat.

2.5.1 Definition. A computation theory $\langle\Theta, <\rangle$ on $\mathfrak{A}$ is called *regular* if

(a) $C = A$.
(b) Θ has selection operators over N.

2.5.2 Remarks. First observe that 2.5.1 (a), *via* definition by cases, implies that the equality relation on A is Θ-computable. Finite theories on two types will not satisfy this condition, but will still admit a reasonable theory of finiteness, see Chapter 4. The infinite theories of Chapter 5 will be regular in the above sense, as will be the finite theories of Chapter 3.

2.5.3 Definition. Let Θ be a computation theory on $\mathfrak{A}$. Let $B \subseteq A$, by the *B-quantifier* on $\mathfrak{A}$ we understand the following consistent functional $\mathbf{E}_B(f)$ defined as

$$\mathbf{E}_B(f) \simeq \begin{cases} 0, & \text{if } \exists x \in B[f(x) \simeq 0] \\ 1, & \text{if } \forall x \in B[f(x) \simeq 1]. \end{cases}$$

The set $B \subseteq A$ is called Θ-*finite* with Θ-canonical code e, if the B-quantifier $\mathbf{E}_B$ is weakly Θ-computable with Θ-code e.

Some of the basic properties of finiteness are given in the next theorem.

2.5.4 Theorem. *Let Θ be a regular computation theory on $\mathfrak{A}$:*

1. *If B is a finite subset of A, then B is Θ-finite.*
2. *If B is Θ-finite, then B is Θ-computable.*
3. *If B is Θ-finite, $D \subseteq B$ and D is Θ-computable, then D is Θ-finite.*
4. *If B is Θ-finite and f is a Θ-computable mapping, then $f[B]$ is Θ-finite.*
5. *If B is Θ-finite and f is a Θ-computable mapping such that for all $x \in B$, $f(x)$ is a Θ-canonical code for a Θ-finite set B_x, then $\bigcup_{x\in B} B_x$ and $\bigcap_{x\in B} B_x$ are Θ-finite.*

We shall briefly indicate the proofs.

1. For simplicity assume that $B = \{y_1, y_2\} \subseteq A$. First find a Θ-code a^* (as a Θ-computable mapping of y_1, y_2) of the Θ-semicomputable relation

$$\{a^*(y_1, y_2)\}_\Theta(\hat{f}, z) \simeq 0 \quad \text{iff} \quad (\{\hat{f}\}(y_1) \simeq 0 \vee \{\hat{f}\}(y_2) \simeq 0) \wedge z = 0 \cdot \vee \cdot$$
$$(\{\hat{f}\}(y_1) \simeq 1 \wedge \{\hat{f}\}(y_2) \simeq 1) \wedge z = 1.$$

(We here use the results of Section 2.4.) Let q be the selection operator on N. We set $\{e\}_\Theta(\hat{f}) = q(a^*(y_1, y_2), \hat{f})$. e will be a Θ-canonical code for $B = \{y_1, y_2\}$ as a Θ-finite set. We first see that

$$\begin{aligned} \mathbf{E}_B(\{\hat{f}\}) \simeq 0 \quad &\text{iff} \quad \exists y \in B[\{\hat{f}\}(y) \simeq 0]. \\ &\text{iff} \quad q(a^*(y_1, y_2), \hat{f}) \simeq 0 \\ &\text{iff} \quad \{e\}_\Theta(\hat{f}) \simeq 0 \end{aligned}$$

Similarly we see that $\mathbf{E}_B(\{\hat{f}\}) \simeq 1$ iff $\{e\}_\Theta.\ (\hat{f}) \simeq 1$.

It remains to verify the subcomputation condition on $\mathbf{E}_B$: Suppose $\mathbf{E}_B(\{\hat{f}\}) \simeq z$, where $z = 0$ or $z = 1$. Let g be the subfunction of $\{\hat{f}\}$ which gives the values of $\{\hat{f}\}$ at the arguments y_1, y_2. By construction $(a^*, \hat{f}, z, 0) <_\Theta (\hat{q}, a^*, \hat{f}, z)$ and $(\hat{q}, a^*, \hat{f}, z) <_\Theta (e, \hat{f}, z)$. By construction of a^* we also see that $(\hat{f}, y_i, v) <_\Theta (a^*, \hat{f}, z, 0)$, whenever $g(y_i) \simeq v$. Combining the inequalities, the result follows.

2. Construct a Θ-computable function $g(x)$ such that

$$\{g(x)\}_\Theta(u) = \begin{cases} 0, & \text{if } u = x \\ 1, & \text{if } u \neq x. \end{cases}$$

Then $f(x) = \mathbf{E}_B(\{g(x)\})$ is a Θ-computable mapping such that $x \in B$ iff $f(x) = 0$.

3. Let f_D be defined as follows

$$f_D(x) \simeq \begin{cases} f(x), & \text{if} \quad x \in D \\ 1, & \text{if} \quad x \notin D. \end{cases}$$

We then observe that $\mathbf{E}_D(f) \simeq \mathbf{E}_B(f_D)$.

4. In this case let

$$\mathbf{E}_{f[B]}(g) = \mathbf{E}_B(g \circ f),$$

where $(g \circ f)(x) \simeq z$ iff $\exists y[f(x) \simeq y \wedge g(y) \simeq z]$.

5. We consider the case $B_0 = \bigcup_{x \in B} B_x$. We have the following equivalences:

$$\begin{aligned} \mathbf{E}_{B_0}(g) \simeq 0 \quad &\text{iff} \quad \exists y \in B_0[g(y) \simeq 0] \\ &\text{iff} \quad \exists x \in B[\exists y \in B_x[g(y) \simeq 0]] \\ &\text{iff} \quad \exists x \in B[\mathbf{E}_{B_x}(g) \simeq 0] \\ &\text{iff} \quad \mathbf{E}_B(\lambda x \cdot \mathbf{E}_{B_x}(g)) \simeq 0. \\ \mathbf{E}_{B_0}(g) \simeq 1 \quad &\text{iff} \quad \forall y \in B_0[g(y) \simeq 1] \\ &\text{iff} \quad \forall x \in B[\forall y \in B_x[g(y) \simeq 1]] \end{aligned}$$

$$\text{iff} \quad \forall x \in B[\mathbf{E}_{B_x}(g) \simeq 1]$$
$$\text{iff} \quad \mathbf{E}_B(\lambda x \cdot \mathbf{E}_{B_x}(g)) \simeq 1.$$

The assumption that we have a Θ-computable mapping f such that $f(x)$ is a Θ-canonical code for B_x, for all $x \in B$, shows that $\mathbf{E}_{B0}$ is Θ-computable.

2.6 Extension of Theories

In Definition 2.2.7 we introduced a notion of Extension, $\Theta \leqslant H$, for computation theories. We remarked in Proposition 2.2.8 that if $\mathbf{f}$ and $\boldsymbol{\varphi}$ are Θ-computable, then $\Theta[\mathbf{f}, \boldsymbol{\varphi}]$, as constructed in 2.2.6, is an extension of Θ. We left open the problem whether $\Theta[\mathbf{f}, \boldsymbol{\varphi}]$ is the least extension of Θ in which $\mathbf{f}$ and $\boldsymbol{\varphi}$ are computable, i.e. if whenever $\mathbf{f}$ and $\boldsymbol{\varphi}$ are H-computable, and H extends Θ, then $\Theta[\mathbf{f}, \boldsymbol{\varphi}] \leqslant H$.

This question has no simple answer for arbitrary consistent partial functionals due to the fact that there is no canonical choice of *subcomputations* for partial objects of higher types. The situation is not problematic if we compute relative to functionals acting only on total objects.

2.6.1 Proposition. *Let $\langle \Theta, <_\Theta \rangle$ be a computation theory on a domain $\mathfrak{A}$ and let* $\mathbf{f}$ *and* $\boldsymbol{\varphi}$ *be given lists of functions and* total *functionals on $\mathfrak{A}$. Let $\langle H, <_H \rangle$ be any extension of Θ in which* $\mathbf{f}$ *and* $\boldsymbol{\varphi}$ *are computable. Let H further satisfy the following condition*

(*) *If φ_i is in the list $\boldsymbol{\varphi}$ and if $\varphi_i(\{\hat{g}\}) \simeq z$ is an H-computation, then $\{\hat{g}\}$ is total and $(\hat{g}, u, v)$ is an H-subcomputation of $(\hat{\varphi}_i, \hat{g}, z)$ for all $u, v \in A$.*

Then there exists an imbedding of $\Theta[\mathbf{f}, \boldsymbol{\varphi}]$ into H.

We shall give the proof in some detail. It is a typical and reasonably complex example of an index transfer theorem based on the second recursion theorem. We have given:

1. $\Theta \leqslant H$ *via* an H-computable mapping $q(a, n)$.
2. $\mathbf{f} = f$ (for simplicity) is H-computable with code $\hat{f}$.
3. $\boldsymbol{\varphi} = \varphi$ is Θ-computable with code $\hat{\varphi}$, and assume for simplicity that φ is unary, i.e. of the form $\varphi(g)$.

By 2.2.6 $\Theta[f, \varphi]$ is inductively constructed using an operator $\Gamma_{f,\varphi}$. We construct the reduction map $p(a, n)$ of $\Theta[f, \varphi]$ to H by cases according to the form of the index a:

(i) If $a = \langle 1, 0\rangle$, $\langle 2, 0\rangle$, $\langle 3, 0\rangle$, $\langle 4, 0\rangle$, or $\langle 5, 0\rangle$, then we set $p(a, n) = a'$, where a' is the corresponding code in H.

(ii) Let $a = \langle 6, 0\rangle$. This is the case of substitution $\varphi(g, h, \sigma) = g(h(\sigma), \sigma)$. We want to have

$$(p(a, n + 2), \hat{g}, \hat{h}, \sigma, z) \in H \quad \text{iff} \quad (a, \hat{g}, \hat{h}, \sigma, z) \in \Theta[f, \varphi].$$

In order to do this we must go through an intermediate stage $(a', p(\hat{g}, n + 1), p(\hat{h}, n), \sigma, z) \in H$, where a' is the fixed H-code for the substitution functional in H.

We first construct in H a mapping $t(a')$ such that

$$\{t(a')\}_H(\hat{p}, \hat{g}, \hat{h}, \sigma) \simeq z \quad \text{iff} \quad \{a'\}_H(\{\hat{p}\}(\hat{g}, n + 1), \{\hat{p}\}(\hat{h}, n), \sigma) \simeq z$$

where $\hat{p}$ is a H-code for the map $p(a, n)$. We can then take

$$p(a, n + 2) = S^1_{n+2}(t(a'), \hat{p}).$$

The construction of $t(a')$ must be such that whenever u is an element such that $\{\{\hat{p}\}(\hat{h}, n)\}_H(\sigma) \simeq u$ and $\{\{\hat{p}\}(\hat{g}, n + 1)\}_H(u, \sigma) \simeq z$, then

$$(\{\hat{p}\}(\hat{h}, n), \sigma, u) \text{ and } (\{\hat{p}\}(\hat{g}, n + 1), u, \sigma, z) <_H (t(a'), \hat{p}, \hat{g}, \hat{h}, \sigma, z).$$

(iii) If $a = \langle 7, 0\rangle$, we proceed as in case (ii).

(iv) The case $a = \langle 8, 0\rangle$ is similar to case (i).

(v) If $a = \langle 9, 0\rangle$, we set $p(a, n) = \hat{f}$, the given H-code for f.

(vi) Let $a = \langle 10, 0\rangle$, here we are in the case of introducing the functional φ. If we have a tuple $(\langle 10, 0\rangle, \hat{g}, z) \in \Theta[f, \varphi]$, this means that for some total function $\{\hat{g}\}$, $\varphi(\{\hat{g}\}) \simeq z$. This is a statement in the theory $\Theta[f, \varphi]$, where the function $\{\hat{g}\}$ occurs "extensionally".

Considered as a statement inside H, we must have $\varphi(\{p(\hat{g}, 1)\}) \simeq z$, since extensionally $\{\hat{g}\}_\Theta = \{p(\hat{g}, 1)\}_H$. This means that $(\hat{\varphi}, p(\hat{g}, 1), z) \in H$ where $\hat{\varphi}$ is the given H-code for φ. We must define $p(\langle 10, 0\rangle, 1)$ such that

$$\begin{aligned} (p(\langle 10, 0\rangle, 1), \hat{g}, z) \in H \quad &\text{iff} \quad (\varphi, p(\hat{g}, 1), z) \in H \\ &\text{iff} \quad (\langle 10, 0\rangle, \hat{g}, z) \in \Theta[f, \varphi]. \end{aligned}$$

It is indeed possible to construct $p(\langle 10, 0\rangle, 1)$ in such a fashion, and moreover, such that

$$(\hat{\varphi}, p(\hat{g}, 1), z) <_H (p(\langle 10, 0\rangle, 1), \hat{g}, z).$$

(This is completely analogous to the construction in case (ii).)

(vii) If $a = \langle 11, a_1\rangle$, we simply set $p(a, n) = q(a_1, n)$, q being the mapping which imbeds Θ into H.

If a is not of one of the forms considered in (i)–(vii), we set $p(a, n) = 0$. An application of the second recursion theorem gives us $p(a, n)$ as an H-computable mapping. It remains to verify that $\Theta[f, \varphi] \leqslant H$ *via* $p(a, n)$.

We have to prove:

(1) If $(a, \sigma, z) \in \Theta[f, \varphi]$, then $(p(a, n), \sigma, z) \in H$.

(2) If $(a_0, \sigma, z) \in H$ and $a_0 = p(a, n)$, then $(a, \sigma, z) \in \Theta[f, \varphi]$.

(3) The subcomputation condition of Definition 2.2.7.

We take case (ii) in the construction of $p(a, n)$ as representative for (1) and (2).

(1) Let $a = \langle 6, 0\rangle$: If some $(\langle 6, 0\rangle, \hat{g}, \hat{h}, \sigma, z) \in \Theta[f, \varphi]$, then there exists a unique u such that $(\hat{h}, \sigma, u), (\hat{g}, u, \sigma, z) \in \Theta[f, \varphi]$. By the induction hypothesis $(p(\hat{h}, n), \sigma, u), (p(\hat{g}, n + 1)\, u, \sigma, z) \in H$. Hence $(a', p(\hat{g}, n + 1), p(\hat{h}, n), \sigma, z) \in H$, which by construction of p gives that $(p(a, n + 2), \hat{g}, \hat{h}, \sigma, z) \in H$.

(2) Conversely, we proceed by induction on the subcomputation relation $<_H$: Assume that we have the premiss of (2) above with $a = \langle 6, 0\rangle$. From the construction of t and the transitivity of $<_H$ we conclude that

$$(p(\hat{h}, n), \sigma, u) \quad \text{and} \quad (p(\hat{g}, n + 1), u, \sigma, z) <_H (a_0, \sigma, z),$$

for a suitable u. The induction hypothesis yields that $(\hat{g}, u, \sigma, z), (\hat{h}, \sigma, u) \in \Theta[f, \varphi]$, hence $(a, \hat{g}, \hat{h}, \sigma, z) \in \Theta[f, \varphi]$.

In order to verify (3) we have to show that if (b, τ, w) and (a, σ, z) are $\Theta[f, \varphi]$-computations and $(b, \tau, u) <_{\Theta[f,\varphi]} (a, \sigma, z)$, then

$$(p(b, m), \tau, w) <_H (p(a, n), \sigma, z).$$

This is proved by induction on the subcomputation relation $<_{\Theta[f,\varphi]}$. We treat case (vi) in the construction of $p(a, n)$ as an example.

Thus let $(\langle 10, 0\rangle, \hat{g}, z) \in \Theta[f, \varphi]$. $\{\hat{g}\}$ is then total and has H-code $p(\hat{g}, 1)$. In this case it suffices to verify that

$$(+) \qquad (p(\hat{g}, 1), u, v) <_H (\hat{\varphi}, p(\hat{g}, 1), z),$$

for all u, v. The conclusion then follows from the way in which $p(\langle 10, 0\rangle, 1)$ was constructed.

Since $(\langle 10, 0\rangle, \hat{g}, z) \in \Theta[f, \varphi]$, the induction hypothesis tells us that $(\hat{\varphi}, p(\hat{g}, 1), z) \in H$. We now invoke condition (*) to conclude that $(p(\hat{g}, 1)\, u, v)$, for all $u \in A$, is an H-subcomputation of $(\hat{\varphi}, p(\hat{g}, 1), z)$, which is the desired conclusion (+).

Remark. There is a stronger notion of imbedding which requires in Definition 2.2.7 that $(b, \tau, w) <_\Theta (a, \sigma, z)$ iff $(p(b, m), \tau, w) <_H (p(a, n), \sigma, z)$. If H were a well-behaved theory, e.g. of the type $\mathrm{PR}[\mathbf{f}, \boldsymbol{\varphi}]$, we could prove Proposition 2.6.1 using this strengthened notion of imbedding. In a general H we lack enough information about $<_H$ to do so.

An immediate corollary is that $\mathrm{PR}[\mathbf{f}, \boldsymbol{\varphi}] \leqslant H$ if every item in the lists $\mathbf{f}$ and $\boldsymbol{\varphi}$ are H-computable in the appropriate sense, in particular, every φ_i in $\boldsymbol{\varphi}$ is total. We further note the following:

2.6.2 Proposition. *Let $\langle \Theta, <_\Theta\rangle$ and $\langle H, <_H\rangle$ be computation theories on the same domain $\mathfrak{A}$. Let $\mathbf{f}$ and $\boldsymbol{\varphi}$ be given lists, where every φ_i in $\boldsymbol{\varphi}$ is total. If $\Theta \leqslant H$, then*

$$\Theta[\mathbf{f}, \boldsymbol{\varphi}] \leqslant H[\mathbf{f}, \boldsymbol{\varphi}].$$

Finally we state the following transitivity (or cut-elimination) lemma.

2.6.3 Proposition. *Let G, H be either partial functions or total functionals on the domain $\mathfrak{A}$, and $\boldsymbol{\varphi}$ a list of total functionals on the same domain. If G is prime computable in H, $\mathbf{f}$, $\boldsymbol{\varphi}$ (i.e. G is* PR$[H, \mathbf{f}, \boldsymbol{\varphi}]$*-computable in the appropriate sense) and H is prime computable in $\mathbf{f}$, $\boldsymbol{\varphi}$, then G is prime computable in $\mathbf{f}$, $\boldsymbol{\varphi}$.*

The reader is invited to construct a proof for him or herself proceeding along the lines of the proof of 2.6.1. Let us also repeat in order to avoid a possible ambiguity that by "total" we mean the same as in the introduction to this section, i.e. we compute relative to functionals acting only on total objects.

2.7 Faithful Representation

We have so far restricted ourselves to functionals acting only on total functions over the domain. The definitions in Section 2.1 do, however, make sense in the more general situation of consistent partial functionals. And the construction in Definition 2.2.1 with clause 10′ replacing 10, as well as Definition 2.2.3 work equally well in the partial case.

It is with the notion of *subcomputations* that difficulties arise. And this is a problem we cannot gloss over by suppressing subcomputations in favor of *length of computations*. In computing in partial objects of higher types there is in general no canonical choice of subcomputations—incompatible subcomputations may serve the same purpose.

The difference between total and partial functionals is brought out clearly by contrasting clauses 10 and 10′ in Definition 2.2.1. At the same time we note that in clauses 6 and 7 there are canonical choices.

One situation where a partial φ leads to problems is when we want to compare its "computability" in different theories. Proposition 2.6.1 is a case in point. Let Θ and H be two theories on a common domain and assume that $\Theta \leqslant H$. From this we would like to conclude that if φ is H-computable, then $\Theta[\varphi] \leqslant H$. But there are difficulties, the subfunction g_H "picked" by the relation $<_H$ to witness a computation $\varphi(g) \simeq z$ in H need not be an extension of the subfunction $g_{\Theta[\varphi]}$ used to secure the computation $\varphi(g) \simeq z$ inside $\Theta[\varphi]$. (Referring to the proof of 2.6.1, it is not necessary that $(\hat{g}, u, v) <_{\Theta[f,\varphi]} (\langle 10, 0\rangle, \hat{g}, z)$ implies that $(p(\hat{g}, 1), u, v) <_H (p(\langle 10, 0\rangle, 1), \hat{g}, z)$.)

Computing in total objects is in an obvious sense *deterministic*, computing in partial objects need not be. In the latter case there is no unique subcomputation tree to verify a computation $\{a\}_\Theta(\sigma) \simeq z$. There is in general a family of subcomputation trees which act as possible *derivations* in Θ of the "theorem" $\{a\}_\Theta(\sigma) \simeq z$.

As this section and other parts of this work will show (see e.g. (4) of Example 3.1.3), there are many cases where we can handle the more general situation, but we have no abstract axiomatic analysis to offer.

2.7.1 Remark. It should be clear that we do not expect a general version of Proposition 2.6.1 for partial functionals. However, a suitable version of the cut-elimination Lemma 2.6.3 is valid for partial functionals and their "derivations".

We now turn to the main topic of this section, a representation theorem for computation theories which is faithful to the notion of subcomputation, i.e. which preserves the complexity of subcomputations.

As always let $\mathfrak{A}$ be a computation domain and $\langle \Theta, < \rangle$ a computation theory on $\mathfrak{A}$. Let $x \in \Theta$. In connection with Definition 2.1.1 we introduced the set of subcomputations

$$\mathbf{S}_x = \{(a, \sigma, z) : (a, \sigma, z) < x\}.$$

We impose the following regularity condition on Θ.

2.7.2 Definition. The theory Θ is called *s-normal* (subcomputation normal) if the sets $\mathbf{S}_x$ are uniformly Θ-finite for $x \in \Theta$.

We spell this out in a bit more detail. Let $\langle x \rangle$ denote the usual code of the finite sequence x. The sets $\mathbf{S}_x$ are uniformly Θ-finite if there exists a Θ-computable mapping p such that whenever $x \in \Theta$, then $p(\langle x \rangle)$ is a Θ-canonical code for $\mathbf{S}_x$ as a Θ-finite set.

We can now state the main result of this section.

2.7.3 Theorem. Faithful Representation. *Let* $\langle \Theta, < \rangle$ *be an s-normal theory on the domain* $\mathfrak{A}$. *There exists a partial weakly* Θ*-computable functional* φ *such that* Θ *is equivalent to* PR$[\varphi]$.

The idea of the proof is rather straightforward. We want to construct φ such that

a. $$\varphi(f, \langle a, \sigma \rangle) \simeq z \quad \text{iff} \quad (a, \sigma, z) \in \Theta \quad \text{and} \quad \forall (b, \tau, w) \in \mathbf{S}_{(a,\sigma,z)}[f(\langle b, \tau \rangle) \simeq w].$$

This φ is seen to be consistent, and if it is weakly Θ-computable, it has, by Theorem 2.3.1 a least fixed-point f_0. We would like to show that f_0 reduces Θ to PR$[\varphi]$. If we construct PR$[\varphi]$ in the correct way, i.e. are careful in introducing the notion of subcomputation for φ, the converse reduction will also follow.

Before going into the details of the proof, we indicate the following part of the reduction $\Theta \leqslant$ PR$[\varphi]$ by f_0.

1. $$(a, \sigma, z) \in \Theta \quad \text{iff} \quad (\hat{f}_0, \langle a, \sigma \rangle, z) \in \text{PR}[\varphi].$$

First, assume that $(a, \sigma, z) \in \Theta$ and that **1** is true for all $(b, \tau, w) < (a, \sigma, z)$. From **a** it then follows that $\varphi(f_0, \langle a, \sigma \rangle) \simeq z$, which by the fixed-point property of f_0, gives $(\hat{f}_0, \langle a, \sigma \rangle, z) \in \text{PR}[\varphi]$. Conversely, if $(\hat{f}_0, \langle a, \sigma \rangle, z) \in \text{PR}[\varphi]$, then $\varphi(f_0, \langle a, \sigma \rangle) \simeq z$, hence $(a, \sigma, z) \in \Theta$.

For the proof we need the following auxiliary construction.

2.7.4 WDC (**Weak Definition by Cases).** Let f and g be functions of the same number of arguments.

$$\mathrm{WDC}(f, g, x, y, \sigma) \simeq \begin{cases} f(\sigma) & \text{if} \quad x = y \\ g(\sigma) & \text{if} \quad x \neq y. \end{cases}$$

This is to be understood in the following sense: If $x = y$, then $\mathrm{WDC}(f, g, x, y, \sigma)\downarrow$ and equal to $f(\sigma)$ iff $f(\sigma)\downarrow$; the question whether $g(\sigma)\downarrow$ is irrelevant. Similarly if $x \neq y$, then $\mathrm{WDC}(f, g, x, y, \sigma)\downarrow$ iff $g(\sigma)\downarrow$.

We construct an index for WDC uniformly in indices for f and g. First let

$$\pi(\hat{f}, \hat{g}, i) = \begin{cases} \hat{f} & \text{if} \quad i = 0 \\ \hat{g} & \text{if} \quad i \neq 0. \end{cases}$$

Next, let $\mathrm{eq}(x, y)$ test equality on the code set C, i.e.

$$\mathrm{eq}(x, y) = \begin{cases} 0 & \text{if} \quad x = y \\ 1 & \text{if} \quad x \neq y. \end{cases}$$

We now see that

$$\mathrm{WDC}(\{\hat{f}\}, \{\hat{g}\}, x, y, \sigma) \simeq \{\pi(\hat{f}, \hat{g}, \mathrm{eq}(x, y))\}(\sigma),$$

and hence we have an index $\hat{w}$ for WDC as a weakly Θ-computable functional.

We now proceed to the proof of Theorem 2.7.3. First let $\hat{g}_0$ be an index for the totally undefined function, i.e. $\{\hat{g}_0\}(\sigma)\uparrow$ for all σ. Next, introduce the following functional

$$(*) \qquad \kappa(f, \langle a, \sigma\rangle) \simeq \begin{cases} 0, & \text{if } \forall\langle b, \tau, z\rangle \in \mathbf{S}_{\langle a,\sigma,-\rangle}(f(\langle b, \tau\rangle) \simeq z) \\ \uparrow, & ow. \end{cases}$$

This definition requires a comment. We compute $\kappa(\{\hat{f}\}, \langle a, \sigma\rangle)$ by starting to compute $\{a\}(\sigma)$. If $\{a\}(\sigma)\downarrow$ and $\simeq z$, we then decide, using s-normality, for each $(b, \tau, w) < (a, \sigma, z)$ whether $\{\hat{f}\}(\langle b, \tau\rangle) \simeq w$.

We construct a code in the following way. First recall that if $(a, \sigma, z) \in \Theta$, then $p(\langle a, \sigma, z\rangle)$ is a Θ-canonical code for the set $\mathbf{S}_{\langle a,\sigma,z\rangle}$. Next let $\hat{f}_1$ be a code computable in $\hat{f}$ such that $\{\hat{f}_1\}(\langle b, \tau\rangle, w) \simeq 1$ iff $\{\hat{f}\}(\langle b, \tau\rangle) \simeq w$. Then we see that

$$\kappa(\{\hat{f}\}, \langle a, \sigma\rangle) \simeq 0 \quad \text{iff} \quad \{p(\langle a, \sigma, \{a\}(\sigma)\rangle)\}(\hat{f}_1) \simeq 1,$$

from which an appropriate code $\hat{\kappa}$ for κ is easily extracted. But $\hat{\kappa}$ constructed above may also make κ defined in cases other than those indicated in (*). This we get rid of by using the totally undefined $\{\hat{g}_0\}$ in WDC as follows,

(**) $\varphi(f, \langle a, \sigma\rangle) \simeq \text{WDC}(\{a\}, \{\hat{g}_0\}, \kappa(f, \langle a, \sigma\rangle), 0, \sigma).$

Using the codes $\hat{w}$ and $\hat{\kappa}$ it is not difficult to construct a code $\hat{\varphi}_1$ such that if f is Θ-computable with code $\hat{f}$, then

(***) $\varphi(\{\hat{f}\}, \langle a, \sigma\rangle) \simeq \{\hat{\varphi}_1\}(\hat{f}, \langle a, \sigma\rangle).$

from which we can conclude that φ is a weakly Θ-computable functional. We may also now prove **a** above. Then apply the first recursion theorem to get a least Θ-computable fixed-point f_0 for φ.

It remains to construct the "correct" version of PR[φ] for proving that $\Theta \sim$ PR[φ], i.e. to define for the set PR[φ] a "correct" notion of immediate subcomputation. Let us use $\hat{\varphi}$ as PR[φ]-code for φ and $\hat{f}_0$ as the PR-code for the least fixed-point for φ.

The notion of subcomputation for φ in PR[φ] can then be determined from condition **a** above. We want an interaction of $\hat{f}_0$ and $\hat{\varphi}$ such that if $(\hat{f}_0, \langle a, \sigma\rangle, z) \in$ PR[φ], then $(\hat{\varphi}, \hat{f}_0, \langle a, \sigma\rangle, z) <_{\text{PR}[\varphi]} (\hat{f}_0, \langle a, \sigma\rangle, z)$. Further, the subfunction h used in the calculation $\varphi(h, \langle a, \sigma\rangle) \simeq z$ shall consist of all tuples $(\hat{f}_0, \langle b, \tau\rangle, w)$ such that $(b, \tau, w) <_\Theta (a, \sigma, z)$.

Remark. The precise clause corresponding to clause 10 of Definition 2.2.1 is

$$\text{If } (\hat{f}, \langle b, \tau\rangle, w) \in \Theta \quad \text{for all} \quad (b, \tau, w) <_\Theta (a, \sigma, z) \quad \text{and} \quad (a, \sigma, z) \in \Theta, \quad \text{then} \quad (\hat{\varphi}, \hat{f}, \langle a, \sigma\rangle, z) \in \Gamma(\Theta^*).$$

From this we determine the "correct" notion of immediate subcomputation replacing clause (v) of Definition 2.2.4.

With this version of PR[φ] we can prove the reduction $\Theta \leqslant$ PR[φ] by f_0.

2. $(b, \tau, w) <_\Theta (a, \sigma, z)$ iff $(\hat{f}_0, \langle b, \tau\rangle, w) <_{\text{PR}[\varphi]} (\hat{f}_0, \langle a, \sigma\rangle, z).$

Assume that $(b, \tau, w) <_\Theta (a, \sigma, z)$. Using the subfunction h above we see that $h(\langle b, \tau\rangle) \simeq w$, hence

$$(\hat{f}_0, \langle b, \tau\rangle, w) <_{\text{PR}[\varphi]} (\hat{\varphi}, \hat{f}_0, \langle a, \sigma\rangle, z) <_{\text{PR}[\varphi]} (\hat{f}_0, \langle a, \sigma\rangle, z).$$

Conversely, we note that if $(\hat{f}_0, \langle b, \tau\rangle, w) <_{\text{PR}[\varphi]} (\hat{f}_0, \langle a, \sigma\rangle, z)$, then also $(\hat{f}_0, \langle b, \tau\rangle, w) <_{\text{PR}[\varphi]} (\hat{\varphi}, \hat{f}_0, \langle a, \sigma\rangle, w)$ by the way things are constructed. Again, by property **a** of φ and the minimality of the fixed-point f_0, this means that $(a, \sigma, z) \in \Theta$ and $(b, \tau, w) <_\Theta (a, \sigma, z)$.

2.7.5 Remark. A weak version of Theorem 2.7.3 was proved by J. Moldestad (unpublished), viz. he obtained a reduction

1′ $(a, \sigma, z) \in \Theta$ iff $(\hat{f}_0, \langle a, \sigma, z\rangle, 0) \in$ PR[φ].

The difference between this result and Theorem 2.7.3 is a bit subtle and merits a comment, it is the difference between a function and its graph. Going from the graph to the function we need some sort of selection operator, and, indeed, in the presence of a selection operator we immediately get 1 from 1′. We were able in 2.7.3 to prove the strong version without assuming a selection operator in the theory.

2.7.6 Remark. A computation theory $\langle \Theta, < \rangle$ on a domain $\mathfrak{A}$ is also a precomputation theory on $\mathfrak{A}$. If we are only interested in the Θ-semicomputable relations on $\mathfrak{A}$, this is adequately represented by the semicomputable relations in some theory PR[**f**], where **f** is a list of partial functions on $\mathfrak{A}$, see Theorem 1.6.3. But this reduction trivializes the computation structure, everything is coded into the functions **f**. What is added in the representation theorem of this section is that by moving up one type, i.e. by considering suitable theories of the type PR[φ], where φ is a consistent partial functional on $\mathfrak{A}$, then we can preserve the structure of subcomputations, in particular, the length function associated with subcomputations.

Part B

Finite Theories

Chapter 3

Finite Theories on One Type

This chapter develops a general theory abstracting the central features of hyperarithmetic theory or, equivalently, recursion in the functional ${}^2E$. "Higher" recursion theory started in the mid 1950's with S. C. Kleene's work on the analytic hierarchy, constructive ordinals, and the hyperarithmetic sets ([79–81]). A number of important contributions were at that time made by C. Spector, to which we shall return later. The theory of functionals and the connection between hyperarithmetic theory and recursion in ${}^2E$ was developed by S. C. Kleene in 1959 ([83]), and the technique of comparing length of computations, the prewellordering property, was introduced by R. Gandy in 1962 ([38]) and used by him to prove the basic selection principle for recursion in a normal type-2 functional.

The early work did not pay proper attention to the notion of finiteness. It was G. Kreisel in [89] and [91] who stressed the importance of this notion for generalized recursion theory (in particular, in the context of meta-recursion theory, see Chapter 5). Y. Moschovakis [113] gave the definition of finiteness which is now considered to be the appropriate one for the general theory (see Definition 2.5.3).

Prewellordering and *finiteness* are the two key concepts in the general theory. In the first section of this chapter we make some general remarks on notions of finiteness, but concentrate mainly on the effect of the prewellordering property for computations. In the second section we single out the important class of Spector theories, which is the "correct" generalization of hyperarithmetic theory. In a final section we tie up the present study with current work on inductive definability.

3.1 The Prewellordering Property

Let $\langle \Theta, < \rangle$ be a computation theory on a domain $\mathfrak{A} = \langle A, C, N; s, M, K, L \rangle$. We shall not in the sequel distinguish between the set of computations Θ and the coded set $\{\langle a, \sigma, z \rangle : (a, \sigma, z) \in \Theta\}$. For $x \in \Theta$ let $|x|_\Theta$ be the ordinal of the computation x, i.e. the ordinal of the set $\mathrm{S}_x = \{y \in \Theta : y < x\}$.

In Section 2.7 we called a theory Θ *s*-normal (subcomputation normal) if the sets S_x are uniformly Θ-finite for $x \in \Theta$. In many cases the related notion of *p*-normality ("*p*" for prewellordering) turns out to be more useful.

3.1.1 Definition. The theory Θ is called *p-normal* if there is a Θ-computable function p such that $p(x, y)\downarrow$ if $x \in \Theta$ or $y \in \Theta$ and

$$\begin{aligned} x \in \Theta \quad \text{and} \quad & |x|_\Theta \leqslant |y|_\Theta \Rightarrow p(x, y) = 0 \\ & |x|_\Theta > |y|_\Theta \Rightarrow p(x, y) = 1. \end{aligned}$$

This means that if $y \in \Theta$, then the set $\{x \in \Theta : |x|_\Theta \leqslant |y|_\Theta\}$ is Θ-computable. It is easy to organize ordinary recursion theory over ω in such a way that the resulting theory is p-normal but not s-normal.

Conversely, s-normality always implies a "weak" form of the prewellordering property: There is a Θ-computable function $q(x, y)$ such that if $x, y \in \Theta$, then

$$q(x, y) = 0 \quad \text{iff} \quad |x|_\Theta \leqslant |y|_\Theta.$$

Since we can computably quantify over finite sets we have the following recursion equation for q

$$|x|_\Theta \leqslant |y|_\Theta \quad \text{iff} \quad \forall x' \in \mathbf{S}_x \, \exists y' \in \mathbf{S}_y |x'|_\Theta \leqslant |y'|_\Theta.$$

If we assume that Θ has selection operators and that there is a Θ-semicomputable extension of the relation $<$ to all tuples (a, σ, z), s-normality implies p-normality. In this case we have the following recursion equations for the function p:

$$\begin{aligned} p(x, y) = 0 \quad & \text{if} \quad \forall x' \in \mathbf{S}_x \quad \exists y'[y' < y \wedge p(x', y') = 0] \\ p(x, y) = 1 \quad & \text{if} \quad \exists x'[x' < x \wedge \forall y' \in \mathbf{S}_y \cdot p(x', y') = 1]. \end{aligned}$$

These assumptions are satisfied in many cases. But we should note that the usual proofs of the existence of selection operators proceed via the prewellordering property.

To conclude this discussion we remark that if the domain A is Θ-finite then p-normality and s-normality lead to essentially the same class of theories, viz. the Spector theories of the next section.

We return to the general discussion. In Definition 2.5.3 we called a set $S \subseteq A$ Θ-finite if the functional

$$\mathbf{E}_S(f) \simeq \begin{cases} 0 & \text{if} \quad \exists x \in S[f(x) \simeq 0] \\ 1 & \text{if} \quad \forall x \in S[f(x) \simeq 1], \end{cases}$$

is weakly Θ-computable. We now introduce a notion of *weak* Θ-finiteness, and refer to the old notion as *strong* Θ-finiteness.

3.1.2 Definition. A set $S \subseteq A$ is called *weakly Θ-finite* if the functional

$$\mathbf{E}'_S(f) \simeq \begin{cases} 0 & \text{if} \quad \forall x \in S[f(x)\downarrow] \wedge \exists x \in S[f(x) \simeq 0] \\ 1 & \text{if} \quad \forall x \in S \quad \exists y \neq 0[f(x) \simeq y] \end{cases}$$

is weakly Θ-computable.

3.1.3 Examples. We list some familiar examples:

(1) Let Θ = ORT on the integers ω. Then Θ is p-normal and a subset $S \subseteq \omega$ is Θ-finite iff it is finite in the ordinary sense. Hence the domain ω is neither strongly nor weakly Θ-finite.

(2) Let $\Theta = \{(a, \sigma, z) : \{a\}(\sigma, F) \simeq z,\ \sigma$ and z integers$\}$ where F is a normal type-2 functional over ω and $\{a\}(\sigma, F) \simeq z$ is defined by Kleene's schemata S1–S9. (See either Kleene's original treatment in [83] or the more general development of recursion in a normal list in Chapter 4.)

Θ is p-normal and the domain is strongly Θ-finite. The proof of p-normality is based on the following recursion equations: (i) if $x \in \Theta$, then $|x| \leqslant |y|$ iff for all subcomputations x' of x there exists a subcomputation y' of y such that $|x'| \leqslant |y'|$; (ii) $|x| > |y|$ iff there exists a subcomputation x' of x such that $|x'| > |y'|$ for all subcomputations y' of y. Normality of F (i.e. the fact that ${}^2E$ is Kleene recursive in F) allows us to "compute" the quantifiers in (i) and (ii) and hence to define a function $p(x, y)$ with the properties of Definition 3.1.1. (For technical hints see the proof of Proposition 3.1.12.)

(3) Let $A = \bigcup_{i<n} \mathrm{Tp}(i)$, where $\mathrm{Tp}(i)$ is the set of objects of type i. Let F be normal of type $\geqslant n + 2$. Let $\Theta = \{(a, \sigma, z) : \{a\}(\sigma, F) \simeq z,\ z \in \omega$ and σ a list of arguments from $A\}$. $\{a\}(\sigma, F) \simeq z$ is again defined by Kleene's formula S1–S9.

Θ is p-normal and for all $i < n$ $\mathrm{Tp}(i)$ is strongly Θ-finite. If F is of type $n + 2$, then $\mathrm{Tp}(n)$ is weakly but not strongly Θ-finite. If F is of type $> n + 2$, then $\mathrm{Tp}(n)$ is strongly Θ-finite. (Proofs can be extracted from Chapter 4, where the basic references for this example are given.)

(4) As remarked in Section 2.7, computing relative to a *partial* type-2 functional is problematic even over the integers ω. Here is one case where we have p-normality. The example is due to P. Hinman [59], see also Aczel [3].

Let $\mathbf{Q}$ be a monotone quantifier, i.e. $\mathbf{Q} \subseteq 2^\omega$ and $A \in \mathbf{Q}$, $A \subseteq B$ implies $B \in \mathbf{Q}$. The dual of $\mathbf{Q}$ is defined by $\breve{\mathbf{Q}} = \{\omega - A : A \notin \mathbf{Q}\}$. Associated with $\mathbf{Q}$ is a partial type-2 functional $F^{\#}_{\mathbf{Q}}$ defined by

$$F^{\#}_{\mathbf{Q}}(f) \simeq \begin{cases} 0 & \text{if } \{x \mid f(x) = 0\} \in \breve{\mathbf{Q}} \\ 1 & \text{if } \{x \mid f(x) > 0\} \in \mathbf{Q}, \end{cases}$$

where f is a partial function. It is easy to see that $\mathbf{E}_\omega$, i.e. the extension of ${}^2E$ to partial objects, is nothing but $F^{\#}_{\{\omega\}}$, i.e. $\mathbf{Q} = \{\omega\}$.

The basic fact now is that recursion in ${}^2E$, $F^{\#}_{\mathbf{Q}}$ is p-normal. For an application see Example 3.3.7 below.

The choice of terminology "weak" versus "strong" is justified by the following simple proposition.

3.1.4 Proposition. *Let $\langle \Theta, < \rangle$ be a computation theory on $\mathfrak{A}$ and $S \subseteq A$. If S is strongly Θ-finite, then S is weakly Θ-finite.*

Proof. The domain of definition of $\mathbf{E}_S$ may be larger than the domain of $\mathbf{E}'_S$. We cut it down to the right size by the following simple trick. Let t be the constant function 1. Then $\mathbf{E}_S(\lambda x\cdot t(f(x)))$ is defined iff f is defined on all of S. And if $\mathbf{E}_S(\lambda x\cdot t(f(x)))\downarrow$, let $\mathbf{E}'_S(f) = \mathbf{E}_S(f)$.

In Section 2.4 we discussed various *closure properties* for the class of Θ-semicomputable relations. We note that if the domain A is strongly Θ-finite, then the class of Θ-semicomputable relations is closed under $\exists$-quantifier. From Section 2.4 we further conclude that if N is strongly Θ-finite, then the Θ-semicomputable relations are closed under $\vee$. The following simple observation shows that weak Θ-finiteness suffices to prove the closure of the Θ-semicomputable relations under $\forall$.

3.1.5 Proposition. *Let the domain A be weakly Θ-finite. $\forall x R(x, \sigma)$ is Θ-semicomputable if $R(x, \sigma)$ is, and an index for $\forall x R(x, \sigma)$ can be found uniformly from an index for R.*

Proof. Let e be an index such that $R(x, \sigma)$ iff $\{e\}(x, \sigma) = 1$, and $\{e\}(x, \sigma)$ is undefined whenever $\neg R(x, \sigma)$. Then $\forall x R(x, \sigma)$ iff $\mathbf{E}'_A(\lambda x\cdot\{e\}(x, \sigma)) = 1$.

We noted in Section 2.4 that if Θ has a selection operator over N, then a relation R is Θ-computable iff $R, \neg R$ are Θ-semicomputable. We shall now prove that p-normality of Θ gives us selection operators over ω, see Definition 2.4.3.

3.1.6 Theorem. *A p-normal theory Θ admits selection operators over N.*

The following proof should be compared with the definition of the μ-operator in 1.4.1. (See also the discussion in Section 1.8.) Let e' be an index (computable from e) such that $\{e'\}(\tau) \simeq 0$ iff $\{e\}(\tau)\downarrow$ and such that $|e', \tau, 0|_\Theta > |e, \tau, y|_\Theta$, where $y \simeq \{e\}(\tau)$. Use the fixed-point theorem to define a function $\{e\}$ by

$$\{e\}(n, a, \sigma) \simeq \begin{cases} n, & \text{if } |a, n, \sigma, 0|_\Theta < |e', n+1, a, \sigma, 0|_\Theta \\ \{e\}(n+1, a, \sigma), & \text{otherwise,} \end{cases}$$

p-normality is used in stating the condition on $\{e\}$. Let

$$q(a, \sigma) \simeq \{e\}(0, a, \sigma).$$

Then q is an n-ary selection operator over N, $n = \mathrm{lh}(\sigma)$. For the proof of this we first note:

i If $\{a\}(n, \sigma)\downarrow$, then $p(\langle a, n, \sigma, 0\rangle, \langle e', n+1, a, \sigma, 0\rangle)\downarrow$. Hence $\{e\}(n, a, \sigma)\downarrow$.
ii If $\{e\}(n+1, a, \sigma)\downarrow$, then $\{e\}(n, a, \sigma)\downarrow$.

From **i** and **ii** it follows that if $\exists n\cdot\{a\}(n, \sigma)\downarrow$, then $q(a, \sigma)\downarrow$.

Suppose next that $q(a, \sigma) \simeq k$. We will prove that $\{a\}(k, \sigma) \simeq 0$. First observe that for some n, $|a, n, \sigma, 0|_\Theta \leqslant |e', n+1, a, \sigma, 0|$; otherwise $\{e\}(n, a, \sigma) \simeq \{e\}(n+1, a, \sigma) \simeq k$ for all n, hence we would get an infinite descending chain $|e, 0, a, \sigma, k|_\Theta > |e, 1, a, \sigma, k|_\Theta > \ldots$.

Let k_0 = least n such that $|a, n, \sigma, 0|_\Theta \leqslant |e', n+1, a, \sigma, 0|_\Theta$. Then $\{a\}(k_0, \sigma) \simeq$

0 and $\{e\}(k_0, a, \sigma) \simeq k_0$. Working backwards we get $k_0 = \{e\}(k_0, a, \sigma) = \ldots = \{e\}(0, a, \sigma) = k$. Thus $\{a\}(k, \sigma) \simeq 0$.

It is possible to arrange the construction so that if $\{a\}(n, \sigma) \simeq 0$ and $q(a, \sigma) \simeq n$, then $(a, n, \sigma, 0) < (\hat{q}, a, \sigma, n)$. In particular, $|a, n, \sigma, 0|_\Theta < |\hat{q}, a, \sigma, n|_\Theta$. This concludes the proof.

3.1.7 Corollary. *Let Θ be p-normal. Then the Θ-semicomputable relations are closed under disjunction and $\exists$-quantification over N. A relation R is Θ-computable iff R and $\neg R$ are Θ-semicomputable.*

3.1.8 Remark. We will not always spell out the uniformity involved in the various constructions. To be explicit in one example: There is a Θ-computable mapping r such that if $R_1(\sigma)$, $R_2(\sigma)$ are Θ-semicomputable with indices e_1, e_2, respectively, and $l = \mathrm{lh}(\sigma)$, then $r(e_1, e_2, l)$ is an index for $R_1(\sigma) \vee R_2(\sigma)$. And if $\{r(e_1, e_2, l)\}(\sigma) \simeq 0$, then

$$|r(e_1, e_2, l), \sigma, 0|_\Theta \geqslant \inf\{|e_1, \sigma, 0|_\Theta, |e_2, \sigma, 0|_\Theta\}.$$

Such extra information is often necessary when one is doing iterated constructions. But we will seldom make the details so explicit. Writing about recursion theory one must try to strike a proper balance between completeness in notations and exposition versus an attention to the mathematical core of an argument.

We shall digress for a moment to discuss further the relationship between strong and weak finiteness. The equivalence between the two notions is tied up with the existence of some sort of selection principle. Over ω we have as a consequence of p-normality the existence of selection operators. As we shall see in the next chapter, in higher types we have only the following *selection principle*: Let B be a non-empty Θ-semicomputable subset of the domain. We can effectively compute from the index of B an index of a Θ-computable non-empty subset $B_0 \subseteq B$. It is precisely this principle which allows us to go from weak to strong finiteness.

We start by formulating the principle more carefully. Let $S \subseteq A$:

(*) *There is a Θ-computable mapping r such that for all z, τ: If $B = \{x \in A : \{z\}(x, \tau) \simeq 0\}$, $B \subseteq S$, and $B \neq \varnothing$, then $\lambda x \cdot \{r(z, \mathrm{lh}(\tau))\}(x, \tau)$ is the characteristic function of a non-empty subset $B' \subseteq B$. If $B = \varnothing$, then $\lambda x \cdot \{r(z, \mathrm{lh}(\tau))\}(x, \tau)$ is totally undefined.*

3.1.9 Remark. If S satisfies the condition (*) it is possible to choose r such that, whenever $B \neq \varnothing$,

$$\inf\{|r(z, \mathrm{lh}(\tau)), x, \tau, 0|_\Theta : x \in B'\} \geqslant \inf\{|z, x, \tau, 0|_\Theta : x \in B\}.$$

3.1.10 Proposition. *Let Θ be p-normal and assume that $A = C$. Then for all $S \subseteq A$, S is strongly Θ-finite iff S is weakly Θ-finite and satisfies condition (*).*

Proof. (1) Suppose that S is strongly Θ-finite. From 3.1.4 we know that S is weakly Θ-finite. It remains to verify (*): Let B be a non-empty Θ-semicomputable subset of S, $B = \{x : \{z\}(x, \tau) \simeq 0\}$. $\mathbf{E}_S$ is Θ-computable and by assumption

$$\mathbf{E}_S(\lambda x\cdot\{z\}(x, \tau)) \simeq 0.$$

Let e_S be a Θ-index for $\mathbf{E}_S$; we may define

$$B' = \{x : |z, x, \tau, 0|_\Theta < |e_S, z_\tau, 0|_\Theta\},$$

where z_τ is an index for $\lambda x\cdot\{z\}(x, \tau)$ computable from z and τ. p-normality shows that B' is Θ-computable, and we may easily construct the function $r(z, n)$ as required by (*).

(2) Let S be weakly Θ-finite and satisfy (*). The functional $\mathbf{E}'_S$ is Θ-computable and S, being weakly finite, is Θ-computable. The following instructions give a procedure for computing $\mathbf{E}_S(\lambda x\cdot\{e\}(x, \tau))$:

Choose an index e' such that

$$\{e'\}(x, \tau) \simeq \begin{cases}\{e\}(x, \tau) & \text{if } x \in S\\ 1 & \text{if } x \notin S.\end{cases}$$

Let $B = \{x : \{e'\}(x, \tau) \simeq 0\}$; $B \subseteq S$ and if $B \neq \varnothing$, then $\lambda x\cdot\{r(e', n)\}(x, \tau)$ is the characteristic function of a non-empty subset $B' \subseteq B$. We now consider the following Θ-semicomputable relation

$$R(t, e, \tau) \text{ iff } (\mathbf{E}'_S(\lambda x\cdot\{r(e', \mathrm{lh}(\tau))\}(x, \tau)) = 0 \wedge t = 0)$$
$$\vee\ (\mathbf{E}'_S(\lambda x\cdot\{e\}(x, \tau)) = 1 \wedge t = 1).$$

By 3.1.6 we have a selection function $q^*(e, \tau)$, and we see that

$$q^*(e, \tau) = 0 \text{ iff } \exists x \in S[\{e\}(x, \tau) \simeq 0],$$

and

$$q^*(e, \tau) = 1 \text{ iff } \forall x \in S\ \exists y \neq 0[\{e\}(x, \tau) \simeq y].$$

We may therefore set $\mathbf{E}_S(\lambda x\cdot\{e\}(x, \tau)) = q^*(e, \tau)$.

3.1.11 Corollary. *If Θ is p-normal, then N is strongly Θ-finite iff it is weakly Θ-finite.*

The proof is immediate since p-normality of Θ gives us selection operators over N, hence the validity of (*).

Let us at this point make the following *methodological remark*: From the above corollary we see that it does not really matter whether we use $\mathbf{E}_N$ or $\mathbf{E}'_N = {}^2E$ in defining recursion over N.

We shall include one more "useful" technical result. Let Θ be a computation theory on a domain $\mathfrak{A}$ and $\mathbf{R} = R_1, \ldots, R_n$ a list of relations on A. As in Definition 2.2.1 we can construct a theory $\Theta[\mathbf{E}_A, \mathbf{R}]$, where $\mathbf{E}_A$ is the strong quantifier on A.

Since $\mathbf{E}_A$ is a consistent partial functional on the domain, it is better to use the length function instead of the subcomputation relation, for reasons discussed in connection with Definition 2.2.3. We therefore assume that we have given a theory $\langle \Theta, |\ \ |_\Theta \rangle$, and we construct the set $\Theta[\mathbf{E}_A, \mathbf{R}]$ with the naturally associated length function.

From the general theory of Chapter 2 we know that the relations $\mathbf{R}$ are $\Theta[\mathbf{E}_A, \mathbf{R}]$-computable and that the domain A is strongly $\Theta[\mathbf{E}_A, \mathbf{R}]$-finite. Θ is imbeddable in $\Theta[\mathbf{E}_A, \mathbf{R}]$, and the imbedding function $r(a, n)$ can be chosen such that

$$|a, \sigma, z|_\Theta = |r(a, \mathrm{lh}(\sigma)), \sigma, z|_{\Theta[\mathbf{E}_A, \mathbf{R}]},$$

whenever $(a, \sigma, z) \in \Theta$.

We add the following complement to these results.

3.1.12 Proposition. *If Θ is p-normal, then so is $\Theta[\mathbf{E}_A, \mathbf{R}]$.*

Note that if $\Theta = \mathrm{ORT}$ over ω, then this result shows that Kleene-recursion in ${}^2E$ over ω is a p-normal theory, hence by Theorem 3.1.6 has selection operators.

The idea behind the proof is described in example (2) of 3.1.3. We must analyze the construction of $\Theta[\mathbf{E}_A, \mathbf{R}]$ from Θ. The function p will be defined by cases via the fixed-point theorem. It will be convenient to omit the argument z from a computation tuple (a, σ, z). This implies no loss of information since our theories are single-valued. We comment below on why the omission of z is convenient, even necessary.

As a typical example let $x = (x_0, \hat{f}, \hat{g}, \sigma)$ be an "abbreviated" substitution and $y = (y_0, h)$ an application of $\mathbf{E}_A$.

We define two auxiliary functions for this case

$$\varphi_1(\hat{p}, x, y) \simeq \mathbf{E}_A(\lambda t \cdot \{\hat{p}\}((\hat{g}, \sigma), (\hat{h}, t)))$$
$$\varphi_2(\hat{p}, x, y) \simeq \mathbf{E}_A(\lambda t \cdot \{\hat{p}\}((\hat{f}, \{\hat{g}\}(\sigma), \sigma), (\hat{h}, t))).$$

(Here is one reason for the abbreviated computation tuple. If in φ_1 we had a subpart $(\hat{g}, \sigma, \{\hat{g}\}(\sigma))$, $\varphi_1(\hat{p}, x, y)$ would be undefined if $\{\hat{g}\}(\sigma)\uparrow$. But $\varphi_1(\hat{p}, x, y)$ shall be defined if y is defined.)

Define a function φ_3 by primitive recursion

$$\varphi_3(0, \hat{p}, x, y) \simeq \varphi_2(\hat{p}, x, y)$$
$$\varphi_3(n + 1, \hat{p}, x, t) \simeq 1.$$

Finally, let

$$\psi(\hat{p}, x, y) \simeq \varphi_3(\varphi_1(\hat{p}, x, y), \hat{p}, x, y).$$

ψ must also be defined for all other possibilities. This done we apply the recursion theorem to obtain a function p with $\Theta[\mathbf{E}_A, \mathbf{R}]$-code $\hat{p}$ such that $p(x, y) \simeq \psi(\hat{p}, x, y)$.

By induction on $\min\{|x|, |y|\}$ one shows that p satisfies the requirement in the definition of p-normality. As an example let us verify that

$$x \in \Theta[\mathbf{E}_A, \mathbf{R}] \wedge |x| \leqslant |y| \Rightarrow p(x, y) = 0,$$

in the case considered above.

From the assumption $x \in \Theta[\mathbf{E}_A, \mathbf{R}]$ we conclude that $(\hat{g}, \sigma), (\hat{f}, \{\hat{g}\}(\sigma), \sigma) \in \Theta[\mathbf{E}_A, \mathbf{R}]$ and have lengths less than $|x|$. By the induction hypothesis

$$p((\hat{g}, \sigma), (\hat{h}, t))\downarrow, \qquad p((\hat{f}, \{\hat{g}\}(\sigma), \sigma), (\hat{h}, t))\downarrow$$

for all t. Since $|x| \leqslant |y|$ we must further have

$$\exists t \cdot |\hat{g}, \sigma| \leqslant |\hat{h}, t| \quad \text{and} \quad \exists t \cdot |\hat{f}, \{\hat{g}\}(\sigma), \sigma| \leqslant |\hat{h}, t|,$$

i.e. $\exists t \cdot p((\hat{g}, \sigma), (\hat{h}, t)) \simeq 0$ and $\exists t \cdot p((\hat{f}, \{\hat{g}\}(\sigma), \sigma), (\hat{h}, t)) \simeq 0$. But then $\varphi_1(\hat{p}, x, y) \simeq 0$ and $\varphi_2(\hat{p}, x, y) \simeq 0$, and

$$p(x, y) \simeq \psi(\hat{p}, x, y) \simeq \varphi_3(0, \hat{p}, x, y) \simeq 0.$$

Thus the proposition is verified in this case.

This concludes our general discussion of the prewellordering property. Theorem 3.1.6 is the important result. The rest are necessary and sometimes useful house-cleaning results. But now on to more substantial matters.

3.2 Spector Theories

The prewellordering property and finiteness of the computation domain come together in the notion of a *Spector theory*. This important class of computation theories was introduced by Y. Moschovakis in [113]. The name was chosen as a tribute to Clifford Spector's many and important contributions to hyperarithmetic theory, of which these theories is an appropriate general version. (We shall, as remarked above, return to Spector's work in connection with the imbedding theorem of Chapter 5.)

In this section we develop some basic "internal" results about Spector theories which lead to a general representation Theorem, 3.2.9. This theorem comes as a natural continuation of the Representation Theorems 1.6.3 and 2.7.3 of Part A and represents a theme which will be taken up again at several points in the further development of the theory, see, in particular, the discussions in Sections 5.4, 7.2, 7.3, and 8.3.

As a preliminary we shall introduce some suitable notations and terminology for the Θ-computable and Θ-semicomputable relations.

3.2.1 Definition. Let Θ be a computation theory on $\mathfrak{A}$.

$\mathrm{sc}^*(\Theta) = \{S \subseteq A$: there exists an index e and constants $a_1, \ldots, a_n \in A$ such that $\lambda x \cdot \{e\}_\Theta(x, a_1, \ldots, a_n)$ is the characteristic function for $S\}$.
$\mathrm{sc}(\Theta) = \{S \subseteq A$: there exists an index e such that $\lambda x \cdot \{e\}_\Theta(x)$ is the characteristic function for $S\}$.

Note, that if the constant functions are computable in Θ, then $\mathrm{sc}^*(\Theta) = \mathrm{sc}(\Theta)$.

$\mathrm{en}^*(\Theta) = \{S \subseteq A$: there exists an index e and constants $a_1, \ldots, a_n \in A$ such that $x \in S$ iff $\{e\}_\Theta(x, a_1, \ldots, a_n) \simeq 0\}$.
$\mathrm{en}(\Theta) = \{S \subseteq A$: there exists an index e such that $x \in S$ iff $\{e\}_\Theta(x) \simeq 0\}$.

One question we may ask is to what extent the *section* $\mathrm{sc}^*(\Theta)$ and the *envelope* $\mathrm{en}^*(\Theta)$ of a theory Θ determines the theory. We show in this section that the envelope determines the theory for the following class.

3.2.2 Definition. Let Θ be a computation theory on the domain $\mathfrak{A}$. Θ is called a *Spector theory* if

(1) $A = C$,
(2) $\mathbf{E}_A$ is Θ-computable, i.e. A is strongly Θ-finite,
(3) Θ is p-normal.

There are a number of remarks to make. First, from (1) it follows that $=_A$ is Θ-computable and that all constant functions are Θ-computable. We could, however, for the results of this section, only require that $=_A$ belongs to $\mathrm{sc}(\Theta)$ and that A might differ from C.

Next, the assumption that A is strongly Θ-finite means that $\mathrm{en}(\Theta)$ is closed under $\wedge$, $\vee$, $\exists_A$, and $\forall_A$.

Finally, p-normality implies that Θ has a selection operator over N. This means that we have a "good" notion of Θ-finiteness and that $\mathrm{sc}(\Theta) = \mathrm{en}(\Theta) \cap \neg\mathrm{en}(\Theta)$, where $\neg\mathrm{en}(\Theta)$ consists of the complements of sets in $\mathrm{en}(\Theta)$.

We now spell out in more detail the properties of $\mathrm{en}(\Theta)$:

3.2.3 Definition. A class Γ of relations on A is called a *Spector class* if it satisfies the following six conditions:

1. $=_A \in \Gamma \cap \neg\Gamma$.

2. Γ is closed under substitution, i.e. if $R(x_1, \ldots, x_i, \ldots, x_n) \in \Gamma$ and $a \in A$, then $R(x_1, \ldots, a, \ldots, x_n) \in \Gamma$.

3. Γ is closed under $\wedge$, $\vee$, $\exists_A$, and $\forall_A$.

4. A has a Γ-*coding scheme*. This means that there is a coding scheme $\langle N, \leqslant, \langle\,\rangle\rangle$ such that $\langle N, \leqslant\rangle$ is isomorphic to the natural numbers with the usual ordering and $\langle\,\rangle$ is an injection of $\bigcup_n A^n \to A$. Associated with $\langle N, \leqslant, \langle\,\rangle\rangle$ is a relation Seq which is the range of $\langle\,\rangle$ and functions lh, q where

$$\mathrm{lh}(x) = \begin{cases} 0, & \text{if } x \notin \mathrm{Seq} \\ n, & \text{if } x = \langle x_1, \ldots, x_n\rangle, \end{cases}$$

and

$$q(x, i) = \begin{cases} 0, & \text{if } \neg(x \in \mathrm{Seq} \wedge i \in N \wedge 1 \leqslant i \leqslant \mathrm{lh}(x)) \\ x_i, & \text{if } x = \langle x_1, \ldots, x_n \rangle \wedge 1 \leqslant i \leqslant n. \end{cases}$$

The coding scheme $\langle N, \leqslant, <> \rangle$ is called a Γ-coding scheme if the relations $x \in N$, $x \leqslant y$, $\mathrm{Seq}(x)$, $\mathrm{lh}(x) = y$, and $q(x, i) = y$ all belong to $\Gamma \cap \neg\Gamma$.

5. Γ is *parametrizable*, i.e. for each n there is an $n + 1$-ary relation $U_n \in \Gamma$ such that if $R(x_1, \ldots, x_n) \in \Gamma$, there is an $a \in A$ such that

$$R(x_1, \ldots, x_n) \quad \text{iff} \quad U_n(a, x_1, \ldots, x_n).$$

6. Γ is *normed*, i.e. every relation $R \in \Gamma$ has a Γ-norm, where $\sigma\colon R \to On$ is a Γ-norm on R if the associated prewellorderings:

$$x \leqslant_\sigma y \quad \text{iff} \quad x \in R \wedge (y \in R \Rightarrow \sigma(x) \leqslant \sigma(y))$$
$$x <_\sigma y \quad \text{iff} \quad x \in R \wedge (y \in R \Rightarrow \sigma(x) < \sigma(y))$$

belong to Γ.

Note in connection with **6** that if $y \in R$ and $\neg(x <_\sigma y)$, then $(y \leqslant_\sigma x)$. We also note that the weak substitution property in **2** is sufficient to show, in conjunction with **1** and **3**, that Γ is closed under "trivial combinatorial substitutions" in the sense of Moschovakis [115, p. 165].

3.2.4 Proposition. *If* Θ *is a Spector theory, then* $\mathrm{en}(\Theta)$ *is a Spector class.*

The proof is immediate from the remarks made in connection with Definition 3.2.2.

3.2.5 Remark. The notion of a Spector class is a natural companion to the notion of a Spector theory, and was introduced by Y. Moschovakis in Chapter 9 of [115], to which we refer the reader for the elementary structure theory of these classes.

Here we only list a few of the classic, but elementary, consequences of the prewellordering property **6** of Definition 3.2.3. Assume that Γ is a Spector class on A.

(A) *Reduction Property:* Let P and Q be sets in Γ, there exist P_1, Q_1 in Γ such that $P_1 \subseteq P$, $Q_1 \subseteq Q$, $P_1 \cap Q_1 = \varnothing$, and $P_1 \cup Q_1 = P \cup Q$.

(B) *Separation Property:* For any disjoint pair of sets P, Q in $\neg\Gamma$ there is an $S \in \Delta = \Gamma \cap \neg\Gamma$ such that $P \subseteq S$ and $S \cap Q = \varnothing$.

(C) *Selection Principle:* Let $R(x, y)$ be in Γ. There is a set $R^* \subseteq R$ in Δ such that

$$\exists y R(x, y) \Rightarrow \exists y R^*(x, y).$$

Note that R^* is not necessarily the graph of a function.

We now prove the converse to Proposition 3.2.4.

3.2.6 Theorem. *Let* Γ *be a Spector class on* A. *There exists a Spector theory* Θ *on* A *such that* $\Gamma = \mathrm{en}(\Theta)$.

The proof should by now be entirely standard and we restrict ourselves to some brief remarks. The theory Θ will be defined by the usual kind of inductive clauses, not forgetting to include the appropriate clauses for the functional $\mathbf{E}_A$.

In this inductive definition the class Γ will be built in as follows: For each $n > 0$, let $U_n \in \Gamma$ be an $n+1$-ary relation enumerating all the n-ary relations in Γ. Let σ_n be a Γ-norm on U_n. The appropriate inductive clause is now

$$(*)\qquad \text{If } (a, \sigma) \in U_n \text{ and } \forall a', \sigma'[(a', \sigma') <_{\sigma_n} (a, \sigma) \Rightarrow (\langle n_0, a'\rangle, \sigma', 0) \in X], \text{ then } (\langle n_0, a\rangle, \sigma, 0) \in \Lambda(X),$$

where Λ is the inductive operator being defined, and n_0 is some suitable natural number index for U_n. Note that this clause allows us to prove in the end that

$$\begin{aligned} (a, \sigma) \in U_n &\text{ iff } (\langle n_0, a\rangle, \sigma, 0) \in \Theta \\ (a', \sigma') <_{\sigma_n} (a, \sigma) &\text{ iff } |\langle n_0, a'\rangle, \sigma', 0| < |\langle n_0, a\rangle, \sigma, 0|. \end{aligned}$$

Θ is now the least fixed-point for Λ, and comes with the length function inherited from the inductive definition.

The inclusion $\Gamma \subseteq \mathrm{en}(\Theta)$ is immediate from the construction of Θ. The converse inclusion follows from the first recursion theorem for Spector classes:

3.2.7 First Recursion Theorem for Spector Classes. *Let* Φ *be a monotone operator and assume that* Γ *is uniformly closed under* Φ. *Then* $\Phi^\infty \in \Gamma$.

This is proved in Moschovakis [114]. Γ is *uniformly closed* under the operator Φ if the relation

$$Q(x, y) \text{ iff } x \in \Phi(\{x' : P(x', y)\}),$$

is in Γ whenever P is in Γ. Since from our point of view a Spector class always is the envelope of some Spector theory and we do have the first recursion theorem for Spector theories (Theorem 2.3.1), we omit the proof of 3.2.7.

With 3.2.7 at hand it is easy to see that the set

$$B = \{x : \mathrm{Seq}(x) \wedge \exists n[n = \mathrm{lh}(x) \wedge x = \langle x_1, \ldots, x_n\rangle \wedge (x_1, \ldots, x_n) \in \Theta]\}$$

is in Γ, hence $\mathrm{en}(\Theta) \subseteq \Gamma$.

The proof of p-normality of Θ follows in the same way as in Proposition 3.1.12. There is, however, one new and important point to observe. In clause (*) we seem to refer to the relation $<_\sigma$ negatively. But this can be circumvented by the remark immediately following Definition 3.2.3, i.e. whenever $(a, \sigma) \in U_n$, then $\neg((a', \sigma') <_{\sigma_n} (a, \sigma))$ can be replaced by $(a, \sigma) \leqslant_{\sigma_n} (a', \sigma')$.

A final remark on the proof: The reader may be worried by the infinity of clauses introduced through the scheme (*), one clause for each U_n, $n = 1, 2, 3, \ldots$. But Γ has a coding scheme, so we may in (*) just take U_1.

This ends our remarks on Theorem 3.2.6.

Proposition 3.2.4 and Theorem 3.2.6 show that Spector theories and Spector classes are for most purposes interchangeable. The following result shows that the Spector class determines the theory up to equivalence.

3.2.8 Theorem. *Let* Θ_1 *and* Θ_2 *be Spector theories on* A. *Then* $\mathrm{en}(\Theta_1) = \mathrm{en}(\Theta_2)$ *iff* $\Theta_1 \sim \Theta_2$.

Equivalence obviously implies equality of envelopes. A proof of the converse can be based on the representation Theorem 2.7.3. Adapted to Spector theories with length instead of subcomputations we now have a Θ-computable functional F such that

$$F(f, \langle a, \sigma\rangle) \simeq z \quad \text{iff} \quad (a, \sigma, z) \in \Theta \wedge \forall(b, \tau, w) \in S^*_{(a,\sigma,z)}[f(\langle b, \tau\rangle) \simeq w],$$

where $(b, \tau, w) \in S^*_{(a,\sigma,z)}$ now means that $|b, \tau, w|_\Theta < |a, \sigma, z|_\Theta$. The same proof now shows that $\Theta \sim \mathrm{PR}[F]$.

Let us now start with Θ_1 and construct the corresponding F. One now shows that since Θ_2 is a Spector theory and $\mathrm{en}(\Theta_1) \subseteq \mathrm{en}(\Theta_2)$, F will also be Θ_2-computable. Hence, $\Theta_1 \leqslant \Theta_2$.

We conclude the discussion of this section by developing a general representation theorem for Spector theories. Since a Spector theory is s-normal we know from Theorem 2.7.3 that it can be written in the form $\mathrm{PR}[F^\#]$, where $F^\#$ is a consistent partial functional of type 2 over the domain. But this result is not entirely satisfactory for not every consistent partial $F^\#$ is s-normal. In the Spector case we can go one step further.

Let us start by analyzing the proof of 2.7.3. In the case of domain ω we define the representing $F^\#$ by the equation

$$F^\#(f, \langle a, \sigma, z\rangle) \simeq 0 \quad \text{iff} \quad (a, \sigma, z) \in \Theta \wedge \forall(b, \tau, w)[(b, \tau, w) \in \mathbf{S}_{(a,\sigma z)} \Rightarrow f(\langle b, \tau, w\rangle) \simeq 0].$$

$F^\#$ is an inductive operator Φ, viz.

(1) $\qquad c \in \Phi(X) \quad \text{iff} \quad c \in \Theta \wedge \forall y[y <_\Theta x \Rightarrow y \in X].$

It is well known that Φ has an associated monotone quantifier $\mathbf{Q}$,

(2) $\qquad X \in \mathbf{Q} \quad \text{iff} \quad \exists a[\exists b \cdot \langle a, b\rangle \in X \wedge a \in \Phi(X_a)],$

where $X_a = \{b : \langle a, b\rangle \in X\}$. Conversely, Φ can be recovered from $\mathbf{Q}$,

(3) $\qquad a \in \Phi(X) \quad \text{iff} \quad \langle a, X\rangle \in \mathbf{Q}.$

According to (4) of Example 3.1.3 we can pass from $\mathbf{Q}$ to a consistent partial functional $F^{\#}_{\mathbf{Q}}$:

$$(4)\qquad F^{\#}_{\mathbf{Q}}(f) \simeq \begin{cases} 0 & \text{if } \{x : f(x) = 0\} \in \breve{\mathbf{Q}} \\ 1 & \text{if } \{x : f(x) > 0\} \in \mathbf{Q}. \end{cases}$$

We know that $\mathrm{PR}[{}^2E, F^{\#}_{\mathbf{Q}}]$ is a Spector theory. Could we not obtain a satisfactory representation theorem for Spector theories by starting with the $F^{\#}_{\mathbf{Q}}$ provided by 2.7.3 and going through the construction above? It is, indeed, trivial to see that from a Spector theory Θ we get a $F^{\#}_{\mathbf{Q}}$ such that $\Theta \leqslant \mathrm{PR}[{}^2E, F^{\#}_{\mathbf{Q}}]$, but we have problems in proving the converse reduction, that $F^{\#}_{\mathbf{Q}}$ is Θ-computable. Our trouble stems from the $\breve{\mathbf{Q}}$-clause in the definition of $F^{\#}_{\mathbf{Q}}$.

Using a construction due to L. Harrington it is possible to get around this difficulty. In the construction of 2.7.3 we cared only about the "positive" part, building up the computation set Θ in stages through the inductive operator $F^{\#} = \Phi$. Now we have to be more careful.

Let Θ be a Spector theory, let $\Gamma = \mathrm{en}(\Theta)$ be the associated Spector class, and let P be universal in Γ, e.g. let P be the coded computation tuples. Define a set

$$R = \{\langle 0, a, b\rangle : b \in P \wedge a \leqslant_{\Theta} b\} \cup \{\langle 1, a, b\rangle : b \in P \wedge b <_{\Theta} a\}.$$

As usual let $R_i = \{\langle a, b\rangle : \langle i, a, b\rangle \in R\}$, $i = 0, 1$. From a norm on P we can introduce a norm $|\ \ |$ on R. We now introduce the following inductive operator Φ,

$$(5)\qquad c \in \Phi(X) \text{ iff } [X_0 \cap X_1 \neq \varnothing] \vee [c \in R \wedge \forall c' \in R(|c'| < |c| \rightarrow c' \in X)].$$

Note the similarity to (1).

From Φ we can construct a functional $F^{\#}_{\mathbf{Q}}$, see (2) and (4). The only difficult point in proving that $\Theta \sim \mathrm{PR}[\mathbf{E}_A, F^{\#}_{\mathbf{Q}}]$ is to verify the Θ-computability of $F^{\#}_{\mathbf{Q}}$. This reduces to an analysis of the $\breve{\mathbf{Q}}$-clause of (4), or, equivalently, to an analysis of the dual operator $\breve{\Phi}$ of Φ. By definition

$$(6)\qquad \begin{aligned} c \in \breve{\Phi}(Y) &\text{ iff } c \notin \Phi(A - Y) \\ &\text{ iff } Y_0 \cup Y_1 = A \wedge [c \notin R \vee \exists c' \in R(|c'| < |c| \wedge c' \in Y)]. \end{aligned}$$

(Recall the meaning of Y_0 and Y_1.)

Let $Y \in \Gamma$ be such that $Y_0 \cup Y_1 = A$. If we can in a Θ-computable way obtain from a code of Y an element $d \in P$ such that

$$(7)\qquad c \in R \wedge \forall c' \in R[|c'| < |c| \rightarrow c' \notin Y] \text{ implies } |c| \leqslant |d|,$$

then we can replace the clause $c \notin R$ in (6) by $c \notin R^{|d|}$ and conclude that $\breve{\Phi}(Y)$ is Θ-semicomputable. This will take care of the $\breve{\mathbf{Q}}$-clause in (4) and prove the following result.

3.2.9 Theorem. *Let Θ be a Spector theory on A. Then Θ is of the form* $\mathrm{PR}[\mathbf{E}_A, F^{\#}_{\mathbf{Q}}]$

for some monotone quantifier **Q** *on* A. *Conversely,* PR$[\mathbf{E}_A, F^{\#}_{\mathbf{Q}}]$ *is always a Spector theory.*

The last part follows from (4) of Example 3.1.3. It remains to prove (7).

Let $c = \langle i, a, b\rangle$ be an element of R satisfying the premiss of (7). Introduce the sets

$$T_0^c = \{\langle a', b'\rangle : a' \leqslant_\Theta b' \leqslant_\Theta b\},$$
$$T_1^c = \{\langle a', b'\rangle : b' <_\Theta b, b' <_\Theta a'\}.$$

Remembering that $Y \in \Gamma$ is such that $Y_0 \cup Y_1 = A$, it is not difficult to verify (i) $T_0^c \subseteq Y_1$, (ii) $T_1^c \subseteq Y_0$, and (iii) $T_0^c \cap T_1^c = \varnothing$. We may now use the reduction property, 3.2.5 (A), to obtain sets X_0, X_1 such that $X_0 \cup X_1 = Y_0 \cup Y_1 = A$, $X_0 \cap X_1 = \varnothing$, and such that

(8) $\qquad T_0^c \subseteq X_0 \subseteq Y_1$ and $T_1^c \subseteq X_1 \subseteq Y_0$.

Note that both X_0, X_1 are Θ-computable and that an index for X_0 can be Θ-effectively computed from an index of Y as a Θ-computable set. Further, note that (8) is true of all c satisfying the premiss of (7). It remains to compute an element from X_0 which can serve as a bound for the conclusion of (7).

To this end introduce a set

$$W(X_0, b_1) = \{\langle a', b'\rangle : \langle b', a'\rangle \notin X_0 \ \wedge\ \langle b', b_1\rangle \in X_0\}.$$

$W(X_0, b_1)$ is Θ-computable, and it is easily seen that if $b_1 <_\Theta b$, then $W(X_0, b_1) = \{\langle a', b'\rangle : a' \leqslant_\Theta b' <_\Theta b_1\}$ is a well-founded set. From this we conclude that

(9) $\qquad |b| \leqslant \sup\{|W(X_0, b_1)| : W(X_0, b_1)$ is well-founded$\}$.

Here we seem to run against a serious obstacle, the notion of well-foundedness is not Θ-computable in every Spector theory Θ; recall that over ω well-foundedness means the Θ-computability of the functional E_1 which is the restriction to total arguments of the functional $E_1^{\#}$ discussed in Example 3.3.7 below.

However, independently of Theorem 3.2.9, we will prove in Section 5.4 that a Spector theory is of the form PR$[{}^2G]$ for a total, normal type-2 G iff Θ is not Θ-Mahlo (see Definition 5.4.6). So we may assume in the proof of Theorem 3.2.9 that Θ is Θ-Mahlo. If this is the case the notion of well-foundedness is weakly Θ-computable. Thus we can compute within Θ from the inequality in (9) an element $d \in P$ giving a suitable bound for (7).

This completes the proof of Theorem 3.2.9. We should perhaps add an explanation as to why well-foundedness can be handled in the Θ-Mahlo case. Given a relation S we can easily construct a consistent partial functional $F_S^{\#}$, uniformly in S, such that if f_S is the least fixed-point of $F_S^{\#}$ then S is well-founded iff $f_S(0) \simeq 0$. (*Hint:* Brouwer-Kleene ordering.) If S is Θ-computable, which is the case if $S = W(W_0, b_1)$, then $F_S^{\#}$ is weakly Θ-computable, hence by Θ-Mahloness $F_S^{\#}$ is

Θ'-computable in some theory $\Theta' <_1 \Theta$ (see Definition 5.4.1). But then the graph f_S of the least fixed-point of $F_S^{\#}$ is Θ-computable, since $\mathrm{en}(\Theta') \subseteq \mathrm{sc}(\Theta)$. This is exactly what is needed to continue from (*g*).

3.2.10 Remark. We have regarded Theorem 3.2.9 as a natural extension of Theorem 2.7.3. Historically, this is not correct. Theorem 3.2.9 is due to L. Harrington who, working in the context of inductive definability and Spector classes, proved that *every Spector class is of the form* IND(**Q**). We see from Example 3.3.7 below that this is equivalent to the computation-theoretic version presented in 3.2.9. Harrington did not publish his proof, we have followed the exposition in A. Kechris [75].

3.3 Spector Theories and Inductive Definability

In recent years there have been several proposals to develop definability theory (descriptive set theory) and generalized recursion theory on the basis of a general theory of inductive definability. Strong advocates for this approach have been Robin Gandy, see e.g. his [39], and Yiannis Moschovakis, see in particular his book [115] and the papers [116] and [117]. The reader should also consult the work of Peter Aczel [4, 6], and [7].

We discussed in the introduction *computations* versus *inductive definability* as a foundation for general recursion theory, and shall not repeat that discussion here. Our aim in this section is to make a connection between computation theories, in particular Spector theories, and inductive definability.

3.3.1 Remark. For a general account of descriptive set theory the reader should consult the recent book of Moschovakis [118]. It follows from the results of this and the next chapter that the theory can be developed in the framework of computation theories, i.e. as finite theories on one or two types.

We assume that the reader is familiar with the basic facts of the theory of inductive definability. However, to fix notation we recall some of the definitions.

Let A be a set and Γ an operator on A, i.e. a map from 2^A to 2^A. Γ defines inductively a set $\Gamma_\infty \subseteq A$ by the following equation, where $\alpha \in On$:

$$\Gamma_\alpha = \bigcup_{\xi<\alpha} \Gamma(\Gamma_\xi),$$

so that $\Gamma_\infty = \bigcup_{\xi\in \mathrm{On}} \Gamma_\xi$. The Γ_α's are called the *stages* of the inductive definition, and

$$|\Gamma| = \text{least } \alpha(\Gamma_{\alpha+1} = \Gamma_\alpha),$$

is called *the ordinal of the inductive definition*. The operator Γ is called *monotone* if $X \subseteq Y$ implies that $\Gamma(X) \subseteq \Gamma(Y)$.

An inductive definition is classified by the complexity of the relation $x \in \Gamma(X)$. In order to obtain a reasonable theory over arbitrary structures $\mathfrak{A}$ we assume that $\mathfrak{A}$ has a coding scheme (see clause 4 of Definition 3.2.3) and that the language of $\mathfrak{A}$ includes the relations and functions of the coding scheme. In this way we have the usual Σ^n_m, Π^n_m, and Δ^n_m classification of the relation $x \in \Gamma(X)$.

3.3.2 Definition. Let $\mathbf{C}$ be a class of operators on a structure $\mathfrak{A}$. Let $\mathbf{C}_\infty$ denote the class of fixed-points Γ_∞ of operators $\Gamma \in \mathbf{C}$.

$$\mathrm{IND}(\mathbf{C}) = \{R : \exists \mathbf{a} \exists \Gamma \in \mathbf{C}[x \in R \text{ iff } (\mathbf{a}, x) \in \Gamma_\infty]\}.$$

IND(**C**) is the class of **C**-inductive relations. The associated ordinal is

$$|\mathbf{C}| = \sup\{|\Gamma| : \Gamma \in \mathbf{C}\}.$$

For every operator Γ on A we define $\neg\Gamma$ by $\neg\Gamma(X) = A \backslash \Gamma(X)$ and $\neg\mathbf{C} = \{\neg\Gamma : \Gamma \in \mathbf{C}\}$. We set

$$\mathrm{IND}(\neg\mathbf{C}) = \{A - R : R \in \mathrm{IND}(\mathbf{C})\},$$

and call this the class of **C**-coinductive relations. Finally, set

$$\mathrm{HYP}(\mathbf{C}) = \mathrm{IND}(\mathbf{C}) \cap \mathrm{IND}(\neg\mathbf{C}),$$

this is called the class of **C**-hyperdefinable relations on $\mathfrak{A}$.

We list some of the basic examples from which the general theory has been abstracted and developed.

3.3.3 Example. Positive Σ^0_1 inductive definitions over ω.

This is a well-known way of developing ORT over ω, due to E. Post. We know that recursively enumerable is the same as Σ^0_1. Hence the first recursion theorem gives the equivalence between recursively enumerable and positive Σ^0_1 inductive definability. This is a theme to which we will often return.

3.3.4 Example. Π^0_1 inductive definitions over ω.

General, i.e. non-monotone, Π^0_1 inductive definability gives us already the class of Π^1_1 sets. This was first proved by R. Gandy, but remained unpublished by him (see, however, [39]). A proof in the setting of α-recursion theory was published in W. Richter [134]. A proof is also contained in Grilliot's [49]. Grilliot's work clearly brings out the important role of *reflection principles* in the theory of inductive definability. We shall in many connections return to this theme, e.g. in the study of finite theories on two types (e.g. recursion in ${}^3E$). The present version reads as follows.

3.3.4.1 Σ^0_2-Reflection. *Let B be a Π^1_1 set and $\Gamma(X)$ a Σ^0_2 relation. If $\Gamma(B)$, then there exists a hyperarithmetic $B_0 \subseteq B$ such that $\Gamma(B_0)$.*

We indicate a proof assuming that the reader has some basic knowledge of Π_1^1-theory. First we note that B can be obtained by an approximation B_α, $\alpha < \omega_1$ (= the first non-recursive ordinal), where each B_α is hyperarithmetic. Let $\Gamma(X)$ be of the form $\exists y \forall z \Gamma_0(y, z, X)$. Since we have $\Gamma(B)$, fix a parameter b such that $\forall z \Gamma_0(b, z, B)$.

The heart of the argument is a boundedness property: From $\forall z \Gamma_0(b, z, B)$ it follows that $\forall z \exists \beta < \omega_1 \Gamma_0(b, z, B_\beta)$. We want to conclude that there is an $\alpha < \omega_1$ such that $\forall z \Gamma_0(b, z, B_\alpha)$.

The details are as follows. Naively we would like to set $\alpha = \sup_z \mu\beta[\Gamma_0(b, z, B_\beta)]$. But $\Gamma_0(b, z, B_\beta)$ for some $\beta < \alpha$ does not necessarily imply $\Gamma_0(b, z, B_\alpha)$. Hence, we have to complicate the construction a bit. Choose the α_n's such that

$$\alpha_{n+1} \geqslant \sup_z \mu\beta[\beta \geqslant \alpha_n \wedge \Gamma_0(b, z, B_\beta)].$$

Then for every z there is a sequence β_n such that $\alpha_n \leqslant \beta_n \leqslant \alpha_{n+1}$ and $\Gamma_0(b, z, B_{\beta_n})$. Let $\alpha = \sup \alpha_n$. Now we get $\Gamma_0(b, z, B_\alpha)$ for *all* z, since $B_\alpha = \bigcup B_{\beta_n}$.

A recursion-theoretic analysis of the construction (including notations for ordinals etc.) and an application of the Spector boundedness principle, shows that $\alpha < \omega_1$.

The application to IND(Π_1^0) is immediate. Let Γ_∞ be a fixed-point for a Π_1^0 operator Γ. It is not difficult to show that $\Gamma_{\omega_1} \in \Pi_1^1$. Consider the relation $x \in \Gamma_{\omega_1+1} = \Gamma(\Gamma_{\omega_1})$. By Lemma 3.3.5 there is some $\alpha < \omega_1$ such that $x \in \Gamma(\Gamma_\alpha) = \Gamma_{\alpha+1}$. Hence $\Gamma_\infty = \Gamma_{\omega_1}$, from which we conclude that IND(Π_1^0) $= \Pi_1^1$ and $|\Pi_1^0| = \omega_1$.

3.3.5 Example. Π_1^1 monotone inductive definitions over ω.

The basic analysis of this case is due to C. Spector [161]. Today this part of the theory is best viewed from the standpoint of admissibility theory. Monotone Π_1^1 is the same as positive. And Π_1^1 on ω corresponds to Σ_1 on the next admissible set, in this case L_{ω_1}. Σ_1 positive inductive operators have fixed-points and the length of the inductive definition is at most the ordinal of the admissible set. Translating back one sees that if Γ is Π_1^1-monotone, then $\Gamma_\infty \in \Pi_1^1$ and $|\Gamma| \leqslant \omega_1$.

If one wants to prove this in the setting of hyperarithmetic theory, the basic fact to verify is that if Γ is Π_1^1-monotone and $B \in \Pi_1^1$, then $x \in \Gamma(B)$ iff there is some hyperarithmetic (i.e. "finite") $B_0 \subseteq B$ such that $x \in \Gamma(B_0)$.

3.3.6 Example. Positive elementary inductive definitions on a structure $\mathfrak{A}$.

Over ω we know that Π_1^0 positive and Π_1^1 monotone give the same class of relations, viz. the Π_1^1 relations. Hence the class of positive elementary operators also gives the same inductively defined relations. (*Elementary* here means first-order in the language of the structure.)

Moschovakis developed in his book *Elementary Induction on Abstract Structures* [115] the theory of positive elementary inductive definability over arbitrary structures $\mathfrak{A}$ (equipped with a suitable coding scheme) as a general approach to definability. Barwise obtained the same theory in his book *Admissible Sets and*

Structures [11] from the standpoint of the theory of HYP$\mathfrak{M}$, the "next admissible" set.

In our approach we note that if **C** = the class of positive elementary operators, then IND(**C**) is the "least" Spector class on the structure. Hence we view the theory as part of the development of Spector theories, see Section 3.2 of this chapter.

The HYP$\mathfrak{M}$—or *imbedding* aspect will be treated in full in Chapter 5.

3.3.7 Example. Σ_1^1 monotone inductive definitions and generalized quantifiers.

This example is due to P. Aczel [3] and has been the source of much of the work on generalized quantifiers in the context of general recursion theory; see Aczel [4, 6], Moschovakis [115], Barwise [13], and Kolaitis [87].

Let $E_1^\#$ be the partial functional given by the equation

$$E_1^\#(f) \simeq \begin{cases} 0 & \text{if } \forall\alpha\exists n \cdot f(\bar{\alpha}(n)) = 0 \\ 1 & \text{if } \exists\alpha\forall n \cdot f(\bar{\alpha}(n)) > 0, \end{cases}$$

where f is a partial function from ω to ω. Aczel's main result in [3] states that if $A \subseteq \omega$, then A is semicomputable in $E_1^\#$ iff $A \in \mathrm{IND}(\Sigma_1^1\text{-mon})$.

The proof is not difficult. An analysis of recursion in $E_1^\#$ shows that it is given by a Σ_1^1 monotone operator. Conversely, if Γ is a Σ_1^1 monotone operator, i.e.

$$x \in \Gamma(\{n : \alpha(n) = 0\}) \quad \text{iff} \quad R(\alpha, n),$$

where R is Σ_1^1, then we can write, using the monotonicity of Γ,

$$x \in \Gamma(X) \quad \text{iff} \quad \exists\alpha\forall n[(\alpha(n) = 0 \to n \in X) \wedge R(\alpha, x)],$$

which has the form

$$x \in \Gamma(X) \quad \text{iff} \quad \exists\alpha\forall n[(R_1(\bar{\alpha}(n)) \vee g_1(\bar{\alpha}(n)) \in X) \wedge S_1(\bar{\alpha}(n), x)],$$

where R_1, S_1, g_1 are recursive.

Introduce the functional F by

$$F(f, x) \simeq 1 \quad \text{iff} \quad \exists\alpha\forall n[(R_1(\bar{\alpha}(n)) \vee f(g_1(\bar{\alpha}(n))) = 1) \wedge S_1(\bar{\alpha}(n), x)].$$

Then one sees that F is $E_1^\#$-computable and monotone. Let g be the least fixed-point of F, g is semicomputable in $E_1^\#$ by the first recursion theorem. Hence Γ_∞ is also semicomputable in $E_1^\#$, since evidently $x \in \Gamma_\infty$ iff $g(x) \simeq 1$.

This example can be generalized to arbitrary monotone quantifiers **Q**. A *monotone quantifier* **Q** is a family of subsets of ω (or some other suitable domain in the generalized versions) such that $A \in \mathbf{Q}$ and $A \subseteq B$ implies $B \in \mathbf{Q}$. The dual of **Q** is defined as $\breve{\mathbf{Q}} = \{\omega - X; X \in \mathbf{Q}\}$. In the example above $\mathbf{Q} = \{A \subseteq \omega;$

$\exists\alpha\forall n\cdot\bar{\alpha}(n) \in A\}$. Analogous to $E_1^{\#}$ we have an associated functional $F_{\mathbf{Q}}^{\#}$ defined by

$$F_{\mathbf{Q}}^{\#}(f) \simeq \begin{cases} 0 & \text{if } \{x : f(x) = 0\} \in \breve{\mathbf{Q}} \\ 1 & \text{if } \{x : f(x) > 0\} \in \mathbf{Q}. \end{cases}$$

The main result above generalizes: Given $A \subseteq \omega$, then A is semicomputable in $F_{\mathbf{Q}}^{\#}$ iff $A \in \mathrm{IND}(\mathbf{Q})$, see Aczel [3].

The reader will have noticed the dramatic difference between the strength of non-monotone versus monotone inductive operators, e.g. both $\mathrm{IND}(\Pi_1^1\text{-mon})$ and $\mathrm{IND}(\Pi_0^1)$ leads to the same class, viz. the Π_1^1 sets over ω.

We shall present a few basic results on the connection between Spector classes/theories and non-monotone inductive definability. Some of the first and basic results are due to Grilliot [49]. Aczel and Richter in their joint paper [135] emphasized the importance of reflection principles and developed a general theory over ω. Aanderaa [1] solved a fundamental and long-standing open problem about the size of ordinals of inductive definitions. And in the paper [116] Moschovakis brought the various developments together and presented a unified approach. The reader should also consult Cenzer [18]. Finally, we should mention the work of Harrington and Kechris [57] on the relationship between monotone and non-monotone induction.

Here is the appropriate definition to get the theory off the ground (see Moschovakis [116]).

3.3.8 Definition. **C** is a *typical non-monotone class* of operators if it satisfies the following six conditions:

A. **C** contains all second-order relations on A which are definable by (first-order) universal formulas of the trivial structure $\langle A \rangle$.

B. **C** is closed under $\wedge$ and $\vee$.

C. **C** is closed under trivial, combinatorial substitutions.

Remark. From these conditions we already have a large part of the structure theory for Spector classes, see Moschovakis [116, §3].

D. **C** contains all second-order relations definable by existential formulas of the trivial structure $\langle A \rangle$.

E. There is an ordering $\leqslant \subseteq A \times A$ isomorphic to the ordering on ω and a 1-1 function $f: A \times A \to A$ which belong to HYP(**C**).

F. For each $n \geqslant 1$ the n-ary IND(**C**) relations are parametrized by an $n + 1$-ary IND(**C**) relation.

From these conditions on **C** it is not surprising that the following result holds.

3.3.9 Theorem. *If* **C** *is a typical non-monotone class of operators on* A, *then* IND(**C**) *is a Spector-class.*

We can add various refinements, e.g. $|\mathbf{C}|$ = supremum of the length of computations in the associated Spector theory. And every second-order relation in **C** is "Δ on Δ".

The notion "Δ on Δ" was introduced by Moschovakis in the setting of Spector classes. Working with the associated Spector theory seems to simplify the conceptual set-up.

Let **C** be a typical non-monotone class of operators. Theorem 3.3.9 says that IND(**C**) is a Spector class. Associated with IND(**C**) is a Spector theory $\Theta(\mathbf{C})$, see Theorem 3.2.6. With every operator Γ in **C** there is associated a functional F_Γ defined as follows

$$F_\Gamma(\alpha, x) = \begin{cases} 1 & \text{if} \quad x \in \Gamma(\text{set}_\alpha) \\ 0 & \text{if} \quad x \notin \Gamma(\text{set}_\alpha). \end{cases}$$

α is here supposed to be total, and $\text{set}_\alpha = \{x : \alpha(x) = 0\}$. It is immediate that F_Γ is $\Theta(\mathbf{C})$-computable for every operator $\Gamma \in \mathbf{C}$.

We further notice that in the Spector theory $\Theta(\mathbf{C})$ we have $A = C$ and the relation $\text{set}_\alpha \in \text{sc}(\Theta(\mathbf{C}))$ is $\Theta(\mathbf{C})$-semicomputable. We say that **C** is Θ-computable if F_Γ is Θ-computable for every operator $\Gamma \in \mathbf{C}$. This is our notion "Δ on Δ".

The crucial point in proving that $\text{IND}(\Pi_1^0) = \Pi_1^1$ was to establish the Reflection Property 3.3.4.1. We state a general definition.

3.3.10 Definition. Let **C** be a class of operators and Θ a Spector theory. Θ has the **C**-*reflection property* if whenever $R \in \text{en}(\Theta)$, $\Gamma \in \mathbf{C}$, and $R_0 \subseteq R$ belongs to $\text{sc}(\Theta)$,

$$\Gamma(R) \Rightarrow \text{there exists } R^* \in \text{sc}(\Theta),\ R_0 \subseteq R^* \subseteq R, \text{ and } \Gamma(R^*).$$

We showed in 3.3.4.1 that Π_1^1 as a Spector theory has the Σ_2^0-reflection property.

The following result of Moschovakis [116] is a general statement of these facts and provides a converse to Theorem 3.3.9.

3.3.11 Theorem. *Let Θ be a Spector theory and* **C** *a typical non-monotone class of operators such that*

(i) *Θ is* **C**-*reflecting, and*
(ii) **C** *is Θ-computable.*

Then $\text{IND}(\mathbf{C}) \subseteq \text{en}(\Theta)$.

The idea of the proof is simple. We use the fact that **C** is Θ-computable to carry out the inductive definition inside $\text{en}(\Theta)$, and conclude from the **C**-reflecting property of Θ that the inductive definition closes at a stage $\leqslant \|\Theta\|$.

3.3.12 Remark. We should supplement Theorem 3.3.9 by noting that the Spector theory $\Theta(\mathbf{C})$ associated to a typical non-monotone class **C** is **C**-reflecting: For simplicity set $R_0 = \varnothing$ in 3.3.10 and assume $R \in \mathrm{en}(\Theta) = \mathrm{IND}(\mathbf{C})$, i.e. $x \in R$ iff $(a, z) \in \Gamma_\infty$, for some $\Gamma \in \mathbf{C}$. Let $\Delta \in \mathbf{C}$ and assume that $\Delta(R)$ is valid.

Choose an element b and let

$$\Psi(X) = \Gamma(X),$$
$$\Phi(X) = \{b : \Delta(\{x : (a, x) \in X\})\}.$$

This is a simultaneous inductive definition in the class **C**. Obviously, $\Psi_\alpha = \Gamma_\alpha$; hence, $\{x : (a, x) \in \Psi_\infty\} = R$ and thus $b \in \Phi_\infty$. But then there is some $\xi < |\mathbf{C}|$ such that $b \in \Phi_{\xi+1}$. This means that we have $\Delta(R^*)$ for $R^* = \{x : (a, x) \in \Psi_\xi\}$. But the latter set belongs to $\mathrm{sc}(\Theta) = \mathrm{HYP}(\mathbf{C})$ by the usual boundedness argument.

We mention one result on the relative size of the ordinals of inductive definitions. The reader should recall the notations of Definition 3.3.2. The following general result is due to S. Aanderaa [1].

3.3.13 Theorem. *Let* **C** *be a typical non-monotone class of operators. If* **C** *has the prewellordering property, then*

$$|\mathbf{C}| < |\neg\mathbf{C}|.$$

In particular, $|\Pi_1^1| < |\Sigma_1^1|$ and $|\Sigma_2^1| < |\Pi_2^1|$.

Recall that a class **C** has the prewellordering property if every relation in **C** has a prewellordering in **C**.

Note that beyond the second level of the analytic hierarchy the prewellordering property depends upon the axioms of set theory, see Aanderaa [1] for complete statements.

We shall give a brief outline of the proof, following the original exposition closely, but correcting several minor mistakes along the way.

Fact 1. There exists $\Gamma \in \mathbf{C}$ such that $|\Gamma| = |\mathbf{C}|$.

This is an easy consequence of the ω-parametrization property of the class **C**. So let us start with a $\Gamma \in \mathbf{C}$ such that $|\Gamma| = |\mathbf{C}|$. We must produce a $\Lambda \in \neg\mathbf{C}$ such that $|\Gamma| < |\Lambda|$.

Fact 2. Given $\Gamma \in \mathbf{C}$ and $\check{\Gamma} \in \neg\mathbf{C}$ such that whenever $\Gamma(S) - S \neq \varnothing$, then $\varnothing \neq \check{\Gamma}(S) - S \subseteq \Gamma(S)$. Then there exists $\Lambda \in \neg\mathbf{C}$ such that $|\Gamma| < |\Lambda|$. (In fact, we construct a $\Lambda \in \neg\mathbf{C}$ such that $|\Lambda| = |\Gamma| + 1$.)

Let us postpone the proof of fact 2 for a moment and see how we produce a suitable $\check{\Gamma} \in \neg\mathbf{C}$ from the Γ given in fact 1.

Fact 3. Let $\Gamma \in \mathbf{C}$ and assume $\mathrm{PWO}(\mathbf{C})$. Then there exist $\hat{\Gamma} \in \mathbf{C}$ and $\check{\Gamma} \in \neg\mathbf{C}$ such that whenever $\Gamma(S) - S \neq \varnothing$, then $\varnothing \neq \hat{\Gamma}(S) = \check{\Gamma}(S) \subseteq \Gamma(S)$.

This proves the theorem.

Let us first give the proof of fact 3. Let $\Gamma'(S) = \Gamma(S) - S$, then $\Gamma' \in \mathbf{C}$. From the assumption PWO(C), let $\|\cdot\|$ be a norm on the class $A = \{(x, S) : x \in \Gamma'(S)\}$. Define

$$\hat{\Gamma}(S) = \{x : x \in \Gamma'(S) \wedge \forall y(\|y, S\| \leqslant \|x, S\| \\ \rightarrow (\|x, S\| \leqslant \|y, S\| \wedge x \leqslant y))\}.$$

Clearly $\hat{\Gamma} \in \mathbf{C}$. And if $\Gamma'(S) \neq \varnothing$, let λ_S be the least ordinal of the form $\|x, S\|$, where $x \in \Gamma'(S)$. Finally, let x_S be the least element x of ω such that $\|x, S\| = \lambda_S$. Then we see that $\hat{\Gamma}(S) = \{x_S\}$. Let now

$$x \in \check{\Gamma}(S) \quad \text{iff} \quad \forall y(y \in \hat{\Gamma}(S) \rightarrow y = x).$$

$\check{\Gamma}$ clearly satisfies the requirements of fact 3.

The construction of Λ in fact 2 uses the operator $\check{\Gamma}$ and the fact that $\varnothing \neq \check{\Gamma}(S) - S \subseteq \Gamma(S)$. This is an exercise in constructing an inductive definition. We outline one possible way.

Let

$$H(S) = \{u : \exists v(\langle v + 2, u\rangle \notin S \wedge \langle 0, v\rangle \in S)\},$$

and define

$$\Lambda(S) = \{\langle 0, x\rangle : x \in \check{\Gamma}(H(S))\} \\ \cup \{\langle z + 2, x\rangle : z \in \check{\Gamma}(H(S)) \wedge z \notin H(S) \wedge x \notin \Gamma(H(S)) \\ \wedge x \notin H(S)\} \\ \cup \{\langle 1, 0\rangle : \forall x(x \in \Gamma(H(S)) \rightarrow x \in H(S))\}.$$

Finally let $f(y) = \langle 0, y\rangle$. By simultaneous induction on the ordinal λ one may now prove

(i) $\Gamma^\lambda = H(\Lambda^\lambda)$ and $f^{-1}(\Lambda^\lambda) \subseteq \Gamma^\lambda$
(ii) $\Gamma(H(\Lambda^\lambda)) \subseteq \Gamma^{\lambda+1}$
(iii) $\Gamma^{\lambda+1} - \Gamma^\lambda \neq \varnothing \Rightarrow \varnothing \neq \check{\Gamma}(H(\Lambda^\lambda)) - H(\Lambda^\lambda) \subseteq \Gamma^{\lambda+1}$
(iv) $\Gamma^{\lambda+1} = H(\Lambda^{\lambda+1})$ and $f^{-1}(\Lambda^{\lambda+1}) \subseteq \Gamma^{\lambda+1}$.

Clearly, $\Lambda \in \neg\mathbf{C}$, and $|\Lambda| = |\Gamma| + 1$ since $\langle 1, 0\rangle \in \Lambda^{|\Gamma|+1} - \Lambda^{|\Gamma|}$.

We started our discussion of non-monotone inductive definability by pointing to the striking difference between IND(Π_1^1-mon) = IND(Π_1^0) = Π_1^1 and IND(Π_1^1). The ordinal e.g. of the latter class is enormously larger than $|\Pi_1^1\text{-mon}| = \omega_1$, the first non-recursive ordinal. This is the situation over the integers. There are, however, cases where the difference disappears. If the notion WF of wellfoundedness is elementary over a structure $\mathfrak{A}$ (i.e. is first-order definable in the relations of the structure), then the classes of elementary monotone and elementary non-monotone coincide. This is a corollary of a more general result due to L. Harrington and A. Kechris [57].

To obtain a sufficiently general result they call a class **C** of operators on a domain *A adequate* if it contains all operators defined by a universal formula of the structure, is closed under $\wedge$, $\vee$, $\exists_A$ and trivial combinatorial substitutions and contains a coding scheme. The reader may want to compare this in detail with Definition 3.3.8.

3.3.14 Theorem. *Let* **C** *be an adequate class of operators on A. If* $\neg$WF $\in$ **C** *and* $\neg$**C** $\subseteq$ IND(**C**-mon), *then* IND(**C**) = IND(**C**-mon).

Here $\neg$WF is the negation of the relation of wellfoundedness, and $\neg$**C** $\subseteq$ IND(**C**-mon) means the usual thing: if $\Gamma(\mathbf{a}, \mathbf{S})$ is an operator in **C**, then there exists in **C** an operator $\Gamma^*(\mathbf{e}, \mathbf{a}, R, \mathbf{S})$ which is monotone in R such that for a suitable choice of parameters $\mathbf{e}_0$

$$\neg\Gamma(\mathbf{a}, \mathbf{S}) \quad \text{iff} \quad \Gamma^*_\infty(\mathbf{e}_0, \mathbf{a}, \mathbf{S}).$$

We conclude our excursion into the theory of definability by relating the classes IND(Σ^0_2) and IND(Π^0_1) over an acceptable structure $\mathfrak{A}$ to the notions of *strong* and *weak* finiteness, respectively. (Recall that a structure $\mathfrak{A}$ is acceptable if it has an elementary coding scheme, see Moschovakis [115].)

Let us call a theory Θ a *weak Spector theory* if we relax the condition of strong finiteness of the domain to weak finiteness (see condition (2) of Definition 3.2.2). The following results are essentially due to Grilliot [49].

3.3.15 Theorem. *Let* $\mathfrak{A}$ *be an acceptable structure.*

(1) *Let* Θ *be a Spector theory on* $\mathfrak{A}$:

a IND(Σ^0_2) $\subseteq$ en(Θ),
b $\Theta \sim$ PR[$\mathbf{E}_A$, =] *iff* en(Θ) = IND(Σ^2_0).

(2) *Let* Θ *be a weak Spector theory on* $\mathfrak{A}$*:*

c IND(Π^0_1) $\subseteq$ en(Θ),
d $\Theta \sim$ PR[$\mathbf{E}'_A$, =] *iff* en(Θ) = IND(Π^0_1).

(*Here* $\mathbf{E}'_A$ *is the functional of Definition* 3.1.2.)

The proof of (1) is essentially contained in the proof of 3.3.4.1, rephrasing the argument inside an arbitrary Spector class on $\mathfrak{A}$ rather than in terms of Π^1_1. Recall also from 3.2.7 and 3.2.8 that there is a one-to-one correspondence between Spector theories and Spector classes, and that a Spector theory is determined by its envelope.

The proof for Π^0_1 is a bit more subtle. Obviously, we cannot substitute Π^0_1-

reflection for Σ_2^0-reflection in the proof for (1), since Π_1^0-reflection implies Σ_2^0-reflection and there exist weak Spector theories which are not strong, e.g. Kleene recursion in ${}^3E$ over the reals $\mathbb{R}$.

The difference can be traced to another fact, which we shall elaborate on in later chapters, that whereas Spector theories correspond to admissible structure *weak* Spector theories do not. Any weak Spector theory Θ has an associated family of (non-transitive) admissible sets, Spec(Θ), see Chapter 8 in particular. The reflection argument to follow is really the argument for case (1) localized to components of Spec(Θ).

We give a sketch of the non-trivial part of (2), adapted from Grilliot [49]: Let S be a Π_1^0 fixed-point, defined from an inductive operator

$$x \in \Gamma(X) \quad \text{iff} \quad \forall z P(x, z, X),$$

where P is quantifier-free. Let $\pi(\sigma)$ be the supremum of lengths of computations in Θ with σ as input. We want to show that $x \in S$ implies $x \in S_{\pi(x)}$, which yields that

$$S = \bigcup_x S_{\pi(x)} = S_\kappa,$$

where κ is the ordinal of Θ, i.e. the supremum of all computations in Θ. A recursion-theoretic analysis will show that $S = S_\kappa \in \text{en}(\Theta)$ (similar to the analysis in 3.3.4 showing that $\Gamma_{\omega_1} \in \Pi_1^1$). Thus $\text{IND}(\Pi_1^0) \subseteq \text{en}(\Theta)$, proving c of (2).

It remains to verify that $x \in S$ implies $x \in S_{\pi(x)}$. For $x \in S$ let $|x|$ be the first stage at which x occurs in S. As induction hypothesis we assume that whenever $|y| < |x|$, then $y \in S_{\pi(y)}$. We also assume that $\pi(x) \leqslant |x|$, since otherwise $x \in S_{|x|+1} \subseteq S_{\pi(x)}$.

As in the proof of 3.3.4.1 we can show that $\forall z P(x, z, S_\alpha)$ is true when $\alpha = \sup \alpha_n$, where the α_n's satisfy

$$\alpha_{n+1} \leqslant \sup_z \mu\beta[\beta \geqslant \alpha_n \wedge P(x, z, S_\beta)].$$

It remains to get $\alpha < \pi(x)$. This is where "admissibility" enters the proof in 3.3.4.1. Here we must argue more carefully.

Suppose that we can compute in Θ an ordinal α_n, uniformly in n. We show how to compute in Θ an ordinal uniformly in n and z larger than

$$\mu\beta[\beta \geqslant \alpha_n \wedge P(x, z, S_\beta)].$$

Case 1: $|x| < \pi(x, z)$. This is immediate since $|x| \geqslant \alpha_n$ and $P(x, z, S_{|x|})$.

Case 2: $\pi(x, z) \leqslant |x|$. We first note that if a clause $b \in X$ occurs in $P(x, z, X)$, then $b \in S_{|x|}$ iff $b \in S_{\pi(x,z)}$. (If $b \in S_{|x|}$, then $|b| < |x|$, hence by induction hypothesis $b \in S_{\pi(b)} \subseteq S_{\pi(x,z)}$. The last inclusion follows since b is computable in x, z.)

From the assumption $P(x, z, S_{|x|})$ we may therefore conclude $P(x, z, S_{\pi(x,z)})$, hence also $P(x, z, S_\beta)$ for some β immediately below $\pi(x, z)$. This follows since P is Π_0^0 and $\pi(x, z)$ is a limit number.

In both cases we can compute some γ_{nz} larger than $\mu\beta[\beta \geqslant \alpha_n \wedge P(x, z, S_\beta)]$. Simple recursion theoretic properties give us an $\alpha_{n+1} \geqslant \sup_z \gamma_{nz}$ and finally $\alpha = \sup_n \alpha_n$. Since the constructions are computable in x, $\alpha < \pi(x)$. This completes the proof of c.

We make a remark on how to prove the converse inclusion that $\text{en}(\Theta) \subseteq \text{IND}(\Pi_1^0)$ if $\Theta \sim \text{PR}[\mathbf{E}'_A, =]$. This follows from a careful analysis of the inductive definition of the theory $\text{PR}[\mathbf{E}'_A, =]$.

The clauses of the "usual" inductive definition of the relations $\{e\}_{\text{PR}[\mathbf{E}'_A, =]}(x)\downarrow$ and $\{e\}_{\text{PR}[\mathbf{E}'_A, =]} \simeq z$ are all of the Π_0^1 form except substitution and application of $\mathbf{E}'_A$. The trick is to replace these parts of the inductive definition by non-monotonic clauses. We take first the case of substitution, $\{e\}(x) \simeq \{e_1\}(\{e_2\})x), x)$. We will introduce a new inductive operator Γ such that in the end $(0, e, x) \in \Gamma_\infty$ iff $\{e\}(x)\downarrow$ and $(1, e, x, z) \in \Gamma_\infty$ iff $\{e\}(x) \simeq z$. For substitution this is obtained by the non-monotonic clauses:

$$\begin{aligned}
(0, e, x) \in \Gamma(X) \quad &\text{if} \quad (0, e_2, x) \in X \wedge \forall u[(1, e_2, x, u) \in X \\
&\qquad\qquad \to (0, e_1, u, x) \in X]. \\
(1, e, x, z) \in \Gamma(X) \quad &\text{if} \quad (0, e_2, x) \in X \wedge \forall u[(1, e_2, x, u) \in X \\
&\qquad\qquad \to (1, e_1, u, x, z) \in X].
\end{aligned}$$

Application of $\mathbf{E}'_A$ is a bit more involved. In this case we have to build the function

$$\{e\}(x) \simeq \mathbf{E}'_A(\lambda y \cdot \{e'\}(y, x)),$$

into the operator Γ. This will be done by the following clauses using two auxiliary tuples $(a_1, \ldots)$ and $(a_2, \ldots)$

$$\begin{aligned}
(a_1, e', x) \in \Gamma(X) \quad &\text{if} \quad \forall y \cdot (0, e', y, x) \in X \\
(0, e, x) \in \Gamma(X) \quad &\text{if} \quad (a_1, e', x) \in X. \\
(a_2, e', x) \in \Gamma(X) \quad &\text{if} \quad \forall y \cdot (0, e', y, x) \in X \wedge \forall y \cdot (1, e', y, x, 0) \notin X. \\
(1, e, x, 1) \in \Gamma(X) \quad &\text{if} \quad (a_2, e', x) \in X \\
(1, e, x, 0) \in \Gamma(X) \quad &\text{if} \quad (a_1, e', x) \in X \wedge (a_2, e', x) \notin X.
\end{aligned}$$

(The definition is arranged so that the ordinals of $(0, e, x)$ and $(1, e, x, z)$ are the same. This is necessary for the inductive proof that $(0, e, x) \in \Gamma_\infty$ iff $\{e\}(x)\downarrow$ and $(1 \cdot e \cdot x \cdot z) \in \Gamma_\infty$ iff $\{e\}(x) \simeq z$.)

3.3.16 Remark. In Corollary 3.1.11 we saw that ω is strongly Θ-finite iff it is weakly Θ-finite. Hence on ω $\text{IND}(\Sigma_0^2) = \text{IND}(\Pi_1^0) = \Pi_1^1$.

Chapter 4

Finite Theories on Two Types

Moving up in types over the integers ω the notion of *finiteness* bifurcates. In higher types one must carefully distinguish between a *weak* and a *strong* notion (see Definition 3.1.2 and the connecting remarks). This was a phenomenon first observed by Y. Moschovakis [111] in his study of the hyperanalytic sets. He proved that the set of reals semicomputable (in the sense of Kleene) in ${}^3E$ is *not* closed under $\exists\alpha$, i.e. existential quantification over the reals. A further analysis reveals that higher type theories can be captured as theories on "two types", i.e. as computation theories on domains of the form $A = S \cup \mathrm{Tp}(S)$, where $\mathrm{Tp}(S) = \omega^S$, and where S is *strongly finite*, but $\mathrm{Tp}(S)$ is only *weakly finite*.

The distinction between two types can already be found in Moschovakis [114], viz. the notion of 2-point class. It was adopted by L. Harrington and D. MacQueen [55] in their proof of the Grilliot selection theorem. A general development of computation theories on two types was first presented by J. Moldestad [105]. His primary aim was to provide a "natural" setting for the general "plus-2" and "plus-1" theorems (see Chapter 7). But the theory can also serve as a framework for developing second-order definability results in general. A variant theory is the theory of Spector second-order classes (see Section 4.4 below).

4.1 Computation Theories on Two Types

We will study theories Θ on domains of the form $\mathfrak{A} = \langle A, S, \mathrm{S}\rangle$, where S is some infinite set, $A = S \cup \mathrm{Tp}(S)$, and S is a coding scheme for S, i.e.

$$\mathrm{S} = \langle N, s, M, K, L\rangle,$$

where $N \subseteq S$ and $\langle N, s\rangle$ is isomorphic to the integers ω with the successor function. M, K, L are the usual pairing and projection functions on S. We assume that N is closed under M, K, and L. We also assume that $\mathrm{Tp}(S) = \omega^S$.

As usual we can introduce the functions and predicates associated with coding of finite tuples (see the discussion in Section 1.5). In particular we define a predicate $\mathrm{Seq}(r)$ by

$$\mathrm{Seq}(r) \quad \text{iff} \quad \exists n \exists r_1 \ldots \exists r_n [r = \langle r_1, \ldots, r_n\rangle].$$

In the concrete case of higher types we can "bound" the quantifiers in the definition of Seq(r). In the general case we must hypothesize the computability of $\mathbf{E}'_S$ so as to handle Seq(r).

Let * be the injection $S \to \mathrm{Tp}(S)$ defined by

$$r^*(s) = \begin{cases} 0 & \text{if} \quad s = r \\ 1 & \text{if} \quad s \neq r, \end{cases}$$

and $^-$ a function $A \to S$ defined by

$$x^- = \begin{cases} r & \text{if} \quad x = r^* \\ 0 & \text{otherwise.} \end{cases}$$

With the help of these functions any coding scheme **S** for S can be lifted to A, e.g.

$$\begin{aligned} M(r, \alpha) &= \lambda s \cdot M(M(r^*(s), \alpha(s)), M(0, 1)) \\ M(\alpha, r) &= \lambda s \cdot M(M(\alpha(s), r^*(s)), M(1, 0)) \\ M(\alpha, \beta) &= \lambda s \cdot M(M(\alpha(s), \beta(s)), M(1, 1)), \end{aligned}$$

where $\alpha, \beta \in \mathrm{Tp}(S)$, $r, s \in S$.

$$\begin{aligned} K(\alpha) &= (\lambda s \cdot K(K(\alpha(s))))^- && \text{if} \quad L(\alpha(0)) = M(0, 1), \\ &= \lambda s \cdot K(K(\alpha(s))) && \text{otherwise.} \\ L(\alpha) &= (\lambda s \cdot L(K(\alpha(s))))^- && \text{if} \quad L(\alpha(0)) = M(1, 0), \\ &= \lambda s \cdot L(K(\alpha(s))) && \text{otherwise.} \end{aligned}$$

With these definitions the associated functions and predicates can be extended to A.

4.1.1 Definition. Let $\mathfrak{A} = \langle A, S, \mathbf{S} \rangle$ be a computation domain. The class PRF of *primitive recursive functions* is generated by the following schemes wherein σ is always a finite list of elements from A.

1. We have a base of nine initial functions

(i) $$f(x, \sigma) = \begin{cases} 0 & \text{if} \quad x \in N, \\ 1 & \text{if} \quad x \notin N. \end{cases}$$

(ii) $$f(x, \sigma) = \begin{cases} 0 & \text{if} \quad x \in S, \\ 1 & \text{if} \quad x \notin S. \end{cases}$$

(iii) $$f(x, \sigma) = \begin{cases} x & \text{if} \quad x \in S, \\ 0 & \text{if} \quad x \notin S. \end{cases}$$

(iv) $$f(x, \sigma) = \begin{cases} x + 1 & \text{if} \quad x \in N, \\ 0 & \text{if} \quad x \notin N. \end{cases}$$

(v) $f(\sigma) = m \quad m \in N.$

(vi) $$f(x, y, \sigma) = \begin{cases} M(x, y) & \text{if } x, y \in S, \\ 0 & \text{otherwise.} \end{cases}$$

(vii) $$f(x, \sigma) = \begin{cases} K(x) & \text{if } x \in S, \\ 0 & \text{otherwise.} \end{cases}$$

(viii) $$f(x, \sigma) = \begin{cases} L(x) & \text{if } x \in S, \\ 0 & \text{otherwise.} \end{cases}$$

(ix) $$f(x, y, \sigma) = \begin{cases} x(y) & \text{if } x \in \mathrm{Tp}(s), \quad y \in S, \\ 0 & \text{otherwise.} \end{cases}$$

2. We then have closure under the following schemes:

(a) *Substitution:* $f(\sigma) = g(h(\sigma), \sigma)$.
(b) *Primitive recursion:* $f(0, \sigma) = g(\sigma)$
$$f(n + 1, \sigma) = h(f(n, \sigma), n, \sigma), \quad n \in N$$
$$f(x, \sigma) = 0, \quad \text{if} \quad x \notin N.$$
(c) *Permutation:* $f(\sigma) = g(\sigma')$, σ' permutation of σ.

Let $R_1, \ldots, R_k$ be predicates, $f_1, \ldots, f_l$ total functions with values in S, i.e. $f_i\colon A^{n_i} \to S$, and $F_1, \ldots, F_m$ total functionals, i.e. each F_i is total and has as arguments total functions with values in S.

We can extend Definition 4.1.1 in the following way.

4.1.2. Definition. Let **L** be a list $R_1, \ldots, R_k, f_1, \ldots, f_l, F_1, \ldots, F_m$ as specified above. The class of functions PRF(**L**) is obtained by adding the following initial schemes to Definition 4.4.1:

(x) $$f_i(\sigma, \tau) = \begin{cases} 0 & \text{if } R_i(\sigma), \\ 1 & \text{if } \neg R_i(\sigma) \end{cases} \quad i = 1, \ldots, k.$$

(xi) $f_i(\sigma, \tau) = f_i(\sigma) \quad i = 1, \ldots, l.$
(xii) $f_i(\sigma) = F_i(\lambda x \cdot g(x, \sigma)) \quad i = 1, \ldots, m.$

We also add the following closure scheme (allowing us to substitute a function for an element of Tp(S)).

(d) *Substitution:* $f(\sigma) = h(\lambda r \cdot g(r, \sigma), \sigma)$, where g takes values in N and r ranges over elements from S.

Note the following lemma.

4.1.3 Lemma. *The graphs of the functions* $*$, $^-$, *and of the functions associated with the extended coding scheme, and the extended predicate* Seq *are primitive recursive in the equality relation on* S *and the functional* $\mathbf{E}'_S$.

Note that $\mathbf{E}'_S$ is defined on total functions only, i.e.

$$\mathbf{E}'_S(f) = \begin{cases} 0 & \text{if} \quad \exists s \in S. \quad f(s) = 0 \\ 1 & \text{if} \quad \forall s \in S. \quad f(s) \neq 0, \end{cases}$$

where $f\colon A \to S$ is total. $\mathbf{E}_S$ is the extension of $\mathbf{E}'_S$ to partial functions.

We come now to the main notion of functions partial recursive in some list $\mathbf{L} = R_1, \ldots, R_k, \varphi_1, \ldots, \varphi_l, F_1, \ldots, F_m$, where $R_1, \ldots, R_k, F_1, \ldots, F_m$ are as above, but $\varphi_1, \ldots, \varphi_l$ are partial functions with values in S. There are 17 clauses in the definition of PR(L), which we give, with some reluctance, for the sake of completeness and explicitness.

4.1.4 Definition. Let $\mathfrak{A} = \langle A, S, \mathbf{S}\rangle$ be a computation domain and $\mathbf{L} = R_1, \ldots, R_k, \varphi_1, \ldots, \varphi_l, F_1, \ldots, F_m$ a list on $\mathfrak{A}$. The class PR(L) of functions partial recursive in $\mathbf{L}$ is given by the following inductive definition:

Let Γ be the following inductive operator

I $\quad (\langle 1, n+1\rangle, x, \sigma, 0) \in \Gamma(X)$ if $x \in N$,
$(\langle 1, n+1\rangle, x, \sigma, 1) \in \Gamma(X)$ if $x \notin N$.

II $\quad (\langle 2, n+1\rangle, x, \sigma, 0) \in \Gamma(X)$ if $x \in S$,
$(\langle 2, n+1\rangle, x, \sigma, 1) \in \Gamma(X)$ if $x \notin S$.

III $\quad (\langle 3, n+1\rangle, x, \sigma, x) \in \Gamma(X)$ if $x \in S$,
$(\langle 3, n+1\rangle, x, \sigma, 0) \in \Gamma(X)$ if $x \notin S$.

IV $\quad (\langle 4, n+1\rangle, x, \sigma, x+1) \in \Gamma(X)$ if $x \in N$,
$(\langle 4, n+1\rangle, x, \sigma, 0) \in \Gamma(X)$ if $x \notin N$.

V $\quad (\langle 5, n, m\rangle, \sigma, m) \in \Gamma(X)$ $m \in N$.

VI $\quad (\langle 6, n+2\rangle, x, y, \sigma, M(x, y)) \in \Gamma(X)$ if $x, y \in S$,
$(\langle 6, n+2\rangle, x, y, \sigma, 0) \in \Gamma(X)$ otherwise.

VII $\quad (\langle 7, n+1\rangle, x, \sigma, K(x)) \in \Gamma(X)$ $x \in S$,
$(\langle 7, n+1\rangle, x, \sigma, 0) \in \Gamma(X)$ otherwise.

VIII $\quad (\langle 8, n+1\rangle, x, \sigma, L(x)) \in \Gamma(X)$ $x \in S$,
$(\langle 8, n+1\rangle, x, \sigma, 0) \in \Gamma(X)$ otherwise.

IX $\quad (\langle 9, n+2\rangle, x, y, \sigma, x(y)) \in \Gamma(X)$ $x \in \mathrm{Tp}(S)$, $y \in S$,
$(\langle 9, n+2\rangle, x, y, \sigma, 0) \in \Gamma(X)$ otherwise.

X $\quad$ If $\exists y[(e, \sigma, y) \in X \wedge (e', y, \sigma, x) \in X]$,
then $(\langle 10, n, e, e'\rangle, \sigma, x) \in \Gamma(X)$.

XI $\quad$ If $(e, \sigma, x) \in X$, then $(\langle 11, n+1, e, e'\rangle, 0, \sigma, x) \in \Gamma(X)$.
If $\exists y[(\langle 11, n+1, e, e'\rangle, m, \sigma, y) \in X \wedge (e', y, m, \sigma, x) \in X]$
then $(\langle 11, n+1, e, e'\rangle, m+1, \sigma, x) \in \Gamma(X)$.

XII $\quad$ If $(e, \sigma', x) \in X$, then $(\langle 12, n, e, i\rangle, \sigma, x) \in \Gamma(X)$,
where σ' is obtained from σ by moving the $i+1$-st object in σ to the front of the list.

XIII $\quad$ If $(e, \sigma, x) \in X$, then $(\langle 13, n+1\rangle, e, \sigma, x) \in \Gamma(X)$.

XIV $\quad (\langle 14, j_i + n, i\rangle, \tau, \sigma, 0) \in \Gamma(X)$ if $R_i(\tau)$
$(\langle 14, j_i + n, i\rangle, \tau, \sigma, 1) \in \Gamma(X)$ if $\neg R_i(\tau)$,
where $\mathrm{lh}(\tau) = j_i$.

XV $(\langle 15, j_i + n, i\rangle, \tau, \sigma, \varphi_i(\tau)) \in \Gamma(X)$ if $\tau \in \mathrm{dom}(\varphi_i)$.

XVI If $\forall x \exists y (e, x, \sigma, y) \in X$, then $(\langle 16, n, e, i\rangle, \sigma, F_i(f)) \in \Gamma(X)$, where $f(x) = y$ iff $(e, x, \sigma, y) \in X$.

XVII If $\forall x \in S, \exists y \in N[(e, x, \sigma, y) \in X]$ and $(e', z, \sigma, u) \in X$, then $(\langle 17, n, e, e'\rangle, \sigma, u) \in \Gamma(X)$, where $z \in \mathrm{Tp}(S)$ is defined by $z(x) = y$ iff $(e, x, \sigma, y) \in X$.

We now let $\mathrm{PR}(\mathbf{L}) = \Gamma_\infty$ = the least fixed-point for Γ. $\mathrm{PR}(\mathbf{L})$ *is the set of functions which are partial recursive in* $\mathbf{L}$. By dropping XIII we get $\mathrm{PRF}(\mathbf{L})$, the functions primitive recursive in $\mathbf{L}$.

As usual we define for $(e, \sigma, x) \in \mathrm{PR}(\mathbf{L})$

$$|e, \sigma, x|_{\mathbf{L}} = \text{least } \xi \text{ such that } (e, \sigma, x) \in \Gamma^{\xi+1}.$$

Let $\kappa_{\mathbf{L}} = \sup\{|e, \sigma, x|_{\mathbf{L}} : (e, \sigma, x) \in \mathrm{PR}(\mathbf{L})\}$.

4.1.5 Remark. We see that clause XIII is the reflection scheme which leads from the primitive recursive to the partial recursive functions. As before, we note that the scheme XI for primitive recursion can be removed in presence of XIII.

The scheme XVII is an extended form for substitution, i.e. the scheme S8 of Kleene [83]. If $\lambda x \cdot \{e\}(x, \sigma)$ is total, then it is an element of $\mathrm{Tp}(S)$ and hence can occur as an argument for a function g. If this function g is computable, i.e. $g(\alpha, \sigma) = \{e'\}(\alpha, \sigma)$, this means that given α, σ by some oracle, we then can compute $\{e'\}(\alpha, \sigma)$. If α itself is not given by some oracle, but by a computation procedure, $\alpha = \lambda s \cdot \{e\}(s, \sigma)$, we should be able to compute g on α, σ, *not* by appealing to an oracle for both α, σ, but by using the computation procedure for α and the oracle only for σ, i.e.

$$f(\sigma) = g(\alpha, \sigma) = \{e'\}(\lambda s \cdot \{e\}(s, \sigma), \sigma),$$

should belong to $\mathrm{PR}(\mathbf{L})$ with an *index* e'' simply computable from e and e'. And this is precisely the content of scheme XVII.

Let us fix some further terminology. Our computations are single-valued, hence there is no loss of information in abbreviating $(e, \sigma, x) \in \mathrm{PR}(\mathbf{L})$ by $\langle e, \sigma\rangle$. In general we call (e, σ, x) a computation. It is *convergent* if it belongs to $\mathrm{PR}(\mathbf{L})$, we denote this by $\langle e, \sigma\rangle\downarrow$. Conversely, $\langle e, \sigma\rangle\uparrow$ means that for no x is $(e, \sigma, x) \in \mathrm{PR}(\mathbf{L})$, and we say that the computation $\langle e, \sigma\rangle$ *diverges*.

We note that given any pair $\langle e, \sigma\rangle$ we can start checking whether it is a convergent computation or not. If $\langle e, \sigma\rangle\downarrow$, then we can from Definition 4.1.4 define a notion of *immediate subcomputation* and *subcomputation* exactly as in Definition 1.5.9. If $\langle e, \sigma\rangle\uparrow$, then either e is not an index according to Definition 4.1.4, or σ is not of the form required by e, in this case we let $\langle e, \sigma\rangle$ be an immediate subcomputation of itself. If $\langle e, \sigma\rangle$ "looks" like a convergent computation, we can again define the immediate subcomputations of $\langle e, \sigma\rangle$—in this case at least one of them must be divergent.

In any case, for every $\langle e, \sigma\rangle$ there is a *well-defined notion of immediate subcomputation and of subcomputation.* And we have the following simple, but important theorem.

4.1.6 Theorem. *For all $\langle e, \sigma\rangle$, $\langle e, \sigma\rangle\downarrow$ iff the computation tree (i.e. the set of subcomputations) is well-founded.*

We shall state in the present framework the first and second recursion theorem. First some definitions. Let

$$\mathbf{F}(\varphi_1, \ldots, \varphi_k, \sigma),$$

be a partial functional with values in S, i.e. $\mathbf{F}$ is partial and defined on partial functions. $\mathbf{F}$ is *monotone* if whenever $\mathbf{F}(\varphi_1, \ldots, \varphi_k, \sigma)\downarrow$ and $\varphi_i \subseteq \psi_i$, $i = 1, \ldots, k$, then $\mathbf{F}(\psi_1, \ldots, \psi_k, \sigma)\downarrow$ and $\mathbf{F}(\varphi_1, \ldots, \varphi_k, \sigma) \simeq \mathbf{F}(\psi_1, \ldots, \psi_k, \sigma)$.

4.1.7 Definition. A functional $\mathbf{F}$ is *partial recursive in a list* $\mathbf{L}$ if there is an index e such that for all $\varphi_1, \ldots, \varphi_k, \sigma$

$$\mathbf{F}(\varphi_1, \ldots, \varphi_k, \sigma) \simeq \{e\}_{\mathbf{L},\varphi_1\ldots\varphi_k}(\sigma).$$

$\mathbf{F}$ is *weakly partial recursive* in $\mathbf{L}$ if there is a primitive recursive function $f(n_1, \ldots, n_k)$ such that for all $e_1, \ldots, e_k, \sigma_1, \ldots, \sigma_k$, where $\mathrm{lh}(\sigma_i) = n_i$, $i = 1, \ldots, k$,

$$\mathbf{F}(\lambda\tau_1\cdot\{e_1\}_{\mathbf{L}}(\tau_1, \sigma_1), \ldots, \lambda\tau_k\cdot\{e_k\}_{\mathbf{L}}(\tau_k, \sigma_n), \sigma) \simeq \{f(n_1, \ldots, n_k)\}_{\mathbf{L}}(e_1, \ldots, e_k, \sigma_1, \ldots, \sigma_k, \sigma).$$

(The reader should compare this with the distinction between weak and strong computability in Section 1.1.7.)

4.1.8 First Recursion Theorem. *Let $\mathbf{F}$ be monotone and weakly recursive in $\mathbf{L}$. Then there is a least φ such that for all σ, $F(\varphi, \sigma) \simeq \varphi(\sigma)$, and this φ is partial recursive in $\mathbf{L}$.*

The proof is standard and therefore omitted, see the similar proofs in 1.7 and 2.3.

4.1.9 Second Recursion Theorem. *For all indices e there is some x such that for all lists $\mathbf{L}$ and all σ*

$$\{e\}_{\mathbf{L}}(x, \sigma) \simeq \{x\}_{\mathbf{L}}(\sigma).$$

The proof is again standard, see 1.2.6.

We will also need a fixed-point theorem for the class PRF, which is generated by clauses I–XII of 4.1.4. Let $\{e\}_{\mathrm{PRF}}$ be the primitive recursive function with index e. We have the following result.

4.1.10 PRF **Recursion Theorem.** *Let* $f(e, \sigma)$ *be* PRF. *Then there is an index e such that for all* σ

$$f(e, \sigma) = \{e\}_{\mathrm{PRF}}(\sigma).$$

It remains to verify that every **F** which is partial recursive in **L** also is weakly partial recursive in **L**. This is an immediate consequence of the following lemma. Note that the converse is not true, as a simple cardinality argument shows.

4.1.11 Lemma. *Let* $\varphi = \lambda\tau\cdot\{e\}_{\mathbf{L}}(\tau, \sigma)$. *There is a primitive recursive function f such that for all* x, σ', y:

$$\{x\}_{\mathbf{L},\varphi}(\sigma') \simeq y \quad \textit{iff} \quad \{f(x)\}_{\mathbf{L}}(\sigma', \sigma) \simeq y.$$

We indicate briefly the proof: We define a primitive recursive function g by cases, one for each clause in the inductive definition of $\{x\}_{\mathbf{L},\varphi}(\sigma')$. Let $\mathrm{lh}(\tau) = k$, $\mathrm{lh}(\sigma) = l$, $\mathrm{lh}(\sigma') = n$.

I. $x = \langle 1, n + 1\rangle$: Let $g(x, t) = \langle 1, n + l + 1\rangle$.

Clauses II–IX are treated similarly.

X. $x = \langle 10, n, e, e'\rangle$: Let $g(x, t) = \langle 10, n + l, g(e, t), g(e', t)\rangle$.

We omit XI and XII.

XIII. $x = \langle 13, n + 1\rangle$: There is a primitive recursive function h such that for all $t, r, \sigma'', \mathbf{L}$: $\{h(t)\}_{\mathbf{L}}(r, \sigma'') \simeq \{\{t\}_{\mathrm{PRF}}(r)\}_{\mathbf{L}}(\sigma'')$. Let $g(x, t) = h(t)$.

XV. (Introduction of φ.) $x = \langle 15, k + n, i\rangle$. There is a primitive recursive function s such that for all σ'' of length n: $\{e\}_{\mathbf{L}}(\tau, \sigma) \simeq \{s(e, n)\}_{\mathbf{L}}(\tau, \sigma'', \sigma)$. Let $g(x, t) = s(e, n)$, where $n + k = \mathrm{lh}(\tau, \sigma'')$.

We now use the fixed-point Theorem 4.1.10 for PRF to get a t_0 such that $g(x, t_0) = \{t_0\}_{\mathrm{PRF}}(x)$ for all x. We define $f(x) = g(x, t_0)$, and prove the equivalence of the lemma by inductions on $|\{x\}_{\mathbf{L},\varphi}(\sigma')|_{\mathbf{L},\varphi}$ and $|\{f(x)\}_{\mathbf{L}}(\sigma', \sigma)|_{\mathbf{L}}$, respectively.

4.2 Recursion in a Normal List

In the first section of this chapter we have constructed a computation theory PR(L) on domains of the type $\mathfrak{A} = \langle A, S, \mathbf{S}\rangle$. In this section we shall study computation theories which correspond to Kleene recursion in normal objects of higher types. In Chapter 3 we studied the abstract version of recursion in normal type-2 objects. The prototype for the following theory is Kleene recursion in ${}^3E$ on the integers and the reals.

4.2.1 Definition. A list **L** is called *normal* if the functional $\mathbf{E}'_A$ is weakly recursive in **L**, the equality relation on S is recursive in **L**, and **L** contains no partial functions. Remember that $\mathbf{E}'_A$ is defined on total functions $f: A \to S$.

4.2.2 Remarks. 1. Usually "normal" is defined in a stronger sense, viz. by using "recursive in **L**" in place of "*weakly* recursive in **L**". The proofs show that the weak notion suffices in most cases; we shall note an exception below.

2. In the concrete setting of higher types one need *not* require that the equality relation on S, which in this case is $\mathrm{Tp}(0) \cup \ldots \cup \mathrm{Tp}(n-1)$, is recursive in **L**.

3. We remind the reader that if **L** is normal, then the relations recursive in **L** are closed under $\forall$ and $\exists$.

For normal lists **L** we have the basic prewellordering theorem and the selection theorem as for Spector theories, see Chapter 3. The proofs are the same, hence we only state the relevant results.

4.2.3 Theorem. *Let* PR(L) *be a normal theory on* $\mathfrak{A} = \langle A, S, \mathrm{S}\rangle$. *Then* PR(L) *is p-normal, i.e. there is a function p partial recursive in* PR(L) *such that*

(i) $x \in \mathbf{C_L}$ *or* $y \in \mathbf{C_L}$ *iff* $p(x, y)\downarrow$,
(ii) $x \in \mathbf{C_L}$ *and* $|x|_{\mathbf{L}} \leqslant |y|_{\mathbf{L}} \Rightarrow p(x, y) = 0$,
$|x|_{\mathbf{L}} > |y|_{\mathbf{L}} \Rightarrow p(x, y) = 1$.

Here $\mathbf{C_L} = \{\langle e, \sigma\rangle : \{e\}_{\mathbf{L}}(\sigma)\downarrow\}$ is the set of computations in PR(L).

4.2.4 Theorem. *Let* PR(L) *be normal on* $\mathfrak{A} = \langle A, S, \mathrm{S}\rangle$. *There is a function* $\varphi \in$ PR(L) *such that for all e,* σ*: If*

$$\exists n \in N\{e\}_{\mathbf{L}}(n, \sigma)\downarrow \;\Rightarrow\; \varphi(e, \sigma)\downarrow \quad \text{and} \quad \{e\}_{\mathbf{L}}(\varphi(e, \sigma), \sigma)\downarrow .$$

Moreover if $\varphi(e, \sigma) \simeq n$, *then* $\{e\}_{\mathbf{L}}(n, \sigma)\downarrow$. *The index of* φ *is primitive recursive in* $\mathrm{lh}(\sigma)$.

4.2.4 is obtained by the standard argument from 4.2.3, see the proof of 3.1.6. We have the usual corollaries, i.e. the PR(L)-semicomputable relations are closed under $\vee$ and $\exists n \in N$.

We shall now prove that the PR(L)-semicomputable relations are *not* closed under $\exists x \in \mathrm{Tp}(S)$. An important result about normal theories PR(L) is that they are closed under $\exists s \in S$; this will be the topic of the next section.

We first need more precise information about subcomputations and computation trees.

4.2.5 Lemma. *Let* $=_S$ *be recursive in* **L** *and* $\mathbf{E}'_S$ *weakly recursive in* **L**. *There is a relation* $S(x, y)$ *semicomputable in* PR(L), *such that if* $x \in \mathbf{C_L}$ *then*

$$S(x, y) \quad \textit{iff} \quad y \textit{ is an immediate subcomputation of } x.$$

The set $\{y : S(x, y)\}$ *is recursive in* x, **L** *when* $x \in \mathbf{C_L}$.

The proof of the lemma comes from an analysis of the clauses of the inductive definition of PR(**L**); see Definition 4.1.4. The assumptions on **L** are required so as to decide the form of x, i.e. to decide the form of the index of the tuple. To do this we need e.g. the predicate Seq., see Lemma 4.1.3. Otherwise the proof is routine.

4.2.6 Lemma. *Let* **L** *be normal. If* $\{e\}_{\mathbf{L}}(\sigma)\downarrow$, *then the computation tree of* $\langle e, \sigma\rangle$ *is recursive in* $\mathbf{L}, \sigma$.

Proof. Let q be the least fixed-point for the monotone **L**-recursive functional **F** defined as follows:

$$\begin{aligned}\mathbf{F}(q, x, y) &\simeq 0 \quad \text{if} \quad x \in \mathbf{C_L} \quad \text{and} \quad (S(x, y) \vee \exists z[S(x, z) \\ &\qquad\qquad\qquad\qquad\qquad\qquad\qquad\qquad \wedge\ q(z, y) \simeq 0]) \\ &\simeq 1 \quad \text{if} \quad x \in \mathbf{C_L} \quad \text{and} \quad (\neg S(x, y) \wedge \forall z[S(x, z) \\ &\qquad\qquad\qquad\qquad\qquad\qquad\qquad\qquad \rightarrow q(z, y) \simeq 1]).\end{aligned}$$

Note that the quantifiers in **F** can be handled by $\mathbf{E}'_A$. We now appeal to the first recursion theorem and obtain a solution q such that $q(x, y)\downarrow$ iff $x \in \mathbf{C_L}$.

If $x \in \mathbf{C_L}$, then $\lambda y \cdot q(\langle e, \sigma\rangle, y)$ is the (**L**-recursive) characteristic function of the computation tree of $\{e\}_{\mathbf{L}}(\sigma)$.

In connection with the coding functions on $\mathfrak{A}$ we introduced two mappings $*: S \rightarrow \mathrm{Tp}(S)$ and $^{-}: A \rightarrow S$. We shall need some further coding and translation devices.

First let Y be a set of elements in Tp(S) indexed by S, i.e. $Y = \{\alpha_r : r \in S\}$. Then all elements in Y can be coded by *one* element in Tp(S),

$$\alpha(r) = \alpha_{(r)_1}((r)_2).$$

For all $r \in S$, $\lambda s \cdot \alpha(\langle r, s\rangle) = \alpha_r$.

Further there is a one-one mapping $**: A \rightarrow \mathrm{Tp}(S)$ and a mapping $^{--}: A \rightarrow A$ such that the graphs of $**$ and $^{--}$ are primitive recursive in $=_S$ and $\mathbf{E}'_S$ and such that $(x^{**})^{--} = x$, for all $x \in A$. One way of defining $**$ and $^{--}$ is as follows:

$$\begin{aligned} r^{**} &= \langle r^*, \mathbf{0}\rangle \\ \alpha^{**} &= \langle \alpha, \mathbf{1}\rangle, \end{aligned}$$

where $\mathbf{0}, \mathbf{1} \in \mathrm{Tp}(S)$ are the constant functions. And

$$x^{--} = \begin{cases} (x)_1 & \text{if} \quad (x)_2(0) = 1 \\ (x)_1^{-} & \text{if} \quad (x)_2(0) = 0. \end{cases}$$

4.2.7 Theorem. *Let* $=_S$ *be* **L***-recursive and* $\mathbf{E}'_S$ *weakly* **L***-recursive. There is* PR(**L**)*-semicomputable relation R such that for all e,* σ*:*

$\{e\}_{\mathbf{L}}(\sigma)\uparrow$ *iff* $\exists\alpha R(\alpha, \langle e, \sigma\rangle)$.

Proof: We use the fact of Theorem 4.1.6: $\{e\}_{\mathbf{L}}(\sigma)$ diverges iff the computation tree of $\langle e, \sigma\rangle$ is not well-founded. The condition for this is that we have an infinite descending chain of immediate subcomputations of $\langle e, \sigma\rangle$, i.e.

$$\exists\alpha_0\exists\alpha_1 \ldots \exists\alpha_n \ldots [\alpha_0^{--} = \langle e, \sigma\rangle \wedge \forall i\cdot S(\alpha_i^{--}, \alpha_{i+1}^{--})].$$

By the remark above the chain can be coded by one element of Tp(S). The sought for relation R is then simply

$$R(\alpha, \langle e, \sigma\rangle) \quad \text{iff} \quad \alpha[0]^{--} = \langle e, \sigma\rangle \wedge \forall i\cdot S(\alpha[i]^{--}, \alpha[i+1]^{--}),$$

where S comes from Lemma 4.2.5 and $\alpha[r] = \lambda s\cdot\alpha(\langle r, s\rangle)$.

4.2.8 Corollary. *The relations which are* PR(**L**)*-semicomputable are not closed under* $\exists\alpha \in \mathrm{Tp}(S)$ *when* **L** *is a normal list.*

4.2.9 Remark. So far we have needed only weak **L**-recursiveness. Here is one example where (strong) **L**-recursiveness is necessary. Call a relation $R \subseteq 2^A \times A$ recursive in **L** if there is an index e such that $\lambda X\lambda x\cdot\{e\}_{\mathbf{L},X}$ is the characteristic function of R.

We have the following normal form theorem: Let **L** be normal and assume that $\mathbf{E}'_A$ is recursive in **L**. Let $B \subseteq A$: Then B is **L**-semicomputable iff there exists $R \subseteq 2^A \times A$ which is recursive in **L** and such that for all $x \in A$:

(*) $\quad x \in B \quad \text{iff} \quad \exists X(X \text{ is recursive in } x, \mathbf{L} \wedge R(X, x)).$

Note that this is reminiscent of the Gandy-Spector theorem, but far weaker, however. In the Gandy-Spector theorem the relation R is arithmetic over the domain.

The proof is omitted, but can be found in Moldestad [105], page 40. We note that the X on the RHS of (*) should be the computation tree of the computation $\{e_0\}_{\mathbf{L}}(x) \simeq 0$, which asserts that $x \in B$. This tree is recursive by Lemma 4.2.6. And we need the strong recursiveness of $\mathbf{E}'_A$ since in the analysis of the tree we meet conditions of the form $\forall x(\ldots x, X, \ldots)$. Weak recursiveness cannot handle such clauses.

4.3 Selection in Higher Types

We now come to one of the most important results about PR(**L**), where **L** is a normal list on a domain of two types $\mathfrak{A} = \langle A, S, \mathbf{S}\rangle$. As we remarked in 4.2.4,

PR(**L**) admits selection operators over N, hence the PR(**L**)-semicomputable relations are closed under $\exists n \in N$. In 4.2.8 we saw that the class is not closed under $\exists \alpha \in \mathrm{Tp}(S)$. In this section we prove that the class is closed under $\exists s \in S$.

4.3.1 Theorem. *Let* **L** *be a normal list on* $\mathfrak{A} = \langle A, S, \mathbf{S}\rangle$. *There is a function* φ *partial recursive in* **L** *with index* $\hat{e}$ *such that if*

$$\exists x \in S \cdot \{e\}_{\mathbf{L}}(x, \sigma)\downarrow,$$

then $\varphi(\langle e, \sigma\rangle)\downarrow$, *and in this case*

$$|\{\hat{e}\}_{\mathbf{L}}(\langle e, \sigma\rangle)|_{\mathbf{L}} \geqslant \min_{x \in S}\{|\{e\}_{\mathbf{L}}(x, \sigma)|_{\mathbf{L}}\}.$$

Conversely, if $\varphi(\langle e, \sigma\rangle)\downarrow$, *then* $\exists x \in S \cdot \{e\}_{\mathbf{L}}(x, \sigma)\downarrow$.

This theorem immediately entails the closure of the PR(**L**)-semirecursive relations under $\exists s \in S$. As we have remarked before, the existence of a single-valued selection operator is closely connected to the existence of a computable well-ordering of the domain. The next best thing without going all the way to multiple-valued theories, is to be able to compute something, in 4.3.1 $\varphi(\langle e, \sigma\rangle)$, which effectively gives us a non-empty computable subset of a given semicomputable set. (The reader should refer back to our discussion in connection with 3.1.10.)

We spell out the details: Let B be an **L**-semicomputable subset of S, i.e.

$$B = \{x \in S : \{e\}_{\mathbf{L}}(x, \sigma)\downarrow\}.$$

We note that $\varphi(\langle e, \sigma\rangle)\downarrow$ iff $B \neq \varnothing$. And if $B \neq \varnothing$, then

$$B_0 = \{x \in S : |\{e\}_{\mathbf{L}}(x, \sigma)|_{\mathbf{L}} \leqslant |\varphi(\langle e, \sigma\rangle)_{\mathbf{L}}|\},$$

is a non-empty **L**-computable subset of B. Obviously, an index for B_0 can be computed in PR(**L**) from the given data.

The selection principle involved in 4.3.1 was first stated by Grilliot [48] in the context of a theory of inductive definability. His proof was, however, defective. A first proof in the context of Kleene recursion in higher types was given in the thesis of D. MacQueen [98], and was published in a somewhat abstract version in L. Harrington and D. MacQueen [55]. A proof, based on MacQueen's thesis, was worked out by Moldestad [105] in the framework of computation theories on two types. It is this proof that we will present in this section.

The remainder of this section is devoted to a proof of Theorem 4.3.1. We make one simple reduction. The set $\{\langle e, x, \sigma\rangle^{**} : x \in S\}$ is a family of elements of $\mathrm{Tp}(S)$ indexed by S, and hence can be coded by one element $\alpha \in \mathrm{Tp}(S)$. Then

$$\exists s \in S \cdot \{e\}_{\mathbf{L}}(s, \sigma)\downarrow \quad \text{iff} \quad \exists s \in S \cdot \alpha[s]^{--} \in \mathbf{C}_{\mathbf{L}},$$

where $\mathbf{C}_{\mathbf{L}}$ is the set of convergent computations, and as above $\alpha[s] = \lambda r \cdot \alpha(\langle s, r\rangle)$.

4.3.2 Definition. For any $\beta \in \mathrm{Tp}(S)$ let $\|\beta\| = \min\{|\beta[s]^{--}|_{\mathbf{L}} : s \in S\}$.

We see that in order to prove Theorem 4.3.1 it suffices to prove the following lemma.

4.3.3 Lemma. *There is an index m such that*

(i) $\|\beta\| < \kappa_{\mathbf{L}} \Rightarrow \{m\}_{\mathbf{L}}(\beta)\downarrow \;\wedge\; \|\beta\| < |\{m\}_{\mathbf{L}}(\beta)|$.
(ii) $\{m\}_{\mathbf{L}}(\beta)\downarrow \;\Rightarrow\; \|\beta\| < \kappa_{\mathbf{L}}$.

The proof will use the recursion theorem. We shall, by induction on the norm of β, i.e. on $\|\beta\|$, show how to define $\{m\}_{\mathbf{L}}(\beta)$. We assume as induction hypothesis

$$\|\beta\| < \mu \Rightarrow \{m\}_{\mathbf{L}}(\beta)\downarrow \;\wedge\; \|\beta\| \leqslant |\{m\}_{\mathbf{L}}(\beta)|_{\mathbf{L}}.$$

Assume that $\alpha \in \mathrm{Tp}(S)$ is such that $\|\alpha\| = \mu$. We shall show how to define $\{m\}_{\mathbf{L}}(\alpha)$. Note that α is kept fixed for the rest of the proof.

Intuitively, α is meant to code a family of computations $\{\langle e, x, \sigma\rangle : x \in S\}$. In order to compute something which dominates some computation in this family, we need to go into a detailed analysis of subcomputations. Lemma 4.2.5 tells us that there is an $\mathbf{L}$-semicomputable relation $S(x, y)$ such that if $x \in \mathbf{C}_{\mathbf{L}}$, then y is an immediate subcomputation of x iff $S(x, y)$.

Except for *substitution*, the notion of *immediate subcomputation* is unproblematic. In the following analysis we control the search for immediate subcomputations through the following "truncated" version of $S(x, y)$: Let $R(x, y, w)$ be a relation such that if $w \in \mathbf{C}_{\mathbf{L}}$:

1 If x is not a substitution, i.e. x is not of the form $\langle\langle 10, n, e, e'\rangle, \sigma\rangle$, then $\{y : R(x, y, w)\} = \{y : S(x, y)\}$.
2 If x is a substitution, i.e. $x = \langle\langle 10, n, e, e'\rangle, \sigma\rangle$, then

$$\{y : R(x, y, w)\} = \begin{cases} \{\langle e, \sigma\rangle\} & \text{if} \quad |w|_{\mathbf{L}} < |\{e\}_{\mathbf{L}}(\sigma)|_{\mathbf{L}}, \\ \{\langle e, \sigma\rangle, \langle e', \{e\}_{\mathbf{L}}(\sigma), \sigma\rangle\} & \text{otherwise.} \end{cases}$$

We see how we use a $w \in \mathbf{C}_{\mathbf{L}}$ as a "cut-off" measure in searching for the immediate subcomputations of x. Note that if $x \in \mathbf{C}_{\mathbf{L}}$, then there is some $w \in \mathbf{C}_{\mathbf{L}}$ such that $R(x, y, w)$ iff $S(x, y)$.

For $\sigma < \kappa_{\mathbf{L}}$ we introduce a set

$$T_\sigma = \{\beta : \forall x \in S \cdot R(\alpha[x]^{--}, \beta[x]^{--}, w)\},$$

where $|w|_{\mathbf{L}} = \sigma$.

Note. At this point there is a contradiction in our notation. But don't make a fuss! Even if our reader refuses to go into the details of the proof, he or she should have no difficulties in correctly classifying the occurrences of the letter σ, whether it is an ordinal or an input sequence of a computation.

Any $\beta \in T_\sigma$ is intended to represent a family of computations which up to the order of approximation $\sigma = |w|_{\mathbf{L}}$ are "pointwise" immediate subcomputations of the family α. Note that

$$\beta \in T_\sigma \Rightarrow \|\beta\| < \|\alpha\|$$
$$\sigma < \tau \Rightarrow \mathrm{T}_\sigma \subseteq \mathrm{T}_\tau.$$

We observe that T_σ is recursive in $\mathbf{L}$, α, w where $|w|_{\mathbf{L}} = \sigma$.

It is clear that by searching through "enough" ordinals σ we can decide whether any β codes a family of immediate subcomputations of α. What "enough" means is made precise in the following lemma.

4.3.4 Lemma. *Let λ be an ordinal such that S is not cofinal in λ. Let $\{\sigma(\tau) : \tau < \lambda\}$ be an increasing sequence of ordinals bounded by $\kappa_{\mathbf{L}}$. Then there exists an ordinal $\tau' < \lambda$ such that*

$$T_{\sigma(\tau)} = T_{\sigma(\tau')},$$

for all τ satisfying $\tau' \leqslant \tau < \lambda$.

We postpone the proof of the lemma. We shall apply it in the following situation. Let

$$W = \{\gamma : \mathrm{PWO}(\gamma) \wedge \mathrm{dom}(\gamma) \subseteq S\}.$$

W is recursive in $\mathbf{L}$—we have prewellorderings of S, not of $A = S \cup \mathrm{Tp}(S)$. This is the point where an attempt to extend the result from $\exists x \in S$ to $\exists x \in A$ would fail, and necessarily so as 4.2.8 shows.

For $\delta \in W$, let $\mathrm{or}(\delta) =$ length of the pwo δ. Letting $\lambda = \sup\{\mathrm{or}(\delta) : \delta \in W\}$, we observe that S is not cofinal in λ.

We are now ready for the inductive computation of $\{m\}_{\mathbf{L}}(\alpha)$:

(i) There exists an index m_1 such

$$\{m_1\}_{\mathbf{L}}(m, \alpha, w)\downarrow \quad \text{iff} \quad w \in \mathbf{C}_{\mathbf{L}} \wedge \{m\}_{\mathbf{L}}(\beta)\downarrow, \quad \text{for all } \beta \in T_{|w|}$$

and if $w \in \mathbf{C}_{\mathbf{L}}$, then

$$|\{m_1\}_{\mathbf{L}}(m, \alpha, w)|_{\mathbf{L}} > |\{m\}_{\mathbf{L}}(\beta)|_{\mathbf{L}}, \quad \text{for all } \beta \in T_{|w|}.$$

We note that it follows from the induction hypothesis that $|\{m_1\}_{\mathbf{L}}(m, \alpha, w)| > \|\beta\|$ for all $\beta \in T_{|w|}$.

(ii) There is an index m_2 (using the recursion theorem) such that

$$\{m_2\}_{\mathbf{L}}(m, \alpha, \gamma)\downarrow \quad \text{iff} \quad \gamma \in W \wedge \forall \gamma' \in W[\mathrm{or}(\gamma') < \mathrm{or}(\gamma)$$
$$\Rightarrow \{m_2\}_{\mathbf{L}}(m, \alpha, \gamma')\downarrow \wedge \{m_1\}_{\mathbf{L}}(m, \alpha, \langle m_2, m, \alpha, \gamma'\rangle)\downarrow].$$

m_2 can be chosen such that for all $\gamma' \in W$ with $\text{or}(\gamma') < \text{or}(\gamma)$:

$$|\{m_2\}_{\mathbf{L}}(m, \alpha, \gamma)|_{\mathbf{L}} > |\{m_2\}_{\mathbf{L}}(m, \alpha, \gamma')|_{\mathbf{L}}, |\{m_1\}_{\mathbf{L}}(m, \alpha, \langle m_2, m, \alpha, \gamma'\rangle)|_{\mathbf{L}}.$$

It follows from the induction hypothesis that $|\{m_2\}_{\mathbf{L}}(m, \alpha, \gamma)| > \|\beta\|$ for all $\beta \in T_{|\langle m_2, m, \alpha, \gamma'\rangle|}$, provided $\text{or}(\gamma') < \text{or}(\gamma)$.

(iii) There is an index m_3 such that

$$\{m_3\}_{\mathbf{L}}(m, \alpha)\downarrow \quad \text{iff} \quad \forall\gamma \in W\{m_2\}_{\mathbf{L}}(m, \alpha, \gamma)\downarrow,$$

and such that whenever $\gamma \in W$

$$|\{m_3\}_{\mathbf{L}}(m, \alpha)| > |\{m_2\}_{\mathbf{L}}(m, \alpha, \gamma)|.$$

By the recursion theorem we now find an index m such that

$$\{m\}_{\mathbf{L}}(\alpha) \simeq \{m_3\}_{\mathbf{L}}(m, \alpha),$$

and

$$|\{m\}_{\mathbf{L}}(\alpha)| \geqslant |\{m_3\}_{\mathbf{L}}(m, \alpha)|.$$

Part (i) of Lemma 4.3.3 now follows if we can show that $|\{m_3\}_{\mathbf{L}}(m, \alpha)| \geqslant \|\alpha\|$.

This will follow from Lemma 4.3.4 applied to the ordinal λ derived from the set of prewellorderings W. In detail, define for each $\tau < \lambda$

$$\sigma(\tau) = \inf\{|\{m_2\}_{\mathbf{L}}(m, \alpha, \gamma)|_{\mathbf{L}}; \gamma \in W \wedge \text{or}(\gamma) = \tau\}.$$

Then $\{\sigma(\tau) : \tau < \lambda\}$ is an increasing sequence bounded by $|\{m_3\}_{\mathbf{L}}(m, \alpha)|_{\mathbf{L}} < \kappa_{\mathbf{L}}$. Lemma 4.3.4 applies, i.e. there is an ordinal $\tau' < \lambda$ such that $T_{\sigma(\tau)} = T_{\sigma(\tau')}$ whenever $\tau' \leqslant \tau < \lambda$. Let $\sigma = \sup\{\sigma(\tau); \tau < \lambda\}$. The induction hypothesis gives $|\{m_3\}_{\mathbf{L}}(m, \alpha)|_{\mathbf{L}} \geqslant \sigma$. The ordinal of a computation is determined in terms of the ordinals of the immediate subcomputations. We have controlled the immediate subcomputations of the family α by the sets $T_{\sigma(\tau)}$ and the ordinal σ, hence by the computation $|\{m_3\}_{\mathbf{L}}(m, \alpha)|_{\mathbf{L}}$. So, we expect, and it only formally remains to verify:

4.3.5 Claim. $\sigma \geqslant \|\alpha\|$,

to end the proof of Lemma 4.3.3 (i). Part (ii) of the lemma will then be proved by induction on the length $|\{m\}_{\mathbf{L}}(\alpha)|$.

The reader who is not particularly interested in the fine details of this "cleaning-up-business" may proceed to the next section.

4.3.6 Proof of Lemma 4.3.4. We proceed by contradiction, i.e. assume that $\forall\tau' < \lambda$ $\exists\tau(\tau' \leqslant \tau < \lambda \wedge T_{\sigma(\tau')} \subsetneqq T_{\sigma(\tau)})$, and construct a function $f: S \to \lambda$ such that $\lambda = \sup\{f(x) : x \in S\}$.

Take $\tau' < \lambda$ and choose τ_0 minimal such that $\tau' \leqslant \tau_0$ and $T_{\sigma(\tau')} \subsetneqq T_{\sigma(\tau_0)}$. Obviously $\tau' < \tau_0$. Choose $w, w' \in \mathbf{C_L}$ such that $\tau' = |w'|$ and $\tau_0 = |w|$.

If $\beta \in T_{\sigma(\tau_0)} - T_{\sigma(\tau')}$, then

$$\forall x \in S \cdot R(\alpha[x]^{--}, \beta[x]^{--}, w)$$
$$\exists x \in S \cdot \neg R(\alpha[x]^{--}, \beta[x]^{--}, w').$$

If $\neg R(\alpha[x]^{--}, \beta[x]^{--}, w')$ but $R(\alpha[x]^{--}, \beta[x]^{--}, w)$, then $\alpha[x]^{--}$ is a substitution $\langle\langle 10, n, e, e'\rangle, \sigma\rangle$, $\beta[x]^{--} = \langle e', \{e\}_{\mathbf{L}}(\sigma), \sigma\rangle$, and $|w'| < |\{e\}_{\mathbf{L}}(\sigma)| \leqslant |w|$. Hence $R(\alpha[x]^{--}, \beta[x]^{--}, w'')$ for all $|w''| \geqslant |w|$.

Let

$$P(\tau') = \{x \in S : \exists \beta \in T_{\sigma(\tau_0)} - T_{\sigma(\tau')}, \neg R(\alpha[x]^{--}, \beta[x]^{--}, w')\}.$$

From the remarks above it follows that $P(\tau') \neq \varnothing$, but $P(\tau') = P(\nu)$, $\tau' \leqslant \nu < \tau_0$; and that $P(\tau') \cap P(\nu) = \varnothing$ if $\nu \geqslant \tau_0$.

Let $f\colon S \to A$ be defined by

$$f(x) = \begin{cases} \text{least } \tau' \text{ such that } x \in P(\tau'), & \text{if} \quad x \in \bigcup P(\tau),\ \tau < \lambda \\ 0, \quad \text{otherwise.} \end{cases}$$

Clearly, $\sup\{f(x) : x \in S\} = \lambda$.

4.3.7 Proof of Claim 4.3.5. We argue, once more, by contradiction, i.e. we assume that $\sigma < \|\alpha\|$. We keep the notations from the claim: in particular, the ordinals τ' and σ have their established meaning.

We make a preliminary remark: Let $x \in S$. If $\alpha[x]^{--}$ is a substitution, then either $|\{e\}_{\mathbf{L}}(\sigma)|_{\mathbf{L}} \leqslant \sigma(\tau')$ or $\sigma \leqslant |\{e\}_{\mathbf{L}}(\sigma)|_{\mathbf{L}}$. (For if $\sigma(\tau') < |\{e\}_{\mathbf{L}}(\sigma)|_{\mathbf{L}} < \sigma$, take a $\beta \in T_{\sigma(\tau')}$ and let $\beta'[y] = \beta[y]$ if $y \neq x$, $\beta'[x]^{--} = \langle e', \{e\}_{\mathbf{L}}(\sigma), \sigma\rangle$. Then $\beta' \in T_{\sigma(\tau')}$. But $\beta' \in T_{\sigma(\tau)}$ for $\sigma(\tau) > |\{e\}_{\mathbf{L}}(\sigma)|_{\mathbf{L}}$. This contradicts the fact that $T_{\sigma(\tau)} = T_{\sigma(\tau')}$.)

Using the assumption $\sigma < \|\alpha\|$ we construct a β in the following way:

(i) If $\alpha[x]^{--}$ is not a substitution, let $\beta[x]^{--}$ be such that $S(\alpha[x]^{--}, \beta[x]^{--})$ and such that $|\beta[x]^{--}| \geqslant \sigma$.

The inequality uses the fact that $\sigma < \|\alpha\| = \min\{|\alpha[x]^{--}|_{\mathbf{L}} : x \in S\}$.

(ii) If $\alpha[x]^{--}$ is a substitution $\langle\langle 10, n, e, e'\rangle, \sigma\rangle$ let

$$\beta[x]^{--} = \begin{cases} \langle e', \{e\}_{\mathbf{L}}(\sigma), \sigma\rangle & \text{if} \quad |\{e\}_{\mathbf{L}}(\sigma)| \leqslant \sigma(\tau') \\ \langle e, \sigma\rangle & \text{if} \quad |\{e\}_{\mathbf{L}}(\sigma)| \geqslant \sigma \end{cases}$$

By construction $\beta \in T_{\sigma(\tau')}$.

We now claim that $\|\beta\| \geqslant \sigma$: By definition $\|\beta\| = \min\{|\beta[x]|^{--} : x \in S\}$. From (i) and (ii) we see that $|\beta[x]^{--}| \geqslant \sigma$ in all cases except possibly when $|\{e\}_{\mathbf{L}}(\sigma)|_{\mathbf{L}} \leqslant \sigma(\tau')$. But even in this case $|\beta[x]^{--}| = |\{e'\}_{\mathbf{L}}(\{e\}_{\mathbf{L}}(\sigma), \sigma)| \geqslant \sigma$, because otherwise $|\alpha[x]^{--}| \leqslant \sigma$, which contradicts the assumption $\sigma < \|\alpha\|$.

Combining this claim with the fact that $\beta \in T_{\sigma(\tau')}$ by construction, we derive the final contradiction: Choose τ such that $\tau' < \tau < \lambda$. Choose $\gamma', \gamma \in W$ such that $\mathrm{or}(\gamma') = \tau'$, $\mathrm{or}(\gamma) = \tau$, $\sigma(\tau') = |\{m_2\}_{\mathbf{L}}(m, \alpha, \gamma')|$, and $\sigma(\tau) = |\{m_2\}_{\mathbf{L}}(m, \alpha, \gamma)|$.

By construction of m_2, $|\{m_2\}_{\mathbf{L}}(m, \alpha, \gamma) \geqslant \|\beta'\|$ for all $\beta' \in T_{|\langle m_2, m, \alpha, \gamma'\rangle|}$. In particular, $|\{m_2\}_{\mathbf{L}}(m, \alpha, \gamma)| \geqslant \|\beta\|$, since $\beta \in T_{\sigma(\tau')} = T_{|\langle m_2, m, \alpha, \gamma'\rangle|}$, contradicting the fact that $|\{m_2\}_{\mathbf{L}}(m, \alpha, \gamma)| = \sigma(\tau) < \sigma \leqslant \|\beta\|$.

Note how this proof bears out our remarks in connection with 4.3.5. We have controlled the subcomputations of α by the ordinal σ, hence $\|\alpha\| \leqslant \sigma$. (See, in particular, the verification that $\|\beta\| > \sigma$.)

4.3.8 Proof of (ii) in Lemma 4.3.3. This is a standard inductive argument on the ordinal of $\{m\}_{\mathbf{L}}(\sigma)$.

Let $\{m\}_{\mathbf{L}}(\alpha)\downarrow$ and assume that 4.3.3 (ii) is satisfied for all β such that $|\{m\}_{\mathbf{L}}(\beta)| < |\{m\}_{\mathbf{L}}(\alpha)|$. Since $\{m\}_{\mathbf{L}}(\alpha)\downarrow$, we have $\{m_3\}_{\mathbf{L}}(m, \alpha)\downarrow$ and $|\{m\}_{\mathbf{L}}(\alpha)| > |\{m_3\}_{\mathbf{L}}(m, \alpha)|$. Further $\{m_2\}_{\mathbf{L}}(m, \alpha, \gamma)\downarrow$ for all $\gamma \in W$ and $|\{m_3\}(m, \alpha)| > |\{m_2\}(m, \alpha, \gamma)|$ for all such γ. Let the ordinals $\{\sigma(\tau) : \tau < \lambda\}$ be defined as above, and choose $\tau' < \lambda$ as before.

We recall that if $\alpha[x]^{--}$ is a substitution, then either $|\{e\}_{\mathbf{L}}(\sigma)| \leqslant \sigma(\tau')$ or $\sigma \leqslant |\{e\}_{\mathbf{L}}(\sigma)|$; see the proof of 4.3.5.

We argue once more by contradiction and so assume that $\|\alpha\| = \kappa_{\mathbf{L}}$, i.e. $\alpha[x]^{--}$ codes a divergent computation for all $x \in S$.

Construct a β as follows: If $\alpha[x]^{--}$ is not a substitution, let $\beta[x]^{--}$ be a divergent subcomputation of $\alpha[x]^{--}$. If $\alpha[x]^{--}$ is a substitution let $\beta[x]^{--}$ be constructed as in (ii) of 4.3.7. By construction $\beta \in T_{\sigma(\tau')}$, and we see that $|\beta[x]^{--}| \geqslant \sigma$ for all $x \in S$; hence $\|\beta\| \geqslant \sigma$.

Choose $\gamma' \in W$ such that $\sigma(\tau') = |\{m_2\}_{\mathbf{L}}(m, \alpha, \gamma')|$, and pick a $\gamma \in W$ such that $\mathrm{or}(\gamma') < \mathrm{or}(\gamma)$. By construction of m_2:

$$|\{m_2\}_{\mathbf{L}}(m, \alpha, \gamma)| > |\{m_1\}_{\mathbf{L}}(m, \alpha, \langle m_2, m, \alpha, \gamma'\rangle)|.$$

By construction of m_1:

$$|\{m_1\}_{\mathbf{L}}(m_1, \alpha, \langle m_2, m, \alpha, \gamma')| > |\{m\}_{\mathbf{L}}(\beta)|,$$

since $\beta \in T_{\sigma(\tau')} = T_{|\langle m_2, m, \alpha, \gamma'\rangle|}$. Hence $|\{m\}_{\mathbf{L}}(\alpha)| > |\{m\}_{\mathbf{L}}(\beta)|$.

So by the induction hypothesis $\|\beta\| < \kappa_{\mathbf{L}}$. By part (i) of Lemma 4.3.3 it follows that $\|\beta\| \leqslant |\{m\}_{\mathbf{L}}(\beta)|$. Hence

$$\|\beta\| < |\{m_2\}(m, \alpha, \gamma)|,$$

for all $\gamma \in W$ such that $\mathrm{or}(\gamma') < \mathrm{or}(\gamma)$. By definition of $\sigma(\tau)$, $\|\beta\| < \sigma(\tau)$ when $\tau' < \tau < \lambda$. Hence $\|\beta\| < \sigma$, which contradicts the fact that $\|\beta\| \geqslant \sigma$ by construction.

4.4 Computation Theories and Second Order Definability

We shall make a few brief remarks on second-order definability. But first we draw the basic results of Sections 4.1–4.3 together in the following theorem.

4.4.1 Theorem. *Let* PR(**L**) *be normal on* $A = S \cup \mathrm{Tp}(S)$. *The following is true*

(a) PR(**L**) *is p-normal, hence admits a selection operator over* N,

(b) A *is weakly but not strongly* **L**-*finite, i.e. the* **L**-*semicomputable relations are not closed under* $\exists x \in \mathrm{Tp}(S)$,

(c) S *is strongly* **L**-*finite, i.e. the* **L**-*semicomputable relations are closed under* $\exists s \in S$.

As we shall see in Chapter 7, properties (a)–(c) characterize Kleene recursion in a normal object in higher types.

4.4.2 Remark. Let us be a bit more explicit about the relationship between recursion on two types and Kleene recursion in higher types. If $S = \mathrm{Tp}(0) \cup \ldots \cup \mathrm{Tp}(n-1)$, $n > 0$, then $\mathrm{Tp}(S)$ can be identified with $\mathrm{Tp}(1) \times \ldots \times \mathrm{Tp}(n)$. If F is an object of type $n + 2$, there is a list **L** such that Kleene recursion in F on $\mathrm{Tp}(0), \ldots, \mathrm{Tp}(n)$ is essentially the same as recursion in **L** on the structure $\mathfrak{A} = \langle A, S, \mathbf{S} \rangle$, where **S** is some standard coding scheme. A converse is also true. Hence results about recursion in higher types can be deduced from the corresponding results for PR(**L**) on $\mathfrak{A}$. Thus in a quite precise sense, higher type theories can really be captured as theories on two types. (A detailed exposition of the connection between two types and higher types can be found in the book of J. Moldestad [105].)

So far we have emphasized the connection between computation theories on two types and the specific example of recursion in higher types. But the theory has a wider scope and can serve as a framework for the study of second-order definability in general. This is completely analogous to the case of finite theories on one type where recursion in ${}^2E$ was, in a sense, the paradigm, but the theory had much wider connections with definability theory and inductive definability.

An alternate way of approaching second-order definability is via the notion of a *Spector 2-class*. This notion is introduced in Moschovakis [118], a survey of concepts and applications can be found in the lectures of A. Kechris [76]. *There is the same relationship between computation theories on two types and Spector 2-classes as there is between finite theories on one type* (*Spector theories*) *and Spector classes* (see Section 3.2).

We do not give a formal development of this relationship in this book, since the applications of Spector 2-classes all fall under the scope of computation theories on two types. However, we strongly recommend that the reader study the lectures of Kechris [76]. And may we suggest that he or she tries to lift from one to two types the development presented in Sections 3.2 and 3.3, i.e. explore the relationship between computation theories on two types and Spector 2-classes and see how the applications come out in the context of two types. The groundwork is done in Sections 4.1 to 4.3, and we shall present selected applications in Chapters 7 and 8.

Part C

Infinite Theories

Chapter 5

Admissible Prewellorderings

The precomputation theories of the first chapter were to a large extent patterned on ordinary recursion theory, ORT. In part B we gave an analysis of "higher recursion theory" through the notion of a finite theory which generalizes ORT by moving up in types over the basic domain.

But there are different ways of extending ORT, e.g. from the integers to all or part of the ordinals. Or ORT can be rephrased as a recursion theory on HF, the hereditarily finite sets, and then be extended to other domains of sets, even to the total universe. Both approaches were followed, and we duly got various notions of ordinal and set recursion, leading to theories of primitive recursive functions, to the rudimentary functions, and to admissibility theory.

We shall not in this book retrace in any detail the line of development from ORT to recursion on ordinals and to the notion of an *admissible ordinal*. The literature on this topic is vast, but the reader would do well to consult one of the classics in the field, the 1967 paper of R. Jensen and C. Karp, *Primitive recursive set functions* [72]. From this one could go to the recent survey of R. Shore, *α-recursion theory* [152], and the *Short course on admissible recursion theory* [156] by S. Simpson, the latter being an advertisement—which we endorse—of a book-length exposition to come.

Pure α-recursion theory was soon transformed into a general theory of *admissible structures*. A thorough exposition of this field is given by J. Barwise in his book, *Admissible Sets and Structures* [11].

Remark. The references above are strictly pedagogical and do not imply a history of ordinal recursion theory and admissibility theory in any way—the names of S. Kripke and R. Platek have not even been mentioned.

Barwise and Shore give some historical remarks in their respective introductions. It remains to be seen what Simpson will do with the history of the subject in his book. We shall try to document the sources for those parts of the theory that we discuss. The reader should in particular see Section 5.3.

The line between set theory and recursion theory is sometimes difficult to draw. Admissibility theory on the ordinals is obviously recursion theory, but general admissible structures may be too "short and fat" to support a reasonable recursion theory. (For an example see F. Gregory [46].) We have decided to draw the line at *resolvable* structures, our reasons are as follows.

A basic result about admissible sets is

Gandy's Fixed-point Theorem. *Let A be an admissible set and Γ a Σ_1 positive inductive operator on A. Then the least fixed-point of Γ is Σ_1 on A.*

A discussion and proof can be found in Barwise [11].

For resolvable structures there is a converse. We recall that an admissible set A is *resolvable* if there is a total A-recursive function p such that

(i) $$A = \bigcup_{y \in A} p(y),$$

(ii) the relation $x \leqslant y$ iff $x \in p(y)$ defines a pwo of A.

Note that since a set is A-finite iff it is an element of A, the resolvability of A means that there is an A-computable pwo of A with A-finite initial segments.

The following result was proved by A. Nyberg [132].

Theorem. *Let $\mathfrak{A} = \langle A, \in, R_1, \ldots, R_k \rangle$ be a resolvable structure where A is transitive and closed under pairing. If $\mathfrak{A}$ satisfies Δ_0-separation then the following conditions are equivalent:*

(i) *$\mathfrak{A}$ is admissible.*

(ii) *Every Σ_1 positive inductive operator on $\mathfrak{A}$ has a Σ_1 definable least fixed-point.*

(iii) *The length of a Σ_1 positive induction does not exceed the ordinal of $\mathfrak{A}$.*

(iv) *There exists a positive first-order inductive definition on $\mathfrak{A}$ of length strictly greater than any Σ_1 positive induction on $\mathfrak{A}$.*

To a recursion theorist this result is an excellent conceptual justification for the following general notion of admissible prewellordering.

Definition (see 5.1.9). Let $(\mathfrak{A}, \preccurlyeq)$ be a computation domain with a pwo $\preccurlyeq$ and let $\mathbf{R}$ be a sequence of relations on the domain of $\mathfrak{A}$. The structure $(\mathfrak{A}, \preccurlyeq)$ is called an $\mathbf{R}$-*admissible prewellordering* if every $\Sigma_1(\preccurlyeq, \mathbf{R})$ positive inductive operator on $\mathfrak{A}$ has a $\Sigma_1(\preccurlyeq, \mathbf{R})$ definable least fixed-point.

Remarks. (1) The notion of $\mathbf{R}$-admissible pwo was introduced by Moschovakis [113] but without the conceptual analysis given by Nyberg's theorem. We like to think that Moschovakis had the recursion theorist's natural faith in the first recursion theorem. And it is nice to know that faith sometimes can be vindicated.

(2) Following Moschovakis we shall prove that admissible pwo's correspond to a certain class of infinite computation theories. Previous to this work C. E. Gordon gave a computation-theoretic analysis of admissibility in his thesis [45]. He showed that given an admissible structure $\mathfrak{A}$, Σ_1 definability on $\mathfrak{A}$ corresponds to multiple valued search computability (in the sense of Moschovakis [112]) in the $\in$-relation and the bounded quantifier

$$\mathrm{bE}(x, f) \to \begin{cases} 0 & \text{if } (\exists y \in x)[f(y) \to 0] \\ 1 & \text{if } (\forall y \in x)[f(y) \to 1]. \end{cases}$$

Search computability in $\in$ and bE is easily seen to take care of Σ_1 definability. For the converse one may use Gandy's fixed-point theorem.

We shall now proceed as follows. In Sections 5.1 and 5.2 we characterize admissible pwo's in computation-theoretic terms. The associated computation theory is the natural domain for degree theoretic arguments. In 5.3 we analyze the "next-admissible set" construction and venture to make a few historical remarks on how this idea developed. In the final section we apply the imbedding results of Section 5.3 to study the structure of finite theories over the integers.

5.1 Admissible Prewellorderings and Infinite Theories

Having given our motivation we make a detour via a class of infinite computation theories before developing the general theory of admissible prewellorderings. The class is not entirely arbitrary since we will show in 5.2 that it suffices to characterize the admissible prewellorderings.

So, let us start with a *p-normal* computation theory $\langle\Theta, <\rangle$ on a domain $(\mathfrak{A}, \preccurlyeq)$, where $\mathfrak{A}$ is a computation domain in which $A = C$ (thus equality is Θ-computable) and $\preccurlyeq$ is a prewellordering of A. As always, our theories are single-valued.

5.1.1 Assumption A. The structure $(\mathfrak{A}, \preccurlyeq)$ is Θ-*resolvable*, i.e.

(i) The domain A is not Θ-finite.

(ii) The pwo $\preccurlyeq$ is Θ-computable and the initial segments of $\preccurlyeq$ are uniformly Θ-finite.

The last requirement of 5.1.1 means that the functional $\mathbf{E}^{\preccurlyeq}$ is Θ-computable, where

$$\mathbf{E}^{\preccurlyeq}(f, x) \simeq \begin{cases} 0 & \text{if} \quad (\exists y \prec x)[f(y) \simeq 0] \\ 1 & \text{if} \quad (\forall y \prec x)[f(y) \simeq 1]. \end{cases}$$

For the definition and simple properties of finiteness we refer back to Section 2.5. See also Remark 5.1.4 below.

Our next assumption introduces the admissibility condition $\|\Theta\| = |\preccurlyeq|$, with a suitable effective addition. Recall that

$$\|\Theta\| = \sup\{|a, \sigma, z|_\Theta : (a, \sigma, z) \in \Theta\},$$

and $|\preccurlyeq|$ is the length of the pwo $\preccurlyeq$.

5.1.2 Assumption B. The assumption comes in two parts:

(i) $\|\Theta\| = |\preccurlyeq|$.

(ii) The set $\Theta^{|w|} = \{(a, \sigma, z) \in \Theta : |a, \sigma, z|_\Theta = |w|_\preccurlyeq\}$ is Θ-computable uniformly in $w \in A$.

Note that we can (due to the computability of $\mathbf{E}^{\preccurlyeq}$) replace $=$ by $<$ or $\leqslant$ in the definition of $\Theta^{|w|}$ and still obtain uniform Θ-computability.

We add one more assumption, the *Grilliot selection principle*, in order to close the Θ-semicomputable relations under existential quantification over the domain A. Since we insist on single-valued theories we forsake the option of adding "multiple-valued" search.

5.1.3 Assumption C. There exists a Θ-computable mapping $q(n)$ such that for all a, σ

$$\exists x \cdot \{a\}_\Theta(x, \sigma) \simeq 0 \quad \text{iff} \quad \{q(n)\}_\Theta(a, \sigma) \simeq 0.$$

And if $\exists x \cdot \{a\}_\Theta(x, \sigma) \simeq 0$, then

$$|q(n), a, \sigma, 0|_\Theta > \inf\{|a, x, \sigma, 0| : \{a\}(x, \sigma) \simeq 0\}.$$

This means that the functional

$$\mathbf{E}^{1/2}(f, \sigma) \simeq 0 \quad \text{iff} \quad \exists x \cdot f(x, \sigma) \simeq 0,$$

is Θ-computable, i.e. "half" of the usual $\mathbf{E}$ functional on A is Θ-computable.

5.1.4 Remark. Assumption C together with the p-normality of Θ gives a "nice" theory for en(Θ). From C follows the fact that the Θ-semicomputable relations are closed under $\exists$ and $\vee$. p-normality implies selection over N, which entails that $R \in \text{sc}(\Theta)$ iff $R, \neg R \in \text{en}(\Theta)$. Selection over N also ensures that the Θ-finite sets behave well.

5.1.5 Definition. Let $\mathfrak{A}$ be a computation domain with a pwo $\preccurlyeq$ and Θ a computation theory on $(\mathfrak{A}, \preccurlyeq)$. Θ is called an *infinite theory* on $(\mathfrak{A}, \preccurlyeq)$ iff Θ satisfies axioms A, B and C.

We give a few simple properties of infinite theories.

5.1.6 Proposition. *Let Θ be an infinite theory on $(\mathfrak{A}, \preccurlyeq)$. A set $B \subseteq A$ is Θ-finite iff it is Θ-computable and $\preccurlyeq$-bounded.*

We verify that Θ-finiteness implies $\preccurlyeq$-boundedness. Assume the contrary, then

$$A = \bigcup_{x \in B} \{y \in A : y \prec x\}.$$

But this means that A is a Θ-finite union of Θ-finite sets, hence A is Θ-finite, contradicting (i) of 5.1.1.

5.1.7 Proposition. *For each n there is a Θ-computable relation R_n such that*

$$\{a\}_\Theta(\sigma) \simeq z \quad \textit{iff} \quad \exists w R_n(a, \sigma, z, w).$$

We use the relation $\Theta^{|w|}$ of (ii) in 5.1.2.

Associated with the computation domain $\mathfrak{A}$, the prewellordering $\preccurlyeq$, and a sequence of relations $\mathbf{R}$ on $\mathfrak{A}$, we have a language

$$\mathbf{L} = \mathbf{L}(\mathfrak{A}, \preccurlyeq, \mathbf{R}).$$

As usual define classes $\Delta_0(\preccurlyeq, \mathbf{R})$, $\Sigma_1(\preccurlyeq, \mathbf{R})$, $\Pi_1(\preccurlyeq, \mathbf{R})$, and $\Delta_1(\preccurlyeq, \mathbf{R})$ over the domain A.

5.1.8 Proposition. *There exist Θ-computable relations $\mathbf{R}_\Theta = R_1, R_2$ on A such that* $\text{en}(\Theta) = \Sigma_1(\preccurlyeq, \mathbf{R}_\Theta)$.

The axioms immediately entail that $\Sigma_1(\preccurlyeq, \mathbf{R}) \subseteq \text{en}(\Theta)$ for any sequence $\mathbf{R}$ of Θ-computable relations on A. Conversely, let

$$R_1(a, x, y, w) \ \text{ iff } \ |a, x, y|_\Theta \leqslant |w|_\preccurlyeq$$
$$R_2(a, b, c) \ \text{ iff } \ a = S_1^1(b, c).$$

A simple induction proof using Proposition 5.1.7 shows that $\text{en}(\Theta) \subseteq \Sigma_1(\preccurlyeq, \mathbf{R}_\Theta)$.

Let us note a version of "Δ_0-separation": If S is $\Delta_0(\preccurlyeq, \mathbf{R})$, where $\mathbf{R}$ is Θ-computable, then the set

$$\{y \in A : y \prec x \wedge S(y)\},$$

is Θ-finite, uniformly in x.

We now return to the conceptually important notion of *admissible prewellordering*. First a notational convention, when we use $\Delta_0(\preccurlyeq, X, \mathbf{R})$ and $\Sigma_1(\preccurlyeq, X, \mathbf{R})$, we always require that X occurs *positively* in the formulas. Such formulas $\theta(\sigma, X)$ then define monotone operators

$$\Gamma_\theta(X) = \{\sigma : \theta(\sigma, X)\}.$$

As usual we let Γ_∞ be the least fixed point for Γ_θ and $|\Gamma_\theta|$ the ordinal of the inductive definition.

5.1.9 Definition. Let $(\mathfrak{A}, \preccurlyeq)$ be a computation domain with a pwo and $\mathbf{R}$ a sequence of relations on A. The structure $(\mathfrak{A}, \preccurlyeq)$ is called an $\mathbf{R}$-*admissible prewellordering* if for every $\Sigma_1(\preccurlyeq, X, \mathbf{R})$ formula θ with parameters from A, the fixed-point Γ_∞ of Γ_θ is a $\Sigma_1(\preccurlyeq, \mathbf{R})$ relation.

5.1.10 Proposition. *Let Θ be an infinite theory on $(\mathfrak{A}, \preccurlyeq)$ and* **R** *any sequence of relations extending* $\mathbf{R}_\Theta$. *Then the structure $(\mathfrak{A}, \preccurlyeq)$ is* **R**-*admissible.*

This is an immediate corollary of the first recursion theorem for Θ. Any formula $\theta(\sigma, X) \in \Sigma_1(\preccurlyeq, X, \mathbf{R})$ determines a Θ-computable functional

$$\varphi(f, \tau) \simeq 0 \quad \text{iff} \quad \theta(\tau, \{\sigma | f(\sigma) \simeq 0\}).$$

Let f^* be the fixed-point of φ. Then $\Gamma_\infty = \{\sigma : f^*(\sigma) \simeq 0\}$. Thus Γ_∞ is Θ-semi-computable, hence $\Sigma_1(\preccurlyeq, \mathbf{R})$ by Proposition 5.1.8.

We thus see that infinite theories Θ on a structure $(\mathfrak{A}, \preccurlyeq)$ give rise to admissible pwo's. We promise a converse. As a first step we shall, in a rather crude way, associate a recursion theory H with an **R**-admissible pwo $\preccurlyeq$.

5.1.11. Let $(\mathfrak{A}, \preccurlyeq)$ be **R**-admissible. Define

$$H = \mathrm{PR}[\mathbf{E}^{\preccurlyeq}, \preccurlyeq, \mathbf{R}, \mathbf{E}^{1/2}],$$

where $\mathbf{E}^{1/2}$ is the functional from Assumption C (5.1.3).

The following properties of H are immediate:

1. H is a computation theory on $(\mathfrak{A}, \preccurlyeq)$ satisfying axioms A and C.
2. $\mathrm{en}(H) = \Sigma_1(\preccurlyeq, \mathbf{R})$.

We make just one comment on 1 and 2. By the construction of H we see that $\Sigma_1(\preccurlyeq, \mathbf{R}) \subseteq \mathrm{en}(H)$. To prove the converse we give a $\Sigma_1(\preccurlyeq, X, \mathbf{R})$ inductive definition of $\mathrm{en}(H)$, and use the admissibility of the pwo to conclude that $\mathrm{en}(H)$ is $\Sigma_1(\preccurlyeq, \mathbf{R})$.

5.1.12 Proposition. *Let $(\mathfrak{A}, \preccurlyeq)$ be* **R**-*admissible and assume that* $\Sigma_1(\preccurlyeq, \mathbf{R}) - \Delta_1(\preccurlyeq, \mathbf{R}) \neq \varnothing$. *Let θ be a $\Delta_0(\preccurlyeq, \mathbf{R})$ formula. Then*

$$(\forall x \prec u)(\exists y)\theta(x, y) \Rightarrow (\exists w)(\forall x \prec u)(\exists y \prec w)\theta(x, y)$$

(Σ_1-*collection principle*).

The assumption that $\Sigma_1 - \Delta_1 \neq \varnothing$ is not serious. Indeed, the results of Section 5.2 can be used to show that it can be omitted.

For the proof we need a sublemma:

5.1.13 Sublemma. $\Sigma(\preccurlyeq, \mathbf{R}) = \Sigma_1(\preccurlyeq, \mathbf{R}) = \mathrm{en}(H)$.

We need to show that every Σ relation is in $\mathrm{en}(H)$. As a typical case take

$$\forall x \prec u \exists y R(x, y),$$

where R is Δ_0. By assumption there exists an H-computable function f such that

$$R(x, y) \text{ iff } f(x, y) \simeq 0.$$

Hence $\exists y R(x, y)$ iff $\{\exists_f\}(y) \simeq 0$, where $\exists_f$ is computable from f and the given index for $\mathbf{E}^{1/2}$ in H. Let g be introduced by $g(y) \simeq 1$ iff $\{\exists_f\}(y) \simeq 0$. Then

$$\forall x \prec u \exists y R(x, y) \text{ iff } \mathbf{E}^{\preccurlyeq}(g, u) \simeq 1.$$

The construction is uniform in the parameters, and thus we can proceed by induction.

Back to the proof of 5.1.12: Choose a $\Sigma_1(\preccurlyeq, \mathbf{R})$ relation $U(\sigma)$ such that $\neg U \notin \Sigma(\preccurlyeq, \mathbf{R})$. We note that $|\preccurlyeq|$ is a limit ordinal, otherwise A would be H-finite.

Assume now:

1. $(\forall x \prec u)(\exists y)\theta(x, y)$.
2. $(\forall w)(\exists x \prec u)(\forall y \prec w)\neg\theta(x, y)$.

This means that for all w we must have some $x \prec u$ and some $y \succcurlyeq w$ such that $\theta(x, y)$. This is used in proving the equivalence in 4 below:

3. $U(\sigma)$ iff $\exists z U_0(\sigma, z)$,

where $U_0 \in \Delta_0(\preccurlyeq, \mathbf{R})$.

4. $$\neg U(\sigma) \text{ iff } (\forall z)\neg U_0(\sigma, z)$$ $$\text{iff } (\forall x \prec u)(\exists y)[\theta(x, y) \wedge (\forall t \prec y)\neg U_0(\sigma, t)].$$

From 4 it follows that $\neg U$ is $\Sigma(\preccurlyeq, \mathbf{R})$—a contradiction that proves Proposition 5.1.12.

Above we showed that en(H) is $\Sigma_1(\preccurlyeq, \mathbf{R})$ by constructing a Σ_1-inductive definition for en(H). If we are going to prove axiom B, in particular, the admissibility condition $\|H\| = |\preccurlyeq|$, we need to have an estimate of the ordinals of inductive definitions on $(\mathfrak{A}, \preccurlyeq)$ in terms of the ordinal of the pwo $\preccurlyeq$. We should expect that $|\Gamma_\theta| \leqslant |\preccurlyeq|$ for all $\Sigma_1(\preccurlyeq, X, \mathbf{R})$ inductive operators, which we indeed will prove in 5.1.15 below. In order to prove the equality $\|H\| = |\preccurlyeq|$, we must be able to carry out the construction of $H = \text{PR}[\ldots]$ in sufficiently many steps. And then we must verify that the equality is "effective" in the sense of axiom B, 5.1.2. But that is the topic of the next section. Here we start by proving an auxiliary lemma.

5.1.14 Lemma. *Let $(\mathfrak{A}, \preccurlyeq)$ be* **R**-*admissible and let $\theta(\sigma, X)$ define a monotone $\Sigma_1(\preccurlyeq, X, \mathbf{R})$ inductive operator. The relation*

$$P(\sigma, x) \textit{ iff } \sigma \in \Gamma_{|x|},$$

is $\Sigma_1(\preccurlyeq, \mathbf{R})$. *(Note that* $\Gamma_{|x|}$ *is the x'th stage of the inductively defined set* Γ_∞.) *If* θ *is* $\Delta_0(\preccurlyeq, X, \mathbf{R})$, *the relation P is* $\Delta_1(\preccurlyeq, \mathbf{R})$.

The proof is carried out inside the associated theory $H = \mathrm{PR}[\mathbf{E}^{\preccurlyeq}, \preccurlyeq, \mathbf{R}, \exists^{1/2}]$. First we construct a code $\hat{p}$ such that

$$\{\hat{p}\}(\sigma, x) \simeq 0 \quad \text{iff} \quad P(\sigma, x).$$

If θ is Δ_0, we also get a code $\hat{q}$ such that

$$\{\hat{q}\}(\sigma, x) \simeq 0 \quad \text{iff} \quad \neg P(\sigma, x).$$

We start the construction of $\hat{p}$ (and $\hat{q}$) by rewriting:

$$\begin{aligned} P(\sigma, x) \quad &\text{iff} \quad \sigma \in \Gamma_{|x|} \\ &\text{iff} \quad \sigma \in \Gamma_\theta\Big(\bigcup_{y \prec x} \Gamma_{|y|}\Big) \\ &\text{iff} \quad \theta(\sigma, \{\sigma' : (\exists y \prec x) P(\sigma', y)\}). \end{aligned}$$

Let f be the function defined by

$$f(p, \sigma, x) \simeq 0 \quad \text{iff} \quad \theta(\sigma, \{\sigma' : (\exists y \prec x)\cdot\{p\}(\sigma', y) \simeq 0\}).$$

We see that f is H-computable, since $\{p\}(\sigma', y) \simeq 0$ is H-semicomputable and $\theta(\sigma, X)$ is positive in X.

By the second recursion theorem there exists a code $\hat{p}$ such that $\{\hat{p}\}(\sigma, x) = (\hat{p}, \sigma, x)$. By induction we now verify that

$$\{\hat{p}\}(\sigma, x) \simeq 0 \quad \text{iff} \quad \sigma \in \Gamma_{|x|}.$$

Assume this true for all $y \prec x$:

$$\begin{aligned} \sigma \in \Gamma_{|x|} \quad &\text{iff} \quad \theta(\sigma, \{\sigma' : (\exists y \prec x)\cdot \sigma' \in \Gamma_{|y|}\}) \\ &\text{iff} \quad \theta(\sigma, \{\sigma' : (\exists y \prec x)\cdot\{\hat{p}\}(\sigma', y) \simeq 0\}) \\ &\text{iff} \quad f(\hat{p}, \sigma, x) \simeq 0 \\ &\text{iff} \quad \{\hat{p}\}(\sigma, x) \simeq 0. \end{aligned}$$

If θ is Δ_0, then

$$\neg P(\sigma, X) \quad \text{iff} \quad \neg\theta(\sigma, \{\sigma' : \neg(\forall y \prec x)\neg P(\sigma', y)\}),$$

is by Proposition 5.1.12 a $\Sigma_1(\preccurlyeq, \mathbf{R})$ relation. Hence there exists an H-computable g such that

$$g(q, \sigma, x) \simeq 0 \quad \text{iff} \quad \neg\theta(\sigma, \{\sigma' : \neg(\forall y \prec x)\cdot\{q\}(\sigma, y) \simeq 0\}).$$

We may then proceed as above to produce a code $\hat{q}$ such that $\{\hat{q}\}(\sigma, x) \simeq 0$ iff $\neg P(\sigma, x)$.

5.1.15 Proposition. *Let* $(\mathfrak{A}, \preccurlyeq)$ *be* **R**-*admissible and let* $\theta(\sigma, X)$ *define a monotone* $\Sigma_1(\preccurlyeq, X, \mathbf{R})$ *inductive operator. Then*

$$|\Gamma_\theta| \leqslant |\preccurlyeq|,$$

where Γ_θ *is the inductive operator associated with* θ.

Since $|\preccurlyeq|$ is a limit number, we must be able to show that $\sigma \in \Gamma_\theta(\Gamma_{|\preccurlyeq|})$ implies that $\exists x[\sigma \in \Gamma_\theta(\Gamma_{|x|})]$, i.e.

$$\theta(\sigma, \{\sigma' : \exists x P(\sigma', x)\}) \Rightarrow \exists x \cdot \theta(\sigma, \{\sigma' : P(\sigma', x)\}).$$

Or, more generally, we show that if $\theta'(X)$ is $\Sigma_1(\preccurlyeq, X, \mathbf{R})$ (in parameters from A), then

$$\theta'(\{\sigma : \exists x P(\sigma, x)\}) \Rightarrow \exists x \cdot \theta'(\{\sigma : P(\sigma, x)\}).$$

But this follows since θ' is positive in X, hence the $\exists$-quantifier can be advanced, e.g.

$$\begin{aligned}(\forall y \preccurlyeq z)(\exists x)P(\sigma, x) &\Rightarrow (\exists w)(\forall y \preccurlyeq z)(\exists x \preccurlyeq w)P(\sigma, x)\\ &\Rightarrow (\exists w)(\forall y \preccurlyeq z)P(\sigma, w),\end{aligned}$$

because of the monotone character of P.

5.2 The Characterization Theorem

We saw in Proposition 5.1.10 that if Θ is an infinite theory on a domain $(\mathfrak{A}, \preccurlyeq)$ and **R** is any sequence of relations extending the relations $\mathbf{R}_\theta$ of Proposition 5.1.8, then $(\mathfrak{A}, \preccurlyeq)$ is **R**-admissible. In 5.1.11 we made a few steps toward proving a converse. We constructed a theory $H = \mathrm{PR}[\mathbf{E}^{\preccurlyeq}, \leqslant, \mathbf{R}, \mathbf{E}^{1/2}]$ which by the very construction satisfied Assumption A of 5.1.1 and Assumption C of 5.1.3. We further noted that $\mathrm{en}(H) = \Sigma_1(\preccurlyeq, \mathbf{R})$. And it is a consequence of Proposition 5.1.15 that $\|H\| \leqslant |\preccurlyeq|$, which goes some way toward verifying Assumption B of 5.1.2. Our program is now to make a more refined construction of H, in fact, slowing up the construction of H, so as to obtain the converse inequality $|\preccurlyeq| \leqslant \|H\|$. And by a careful analysis of the construction we shall be able to get the "effective" content of the equality $\|H\| = |\preccurlyeq|$, i.e. Assumption B.

Let $(\mathfrak{A}, \preccurlyeq)$ be a computation domain with a pwo $\preccurlyeq$. We assume that the code set of $\mathfrak{A}$ is equal to the whole domain and that $\mathfrak{A}$ includes a pairing structure. We start from the following basic assumption:

5.2.1 Assumption. Let $\mathbf{R} = R_1, \ldots, R_k$ be relations on A such that $(\mathfrak{A}, \preccurlyeq)$ is $\mathbf{R}$-admissible.

We shall now construct a theory $\Theta = \mathrm{PR}[[\preccurlyeq, \mathbf{R}]]$ which in many respects equals the prime recursion theory $H = \mathrm{PR}[\mathbf{E}^{\preccurlyeq}, \preccurlyeq, \mathbf{R}, \exists^{1/2}]$.

Our main problem in constructing $\mathrm{PR}[[\preccurlyeq, \mathbf{R}]]$ is to "delay" the definition of the inductive operator Γ such as to get in the end the inequality $\|\Theta\| = |\Gamma| \leqslant |\preccurlyeq|$. We use the following trick.

5.2.2 Definition. For every set Δ of tuples on A of length $\geqslant 2$ we set

$$\Delta^+ = \{x : (\forall u \prec x)[(\langle 12, 0\rangle, u, u) \in \Delta]\}.$$

The intention is that $\langle 12, 0\rangle$ will be a special index for the identity function. Δ^+ will always be an initial segment of A, and, in particular, $\emptyset^+ = \{x \in A : |x| = 0\}$, where $|x|$ is the ordinal of x in the pwo $\preccurlyeq$.

5.2.3 Construction of the Inductive Operator Γ. We give a few but typical cases:

1. Successor function: If $\langle 1, 0\rangle, x, s(x) \in \Theta^+$, then

$$(\langle 1, 0\rangle, x, s(x)) \in \Gamma(\Theta).$$

2. Substitution: If $\langle 6, 0\rangle, \hat{f}, \hat{g}, \sigma, z \in \Theta^+$ and $\exists u \in \Theta^+[(\hat{f}, u, \sigma, z) \in \Theta \wedge (\hat{g}, \sigma, u) \in \Theta]$, then

$$(\langle 6, 0\rangle, \hat{f}, \hat{g}, \sigma, z) \in \Gamma(\Theta).$$

(The indices $\langle 2, 0\rangle, \ldots, \langle 5, 0\rangle$ are used for introducing the pairing structure and definition by cases. In the same way $\langle 7, 0\rangle$ is used for $\mathbf{P}$ and $\langle 8, \ldots\rangle$ for the *s-m-n* function.)

3. Introduction of $\preccurlyeq$ *and* $\mathbf{R}$*:* We let

$\preccurlyeq$ have the code $\langle 9, 1\rangle$
R_i have the code $\langle 9, 1 + i\rangle$,

and add the obvious inductive clauses.

4. Closure under $\exists$*-quantification:* If $\langle 10, 0\rangle, \hat{f}, \sigma, 0 \in \Theta^+$ and if $(\exists x \in \Theta^+)\cdot [(\hat{f}, x, \sigma, 0) \in \Theta]$, then

$$(\langle 10, 0\rangle, \hat{f}, \sigma, 0) \in \Gamma(\Theta).$$

5. Introduction of the functional $\mathbf{E}^{\preccurlyeq}$*:* If $\langle 11, 0\rangle, \hat{f}, x, 0 \in \Theta^+$ and $(\exists y \prec x)\cdot [(\hat{f}, y, 0) \in \Theta]$, then

$$(\langle 11, 0\rangle, \hat{f}, x, 0) \in \Gamma(\Theta).$$

If $\langle 11, 0\rangle, \hat{f}, x, 1 \in \Theta^+$ and $(\forall y \prec x)[(\hat{f}, y, 1) \in \Theta]$, then

$$(\langle 11, 0\rangle, \hat{f}, x, 1) \in \Gamma(\Theta).$$

(Recall that Θ^+ is always an initial segment.)

6. The identity function: If $y \in \Theta^+$, then

$$(\langle 12, 0\rangle, y, y) \in \Gamma(\Theta).$$

Note how this clause differs slightly from the previous ones but, as subsequent lemmas will show, the difference is important.

The operator Γ is monotone, we define as usual

$$\Theta^\xi = \Gamma\left(\bigcup_{\eta<\xi} \Theta^\eta\right),$$

and can now introduce the associated theory.

5.2.4 Definition. Let **R** be introduced by Assumption 5.2.1 and Γ constructed by 5.2.3. We define

$$\Theta = \mathrm{PR}[[\preccurlyeq, \mathbf{R}]] = \Theta^\infty = \bigcup_\xi \Theta^\xi.$$

And for $(a, \sigma, z) \in \Theta$ we set

$$|a, \sigma, z|_\Theta = \text{least } \xi \text{ such that } (a, \sigma, z) \in \Theta^\xi.$$

5.2.5 Lemma. $(\langle 12, 0\rangle, y, y) \in \Theta^\xi$ *iff* $|y| \leqslant \xi$.

The proof is by induction, so assume that the lemma is true for all $\eta < \xi$. Let $(\langle 12, 0\rangle, y, y) \in \Theta^\xi - \bigcup_{\eta<\xi} \Theta^\eta$. By 6 of 5.2.3 this means that

$$y \in \left(\bigcup_{\eta<\xi} \Theta^\eta\right)^+,$$

i.e.$(\forall u \prec y)[(\langle 12, 0\rangle, u, u) \in \bigcup_{\eta<\xi} \Theta^\eta$. But this in turn means that $(\forall u \prec y).(\exists n < \xi)[|u| \leqslant \eta]$, i.e. $|y| \leqslant \xi$. The argument also works in reverse which proves the lemma.

5.2.6 Lemma. *If* $a \neq \langle 12, 0\rangle$ *and* $(a, x_1, \ldots, x_n, z) \in \Theta^\xi$, *then* $|a|, |x_1|, \ldots, |x_n|, |z| \leqslant \xi$.

From the assumption that $(a, x_1, \ldots, x_n, z) \in \Gamma(\bigcup_{\eta<\xi} \Theta^\eta)$ and $a \neq \langle 12, 0\rangle$, it follows that $a, x_1, \ldots, x_n, z \in (\bigcup_{\eta<\xi} \Theta^\eta)^+$: But for any $y \in (\bigcup_{\eta<\xi} \Theta^\eta)^+$ we have the truth of

$$(\forall u \prec y)(\exists \eta < \xi)[(\langle 12, 0\rangle, u, u) \in \Theta^\eta].$$

By Lemma 5.2.5 this means that $(\forall u \prec y)(\exists \eta < \xi)[|u| \leqslant \eta]$, i.e. $|y| \leqslant \xi$.

5.2.7 Lemma. PR[[$\preccurlyeq$, **R**]] *with the given length function is a p-normal computation theory on* $(\mathfrak{A}, \preccurlyeq)$ *in which* $\preccurlyeq$, **R**, *and* $\mathbf{E}^{\preccurlyeq}$ *are computable.* en(PR[[$\preccurlyeq$, **R**]]) *is closed under* $\exists$*-quantification.*

This lemma is a direct consequence of the construction. For p-normality we can use the proof of Proposition 3.1.12 of Section 3.1, replacing the functional $\mathbf{E}_A$ of that proof by the functionals $\mathbf{E}^{1/2}$ and $\mathbf{E}^{\preccurlyeq}$ introduced in 4 and 5 of Construction 5.2.3.

5.2.8 Lemma. en(PR[[$\preccurlyeq$, **R**]]) $= \Sigma_1(\preccurlyeq, \mathbf{R})$.

There are two things to verify:

(i) The inductive operator of 5.2.3 is of $\Sigma_1(\preccurlyeq, X, \mathbf{R})$ form, hence by **R**-admissibility of $(\mathfrak{A}, \preccurlyeq)$, has a $\Sigma_1(\preccurlyeq, \mathbf{R})$ least fixed-point; it follows that en(PR[[$\preccurlyeq$, **R**]]) $\subseteq \Sigma_1(\preccurlyeq, \mathbf{R})$.

(ii) A simple analysis of $\Delta_0(\preccurlyeq, R)$ relations shows that they are computable in PR[[$\preccurlyeq$, **R**]]. Closure under $\exists$-quantifier shows that $\Sigma_1(\preccurlyeq, \mathbf{R}) \subseteq$ en(PR[[$\preccurlyeq$, **R**]]).

We also note that sc(PR[[$\preccurlyeq$, **R**]]) $= \Delta_1(\preccurlyeq, \mathbf{R})$, this being a consequence of p-normality (see Remark 5.1.4).

From what we have proved so far we see that PR[[$\preccurlyeq$, **R**]] satisfies Assumptions A and C. And it follows from Lemmas 5.1.15 and 5.2.5 that $\|$PR[[$\preccurlyeq$, **R**]]$\| = |\preccurlyeq|$, which is the first part of Assumption B. We proceed to a more detailed analysis of the construction.

5.2.9 Definition. To the operator $\Gamma(\Theta)$ we associate an operator $\Gamma(w, \Theta)$ obtained by relativizing the quantifiers in $\Gamma(\Theta)$ to "$\preccurlyeq w$" and replacing "$x \in \Theta^+$" by "$x \in \Theta^+ \wedge x \preccurlyeq w$". (Note that $\langle 12, 0\rangle$ is exempted from this restriction.) Θ_w^ξ is defined as usual.

5.2.10 Lemma. *If* $\xi \leqslant |w|$, *then* $\Theta^\xi = \Theta_w^\xi$.

The proof is by induction on subcomputations using Lemma 5.2.6 in an essential way.

5.2.11 Definition. Let Θ be a computation theory on $(\mathfrak{A}, \preccurlyeq)$. Set

$$X(\Theta) = \{(m, x) : m \in N \wedge m \geqslant 2 \wedge ((x)_1, \ldots, (x)_m) \in \Theta\}$$
$$\mathbf{\Theta}(X) = \{((x)_1, \ldots, (x)_m) : (m, x) \in X\}.$$

Introduce the following operators:

$$\bar{\Gamma}(X) = \boldsymbol{X}(\Gamma(\boldsymbol{\Theta}(X))).$$
$$\bar{\Gamma}(w, X) = \boldsymbol{X}(\Gamma(w, \boldsymbol{\Theta}(X))).$$

The corresponding stages X^{ξ} and X^{ξ}_{w} are defined as usual.

Note that we have the following equalities:

$$X^{\xi}_{w} = \boldsymbol{X}(\Theta^{\xi}_{w}) \qquad X^{\xi} = \boldsymbol{X}(\Theta^{\xi})$$
$$\Theta^{\xi}_{w} = \boldsymbol{\Theta}(X^{\xi}_{w}) \qquad \Theta^{\xi} = \boldsymbol{\Theta}(X^{\xi}),$$

i.e. coding and decoding are carried along the stages.

5.2.12 Lemma. PR[[$\preccurlyeq$, **R**]] *satisfies Assumption* B *of* 5.1.2.

We first note that the operator $\bar{\Gamma}(x, X)$ is defined by a $\Delta_0(\preccurlyeq, X, R)$ positive formula, hence by Lemma 5.1.14 the relation

$$(m, x) \in X^{|y|}_{w},$$

is $\Delta_1(\preccurlyeq, \mathbf{R})$.

By Lemma 5.2.10 we have

$$\begin{aligned}(m, x) \in X^{|y|} \quad &\text{iff} \quad (\exists w)[y \preccurlyeq w \ \wedge \ (m, x) \in X^{|y|}_{w}] \\ &\text{iff} \quad (\forall w)[y \preccurlyeq w \rightarrow (m, x) \in X^{|y|}_{w}],\end{aligned}$$

so the relation $(m, x) \in X^{|y|}$ is $\Delta_1(\preccurlyeq, \mathbf{R})$.

Since $\bar{\Gamma}$ is defined by a $\Sigma_1(\preccurlyeq, X, \mathbf{R})$ formula, we get $|\bar{\Gamma}| \leqslant |\preccurlyeq|$ by Proposition 5.1.15. Hence if $(m, x) \in \boldsymbol{X}(\Theta)$, there is a w such that $[m, x] = |w|$, where $[m, x]$ is the ordinal of the coded computation tuple (m, x). But

$$[m, x] = |w| \quad \text{iff} \quad (m, x) \in X^{|w|} \wedge (\forall u \prec w)[(m, x) \notin X^{|u|}],$$

so the relation $[m, x] = |w|$ is $\Delta_1(\preccurlyeq, \mathbf{R})$, i.e. PR[[$\preccurlyeq$, **R**]]-computable, which is the substance of (ii) in Assumption B.

5.2.13 Theorem. *Let* $(\mathfrak{A}, \preccurlyeq)$ *be an* **R**-*admissible prewellordering. There exists a p-normal computation theory* Θ = PR[[$\preccurlyeq$, **R**]] *on* $\mathfrak{A}$ *satisfying A, B, and C such that* $\text{en}(\Theta) = \Sigma_1(\preccurlyeq, \mathbf{R})$ *and* $\text{sc}(\Theta) = \Delta_1(\preccurlyeq, \mathbf{R})$.

The development of Sections 5.1 and 5.2 is patterned on Section 10 of Moschovakis [113]. There are many differences due to the fact that he uses multiple-valued theories whereas we have insisted on single-valued ones. But many of the key technical points are taken from his paper.

We stop short with Theorem 5.2.13. One could go on to investigate to what extent an infinite theory is characterized by its associated pwo: If we start with an infinite theory Θ and construct the associated pwo (5.1.10) with respect to the relations $\mathbf{R}_\Theta$ and then pass to the theory PR[[$\preccurlyeq$, $\mathbf{R}_\Theta$]], are then the two theories equivalent? The answer is "almost", we need to be more careful in the construction of PR[[$\preccurlyeq$, $\mathbf{R}_\Theta$]]; the interested reader may consult theorem (xx) of Section 10 in Moschovakis [113] for technical inspiration and then prove a similar result in the present framework.

5.2.14 Remark. If a multiple-valued theory Θ has a pmv selection operator $q(a, \sigma)$, we get a "nice" theory for en(Θ) in a straightforward way, see Section 1.3.

It could have been the case that multiple-valued selection was necessary for theories associated with an admissible pwo. In a set-theoretic context we assume computability of the union operator, i.e. if $x \in A$, then $f(x) = \bigcup x$ is also an element of A (f is a *rudimentary operation*). And if x is a unit set, then $f(x) = \bigcup x$ is the unique element of x, i.e. we have a special kind of selection operator. This is precisely what we need in proving that $R(\sigma)$ is computable if both $R(\sigma)$ and $\neg R(\sigma)$ are semicomputable. We are led to the construction of a set B_σ (uniformly in σ) which has 0 as its only member if $R(\sigma)$ and 1 as its only member if $\neg R(\sigma)$. $f(B_\sigma) = \bigcup B_\sigma$ would then be the characteristic function of $R(\sigma)$, proving that it is computable.

Over an arbitrary **R**-admissible pwo $(\mathfrak{A}, \preccurlyeq)$ the operation $\bigcup$ makes no sense. So we use multiple valued search: $q(a, \sigma)$ selects a set of elements satisfying a condition $\{a\}(x, \sigma) \simeq 0$. And if for all σ there is a unique x satisfying the condition, then $q(a, \sigma)$ defines a single-valued mapping. (The reader should at this point look back at the proof of Theorem 1.3.4.)

The union operator leads to a new element of the domain, multiple-valued selection leads to a subset. It is an interesting and important technical point that one can extend the formalism of recursion theory to include pmv functions. They can even be made to work in the context of priority arguments, see Stoltenberg-Hansen [163]. But we may have *conceptual* doubts, strong enough to resist a technical point, however ingenious and elegant.

5.3 The Imbedding Theorem

The interplay of recursion theory and set theory has been a rich source of ideas for the general theory. In this section we shall analyze the "next-admissible" set/ordinal construction. We start by tracing some of the history.

"Higher" recursion theory started in the mid 1950's with S. C. Kleene's work on *the analytic hierarchy*, *constructive ordinals*, and *the hyperarithmetic sets* [79–81].

This work was followed by a number of basic contributions by *Clifford Spector* (1930–1961).

(i) In his paper of 1955, *Recursive wellorderings* [158], he introduced and used as a basic tool the boundedness theorem for hyperarithmetic theory (i.e. every Σ_1^1 subset of 0, the set of ordinal notations, is bounded by some recursive ordinal).

(ii) In his 1959 paper *Hyperarithmetic quantifiers* [160] he proved that Π_1^1 equals Σ_1^1 on HYP, i.e. every Π_1^1 predicate $P(a)$ can be written in the form

$$P(a) \quad \text{iff} \quad \exists\alpha(\alpha \in \text{HYP} \wedge R(\alpha, a)),$$

where HYP is the set of hyperarithmetic functions and R is an arithmetic relation.

(iii) In his study of *Ordinals of inductive definitions* in [161] he established a number of basic results, in particular, that

$$|\Pi_1^1\text{-mon}| = |\Pi_1^0| = \omega_1,$$

ω_1 being the first non-recursive ordinal.

(iv) Less known, and published only as an abstract, is a paper from the 1957 Cornell Summer Institute for Symbolic Logic, *Recursive ordinals and predicative set theory* [159]. Restated in modern terminology he proves that L_{ω_1} is the collection of hereditarily hyperarithmetic sets. (See also Hao Wang's contribution [171] to the same meeting where, in particular, the connection with Godel's notion of *constructibility* is emphasized.)

Writing history is not easy. Too often one restates and interprets the past from our present and "correct" point of view. As a result one is often amazed at the lack of insight sometimes shown by our predecessors.

Thus from our "correct" point of view it is remarkable that Spector did not put the pieces together and arrive at our way of looking at general recursion theory. From (ii) and (iv) he "knew" that Π_1^1 corresponds to Σ_1 on L_{ω_1}. From (iii) he "had" the connection with inductive definability. And a basic tool in the proofs was the boundedness theorem of (i), which is the clue to the set-theoretic description of L_{ω_1}, i.e. to admissibility theory. It was all in his hands!

Perhaps, not. The paper referred to in (iv) was related to Hao Wang's program for building a constructive or predicative foundation for mathematics. At the time it could be described as an example of "applied" recursion theory. Today it can be viewed as a central piece of the "pure" theory.

But Spector, who died 31 years old, had some of the basic technical results which we have exploited over and over again in arriving at our present "correct" version of the theory.

Remark. In his paper of 1959, *Quantification of number-theoretic functions* [84], Kleene proves that the ramified analytic hierarchy up to ω_1, R_{ω_1}, equals Δ_1^1. There is no reference to Spector's Cornell paper [159]. Did one not see any connection?

We have briefly focused on the work of Spector. It soon flowered into a rich general theory. Kreisel's and Sacks' 1965 paper, *Metarecursive sets* [91], marked an important stage. It gave both a conceptual analysis of the fundamental notions, in particular, of the notion of *finiteness* (see the introduction to Chapter 3), and contributed significantly to the techniques of ordinal recursion theory. At the same time, S. Kripke [92] developed the more general theory of recursion on an arbitrary *admissible ordinal.* A topic developed independently by R. Platek in his 1966 Stanford thesis [133], where also the theory of *admissible sets* is studied as an important part of the general theory.

In all this work the connection between hyperarithmetic theory and L_{ω_1} was of central importance. The metarecursion theory of Kreisel and Sacks was in a sense "constructed" out of hyperarithmetic theory and notations for recursive ordinals. And what had been a guiding principle was soon made into a precise theory: the "next-admissible" set/ordinal construction.

The basic reference is the 1971 paper of Barwise, Gandy, and Moschovakis, *The next-admissible set* [14]. Let A be a transitive set closed under the formation of unordered pairs. Define

$$A^+ = \bigcap \{M;\ M \text{ is admissible and } A \in M\}.$$

The first basic result in their study is:

(1) A^+ is admissible, in fact,
$A^+ = L_\kappa(A)$,
where $\kappa = \sup\{|\Gamma| : \Gamma \text{ is first-order positive inductive operator on } A\}$.

This result was proved using the theory of hyperprojective sets of Moschovakis [112]. A further result is

(2) A subset S of A is hyperprojective iff $S \in A^+$.

In the context of hyperarithmetic on ω hyperprojective theory is the same as Δ^1_1. In this case (1) constructs L_{ω_1} and (2) asserts that a subset of ω is Δ^1_1 iff it is an element of L_{ω_1}.

Hyperprojective theories are a special class of Spector theories. The above results immediately call for a generalization. This was provided in the context of Spector classes over transitive sets by Moschovakis in Chapter 9 of his book [115] on elementary induction. We discussed the problem in the context of Spector theories over general computation domains in our *On axiomatizing recursion theory* [26]. A proof of these results would use the theory of admissible sets with urelements, which was developed by Barwise [10] at the same time.

The "next-admissible" ordinal construction was carried out by P. Aczel [5] at the same time, formalizing the construction of metarecursion theory from hyperarithmetic theory.

It is time to be technical. Let Θ be a Spector theory on a computation domain $\mathfrak{A}$. There are two basic objects associated with Θ. First, an ordinal $\|\Theta\| = \sup\{|a, \sigma, z|;\ (a, \sigma, z) \in \Theta\}$. And next a relation R defined by

$$R(x, \alpha) \quad \text{iff} \quad x \in \Theta \wedge |x|_\Theta = \alpha.$$

These two objects are the essential ingredients in the "next-admissible" construction.

If we want to construct the next-admissible set, we should look at $L_{\|\Theta\|}[A; R]$, where A, the underlying set of $\mathfrak{A}$, is a set of urelements. If we want to construct the next-admissible ordinal we should look for a two-sorted theory $(A, \|\Theta\|)$, i.e. an admissible ordinal with urelements.

Both approaches are possible. But we shall follow a third way, and look for an R-admissible prewellordering above the given Spector theory Θ on $\mathfrak{A}$. There is an immediate candidate.

Let Θ be a Spector theory on $\mathfrak{A} = \langle A, A, \ldots \rangle$. Define the ordinal $\|\Theta\|$ as above. Let us be a bit more careful with the relation R:

$$R(x, \alpha) \text{ iff } x = \langle n, \langle x_1, \ldots, x_n \rangle \rangle \wedge (x_1, \ldots, x_n) \in \Theta \wedge |x_1, \ldots, x_n|_\Theta = \alpha.$$

5.3.1 Definition. We define a prewellordering $(\mathfrak{A}^*, \preccurlyeq)$ by setting

$$A^* = A \times \|\Theta\|,$$

and

$$(x, \alpha) \preccurlyeq (y, \beta) \text{ iff } \alpha \leqslant \beta.$$

For $(\mathfrak{A}^*, \preccurlyeq)$ we construct an appropriate language $L^*(\preccurlyeq, R)$ and introduce as usual classes $\Delta_0(\preccurlyeq, R)$ and $\Sigma_1(\preccurlyeq, R)$.

5.3.2 Lemma. *Let $B \subseteq A$. Then $B \in \text{en}(\Theta)$ iff B is $\Sigma_1(\preccurlyeq, R)$.*

(We use the simple imbedding $x \to (x, 0)$ of A into A^*.) For the proof assume first that B is Θ-semicomputable, i.e. for some code a,

$$\begin{aligned} x \in B &\text{ iff } (a, x, 0) \in \Theta \\ &\text{ iff } \exists y \in A^*[(y)_0 = \langle 3, \langle a, x, 0 \rangle \rangle \wedge R(y)]. \end{aligned}$$

Remember that we have coding-decoding in a pwo.

Conversely, assume that B is $\Sigma_1(\preccurlyeq, R)$. Then for some $\Delta_0(\preccurlyeq, R)$ formula Φ,

$$\begin{aligned} x \in B &\text{ iff } \exists y \in A \times \|\Theta\| \cdot \Phi((x, 0), y) \\ &\text{ iff } \exists a, z, w \in A[(a, z, 0) \in \Theta \wedge \Phi((x, 0), (w, |a, z, 0|))]. \end{aligned}$$

Here the matrix is Θ-semicomputable. Note that bounded quantification in Φ can be handled by *finiteness* and *prewellordering*, the characteristic properties of a Spector theory.

It remains to show that the pwo $(\mathfrak{A}^*, \preccurlyeq)$ is R-admissible. So let $\theta(x, X)$ be a $\Sigma_1(\preccurlyeq, X, R)$ formula in which X occurs positively. We must show that the least fixed-point of the associated inductive operator Γ_θ is $\Sigma_1(\preccurlyeq, R)$. In view of Lemma 5.3.2 all we need to do is to use the first recursion theorem for the underlying Spector theory.

With Θ we associate a Θ-computable functional φ in the following way (here we think of a as (a_1, a_2)):

$$\varphi(f, a) \simeq 0 \quad \text{iff} \quad a_2 \in \Theta \wedge \theta((a_1, |a_2|), \{(b_1, |b_2|) : \exists b[R((b, |b_2|) \wedge f((b_1, b)) \simeq 0]\}).$$

By necessity there is some coding involved!

Let f^* be the least—and Θ-computable—fixed-point for φ. Define X^* by

$$x \in X^* \quad \text{iff} \quad \exists b[R(b, x_2) \wedge f^*((x_1, b)) \simeq 0].$$

By Lemma 5.3.2 X^* is $\Sigma_1(\preccurlyeq, R)$. Assume now that $\theta(x, X^*)$ is true, we must verify that $x \in X^*$. Now x will be of the form (x_1, α) where $x_1 \in A$ and $\alpha < \|\Theta\|$. Let x' be (x_1, b) where $|b|_\Theta = \alpha$. We then get $\varphi(f^*, (x_1, b)) \simeq 0$, from which we conclude $f^*((x_1, b)) \simeq 0$. But then it follows that $\exists b[R(b, \alpha) \wedge f^*((x_1, b)) \simeq 0]$, i.e. $x \in X^*$, which was what we had to prove. And minimality of X^* follows from the minimality of f^*. QED.

5.3.3 Theorem. *Let Θ be a Spector theory on a computation domain $\mathfrak{A}$. Let $(\mathfrak{A}^*, \preccurlyeq)$ be the* pwo *introduced in Definition* 5.3.1 *and let $R(x, \alpha)$ be the relation: $x \in \Theta \wedge |x|_\Theta = \alpha$. Then*

(i) *$(\mathfrak{A}^*, \preccurlyeq)$ is R-admissible:*

(ii) *A subset $B \subseteq A$ is Θ-semicomputable iff it is $\Sigma_1(\preccurlyeq, R)$ under the imbedding $x \to (x, 0)$.*

This is our version of the "next-admissible" construction. And we feel that our analysis has isolated the crucial recursion-theoretic content: *the first recursion theorem.*

Adding Theorem 5.2.13 to the above construction yields a good infinite theory Θ^* over Θ. And, as we will show in the next chapter, it is this infinite theory Θ^* which will be the setting for "post-Fridberg" recursion theory, i.e. priority arguments and fine structure theory, which by "pull-back" should yield information about the given theory Θ.

5.3.4 Remark. A similar point of view was taken in Chapter 6 of Barwise, *Admissible Sets and Structures* [11]. He used his construction $\text{HYP}_{\mathfrak{M}}$, the "next-admissible" over $\mathfrak{M}$, to develop the theory of inductive definability over $\mathfrak{M}$. (See a remark on this in connection with Example 3.3.7.)

5.4 Spector Theories Over ω

Two important examples of Spector theories over ω are: (i) prime recursion in a total, normal type-2 functional, and (ii) prime recursion in ${}^2E$ and a consistent partial functional $F^{\#}_{\mathbf{Q}}$, derived from a monotone quantifier $\mathbf{Q}$ (see Example 3.1.3). From the representation Theorem 3.2.9 we know that these examples are exhaustive. One question remains: *Can we characterize those Spector theories which are equivalent to prime recursion in some total and normal type-2 functional?*

In discussing this question we shall have more to say about the interplay between set theory and recursion theory, thus continuing the discussion of the last section.

Let Θ and Ψ be Spector theories on ω. Theorem 3.2.8 tells us that $\Theta \sim \Psi$ iff $\mathrm{en}(\Theta) = \mathrm{en}(\Psi)$ and that $\Psi \leqslant \Theta$ iff $\mathrm{en}(\Psi) \subseteq \mathrm{en}(\Theta)$. We shall introduce a special notion for "strictly less than". But first a piece of notation: For any theory Θ we shall use α_Θ for the ordinal $\|\Theta\|$ of the theory. If F is a normal type-2 functional over ω we use α_F to denote the ordinal of the theory $\mathrm{PR}[F]$.

5.4.1 Definition. Let Θ and Ψ be Spector theories over ω. We define

$$\Psi <_1 \Theta \quad \text{iff} \quad \mathrm{en}(\Psi) \subseteq \mathrm{en}(\Theta) \wedge \alpha_\Psi < \alpha_\Theta.$$

5.4.2 Remark. We note the obvious consequences of the definition: If $\Psi \leqslant \Psi'$, $\Psi' <_1 \Theta'$, $\Theta' \leqslant \Theta$, then $\Psi <_1 \Theta$. And $\Psi <_1 \Theta$ implies that $\mathrm{en}(\Psi) \subsetneqq \mathrm{en}(\Theta)$.

We remind the reader that a functional F is Θ-computable if it is weakly Θ-computable, i.e. $F(g, \sigma)$ is Θ-computable if for some primitive recursive $f\colon \omega \to \omega$

$$F(\lambda\tau\cdot\{e\}_\Theta(\tau, \sigma_1), \sigma) \simeq \{f(n)\}_\Theta(e, \sigma_1, \sigma).$$

5.4.3 Remark. The following facts are immediate

(i) If F is Θ-computable, then $\mathrm{PR}[F] \subseteq \Theta$.
(ii) If Θ is Spector and $\Theta \sim \mathrm{PR}[F]$, then F is Θ-computable.

If Θ is a Spector theory and F is a Θ-computable total functional such that $\alpha_F = \alpha_\Theta$ it could still happen that $\mathrm{en}(F)$ (i.e. $\mathrm{en}(\mathrm{PR}[F])$) was strictly contained in $\mathrm{en}(\Theta)$. F could be too "thin" to code up all computations in Θ. What the next result shows is that we can use F to construct another normal and total G such that $\mathrm{PR}[G] \sim \Theta$.

5.4.4 Fattening Lemma. *Let Θ be a Spector theory and F a total Θ-computable functional such that $\alpha_F = \alpha_\Theta$. Then there is a total normal G such that $\Theta \sim$* $\mathrm{PR}[G]$.

For the proof we first pick an index e_1 such that $\{e_1\}_\Theta(e)\downarrow$ iff $\lambda x\{e\}_\Theta(x)$ is total, in which case $|\{e\}_\Theta(x)| < |\{e_1\}_\Theta(e)|$, for all $x \in \omega$. We further define for f a total function from ω to ω:

5.4.5 Definition. $\mathrm{Ord}(f) = \inf\{|\{e_1\}_\Theta(e)|_\Theta : f = \lambda x\cdot\{e\}_\Theta(x)\}$.

Using $\mathrm{Ord}(f)$ as a "cut-off" we construct a functional G_0 as follows:
If $\mathrm{Ord}(f)$ is defined, then

$$G_0(\langle f, e, \sigma\rangle) = \begin{cases} \{e\}_\Theta(\sigma) + 1 & \text{if } |\{e\}_\Theta(\sigma)| \leqslant \text{Ord}(f) \\ 0 & \text{otherwise,} \end{cases}$$

where $\langle f, e, \sigma\rangle = \lambda x \cdot \langle f(x), e, \sigma\rangle$. To make G_0 total we set $G_0(g) = 0$ if g is not of the form $\langle f, e, \sigma\rangle$, or if $g = \langle f, e, \sigma\rangle$ and Ord(f) is undefined.

We note first that G_0 is Θ-computable. We have the following instructions for computing G_0 on a function φ with index e': First use ${}^2E$ to check if φ is total. If φ is total, check if $\varphi = \langle f, e, \sigma\rangle$ for some f, e, σ. If not, let $G_0(\varphi) = 0$. If $\varphi = \langle f, e, \sigma\rangle$, then Ord($f$)$\downarrow$ since φ is Θ-computable. It remains to check if $|\{e\}_\Theta(\sigma)| \leqslant$ Ord(f). But since Θ is Spector, the relation

$$\forall h(f = \lambda x\{h\}_\Theta(x) \Rightarrow |\{e\}_\Theta(\sigma)|_\Theta \leqslant |\{e_1\}_\Theta(h)|_\Theta),$$

is Θ-computable. Hence we have Θ-computability of G_0. To ensure normality, take the join of G_0 and ${}^2E$, which we denote by G_0'. To ensure that G_0' does not close off too soon, take further the join of G_0' and the F we started with, denote the result by G.

One half of the lemma is now immediate, en(G) $\subseteq$ en(Θ). The converse needs more work. First define

$$\kappa = \sup\{\text{Ord}(f) : f \text{ is computable in } G\}.$$

Claim. $\kappa = \alpha_\Theta$.

Granted this claim the "fattening lemma" follows immediately. Suppose $\{e\}_\Theta(\sigma)\downarrow$. By the claim there is an index m such that $f = \lambda t \cdot \{m\}_G(t)$ is total and $|\{e\}_\Theta(\sigma)| \leqslant$ Ord(f). Hence $G_0(\langle \lambda t \cdot \{m\}_G(t), e, \sigma\rangle) \simeq \{e\}_\Theta(\sigma) + 1$. Since PR[$G$] is Spector we have a selection operator $\nu(e, \sigma)$, and we can define

$$\{e\}_\Theta(\sigma) \simeq G(\langle \lambda t \cdot \{\nu(e, \sigma)\}_G(t), e, \sigma\rangle) - 1.$$

Thus en(Θ) $\subseteq$ en(G).

To prove the claim we first note that if f is G-recursive, then f is Θ-computable, hence Ord(f)$\downarrow$ and $< \alpha_\Theta$. Therefore, $\kappa \leqslant \alpha_\Theta$.

To prove the converse we use the fact that $\alpha_F = \alpha_\Theta$. Hence $\alpha_G = \alpha_\Theta$, since obviously $\alpha_F \leqslant \alpha_G \leqslant \alpha_\Theta$. Thus we have to prove that $\alpha_G \leqslant \kappa$. Since α_G is the supremum of the lengths of G-recursive prewellorderings on ω, it suffices to prove that for each such prewellordering there is a G-recursive f such that Ord(f) $\geqslant$ the length of the prewellordering. This is an exercise in the use of the second recursion theorem, which we omit. However, we shall in Chapter 7 return to this point in the context of computation theories on two types.

5.4.6 Definition. A Spector theory Θ on ω is called Θ-*Mahlo* if for all normal Θ-computable F there exists a Spector theory Θ′ such that

(i) $\Theta' <_1 \Theta$,
(ii) F is Θ'computable.

Below we shall comment on the relationship of this notion of Mahloness to the notion of Mahloness in ordinal recursion theory. Here we have the characterization theorem.

5.4.7 Theorem. *Let Θ be a Spector theory on ω. Then Θ is equivalent to* PR[G] *for a normal, total type-2 G iff Θ is not Θ-Mahlo.*

We have here restricted ourselves to ω as domain. A more general result is valid. We return to this in Chapter 7 in the context of recursion theories on two types.

Half of the theorem is immediate: Let $\Theta \sim \text{PR}[G]$, then, using G in Definition 5.4.6, we see that there is no Θ' satisfying (i) and (ii). Conversely, assume that Θ is not Θ-Mahlo. Hence there exists a normal Θ-computable F such that, in particular, PR[F] is *not* $<_1$ than Θ. Since by Remark 5.4.3 $\text{en}(F) \subseteq \text{en}(\Theta)$, this means that $\alpha_F = \alpha_\Theta$. Hence, by the "Fattening Lemma" 5.4.4 there exists a total, normal G such that $\Theta \sim \text{PR}[G]$.

"Mahlo is Mahlo", we shall prove that the definition of Θ-Mahlo in 5.4.6 is the same as the ordinal-theoretic notion of Mahloness. And this will give us an opportunity to elaborate further on the relationship between Spector theories on ω and admissibility theory.

From a Spector theory Θ we derive an ordinal $\alpha_\Theta = \|\Theta\| = \sup\{|a, \sigma, z|_\Theta : (a, \sigma, z) \in \Theta\}$ and a relation $R_\Theta(x, \alpha)$ iff $x \in \Theta \wedge |x|_\Theta = \alpha$. From α_Θ and R_Θ we can construct the admissible set $L_{\alpha_\Theta}[R_\Theta]$. And we know from the imbedding theorem that if Θ is Spector, then α_Θ is R-admissible. (We shall further discuss this in connection with the one-section Theorem 5.4.24.)

As a variant of the standard procedure we shall now look at admissible ordinals from the standpoint of Spector theories. Θ is given. For $\tau \leqslant \alpha_\Theta$ introduce

$$\Theta_\tau = \{(a, \sigma, z) : \{a\}_\Theta(\sigma) \simeq z \wedge |\{a\}_\Theta(\sigma)|_\Theta < \tau\}.$$

5.4.8 Definition. τ is called *Θ-admissible* if Θ_τ is a Spector theory.

5.4.9 Definition. (i) A relation R on ω is called *Θ_τ-semicomputable* if there exists an index e such that

$$R(\sigma) \quad \text{iff} \quad |\{e\}_\Theta(\sigma)| < \tau.$$

(ii) Let $\pi: \tau^n \to \tau$ be a (partial) function. π is called *Θ_τ-computable* if the relation

$$\{(x, \ldots, y) : |x|_\Theta, \ldots, |y|_\Theta < \tau \wedge \pi(|x|_\Theta, \ldots) \simeq |y|_\Theta\},$$

is Θ_τ-semicomputable.

(iii) π is called Θ-*computable* if it is Θ_{α_Θ}-computable.

5.4.10 Definition. An ordinal $\tau \leqslant \alpha_\Theta$ is called Θ-*Mahlo* iff

(i) τ is Θ-admissible.
(ii) Every normal Θ_τ-computable function π has a Θ-admissible fixed-point less than τ.

We use standard set-theoretic terminology: π as a function from ordinals to ordinals is *normal* if it is strictly increasing and continuous at limit ordinals.

We give two simple results to show that the above definitions are the standard ones.

5.4.11 Proposition. *Let Θ be a Spector theory on ω. Let $\nu < \alpha_\Theta$ and π a partial Θ-computable functional. If $\pi(\xi)\downarrow$ for all $\xi < \nu$, then there exists a $\nu' < \alpha_\Theta$ such that $\pi(\xi) < \nu'$, for all $\xi < \nu$.*

For the proof note that the set

$$\{(x, y) : |x|_\Theta, |y|_\Theta < \alpha_\Theta \quad \text{and} \quad \pi(|x|) \simeq |y|\},$$

is Θ-semicomputable. Being in the Spector case, we have a selection function $\nu(x)$ such that if $\exists y[\pi(|x|) \simeq |y|]$, then $\nu(x)\downarrow$ and $\pi(|x|) \simeq |\nu(x)|$. It is not difficult to construct a computation x_0 such that $|\nu(x)|_\Theta < |x_0|_\Theta$ for all x such that $|x|_\Theta < \nu$. Let $\nu' = |x_0|_\Theta$.

5.4.12 Proposition. *Let F be a total, normal type-2 functional on ω. Then α_F is the least F-admissible ordinal.*

F-admissible is, of course, the same as PR[F]-admissible. For the proof let τ be a limit ordinal such that $\omega < \tau < \alpha_F$. We must show that τ is not F-admissible. Since $\tau < \alpha_F$ there must be F-computations of length τ, and since τ is a limit ordinal this computation must be an application of F to some function $\lambda x\{e\}_F(x, \sigma)$, where the function is total and $\tau = \sup\{|\{e\}_F(x, \sigma)|_F + 1 : x \in \omega\}$. Define π as $\pi(n) = |\{e\}_F(n, \sigma)|_F$. Then $\tau = \sup\{\pi(n) : n \in \omega\}$. If τ were F-admissible, then π would be F_τ-computable. 5.4.11 would then tell us that $\sup\{\pi(n) : n \in \omega\} < \tau$, a contradiction.

5.4.13 Remarks. These results should convince the reader that we are just looking at admissibility theory from a different point of view. We can even prove more: If α is admissible and projectible to ω (i.e. there is a one-one mapping π from α into ω which is α-recursive in constants less than α) then there is a Spector theory Θ on ω such that $\alpha_\Theta = \alpha$. We shall return to this point below in connection with the one-section result.

We claimed above that "Mahlo is Mahlo":

5.4.14 Theorem. *Let Θ be a Spector theory on ω. Then Θ is Θ-Mahlo (in the sense of Definition* 5.4.6) *iff α_Θ is Θ-Mahlo (in the sense of Definition* 5.4.10).

The proof is split into two lemmas.

5.4.15 Lemma. *Let Θ be a Spector theory on ω and F a normal Θ-computable type-2 functional. There exists a Θ-computable normal function π with no Θ-admissible fixed-points $< \alpha_F$.*

5.4.16 Lemma. *Let Θ be a Spector theory on ω and π a normal Θ-computable function. There exists a Θ-computable normal F such that α_F is Θ-admissible and a fixed-point for π.*

The theorem is a simple combination of the lemmas. Let first α_Θ be Θ-Mahlo and F a normal Θ-computable functional. By 5.4.15 there is a π with no Θ-admissible fixed-points $< \alpha_F$. But α_Θ is Θ-Mahlo, so π has Θ-admissible fixed points $< \alpha_\Theta$. Thus $\alpha_F < \alpha_\Theta$, and Θ is easily seen to be Θ-Mahlo. Conversely, let Θ be Θ-Mahlo and π a normal Θ-computable function. By 5.4.16 we have a Θ-computable F such that α_F is Θ-admissible and a fixed-point for π. Since Θ is Θ-Mahlo, $\alpha_F < \alpha_\Theta$. Thus α_Θ is Θ-Mahlo.

It remains to prove the lemmas; the reader not interested in the technical details may move on to Remark 5.4.17.

For Lemma 5.4.15 we first note that since F is Θ-computable there is an index t such that

$$x \in \mathbf{C}_F \quad \text{iff} \quad \langle t, x\rangle \in \mathbf{C}_\Theta,$$

where $\mathbf{C}_\Theta$ is, as before, the coded set of convergent computations. Use now Proposition 5.4.11 to prove the following two facts:

If $\nu < \alpha_\Theta$, there exists $\mu < \alpha_\Theta$ such that for all x,

$$|x|_F < \nu \Rightarrow |\langle t, x\rangle|_\Theta < \mu.$$

If $\nu < \alpha_\Theta$, there exists $\mu < \alpha_\Theta$ such that for all x,

$$|\langle t, x\rangle|_\Theta < \nu \Rightarrow |x|_F < \mu.$$

This done, define π as follows: $\pi(0) = 0$ and π is continuous at limit ordinals. $\pi(\nu + 1)$ is the least ordinal μ such that

(i) $\pi(\nu) < \mu$.
(ii) For all x, $|x|_F < \nu \Rightarrow |\langle t, x\rangle|_\Theta < \mu$,
(iii) For all x, $|\langle t, x\rangle|_\Theta < \nu \Rightarrow |x|_F < \mu$.

Using the second-recursion theorem we see that π is Θ-computable. π is normal by construction, and $\pi(\nu) < \alpha_\Theta$ whenever $\nu < \alpha_\Theta$. It remains to verify that π has no Θ-admissible fixed-points $< \alpha_F$.

Assume that $\tau < \alpha_F$, τ a limit ordinal, and $\pi(\tau) = \tau$. As before we have an index e such that the function $\lambda x \cdot \{e\}_F(x, \sigma)$ is total, $|\{e\}_F(x, \sigma)|_F < \tau$ for all x, and

(iv) $\sup\{|\{e\}_F(x, \sigma)|_F : x \in \omega\} = \tau.$

From (ii) we conclude that $|\langle t, \langle e, x, \sigma\rangle\rangle|_\Theta < \pi(\tau) = \tau$, all x. Whence,

(v) $\sup\{|\langle t, \langle e, x, \sigma\rangle\rangle|_\Theta : x \in \omega\} = \tau,$

viz. if the sup in (v) was $\tau' < \tau$, then by (iii) $|\{e\}_F(x, \sigma)|_F < \pi(\tau' + 1) < \pi(\tau) = \tau$, contradicting (iv) above.

If τ was Θ-admissible, then $\rho(x) = |\langle t, \langle \rho, x, \sigma\rangle\rangle|_\Theta$, $x \in \omega$, would be Θ_τ-computable. By 5.4.11 this would give $\sup\{\rho(x) : x \in \omega\} < \tau$, contradicting (v) above. Hence π has no Θ-admissible fixed-points $< \alpha_F$.

The proof of Lemma 5.4.16 necessitates a few preparatory remarks. First of all we need to keep track of how ordinals of computations in Θ grow. If $\nu < \alpha_\Theta$ by virtue of 5.4.11 there exists an ordinal $\mu < \alpha_\Theta$ such that:

(i) (substitution) If there exists an u such that $\{e\}_\Theta(\sigma) \simeq u$ and $\{f\}_\Theta(u, \sigma) \simeq x$ and $|\{e\}_\Theta(\sigma)|_\Theta$, $|\{f\}_\Theta(u, \sigma)|_\Theta < \nu$, then $|\{g_1(e, f, n)\}_\Theta(\sigma)| < \mu$, where $n = \mathrm{lh}(\sigma)$ and g_1 is an index for substitution.

(ii) (prewellordering) If $|x|_\Theta < \nu$ or $|y|_\Theta < \nu$, then $|\{\hat{p}\}_\Theta(x, y)| < \mu$.

(iii) (application of $\mathbf{E}_\omega$) If for some x, $\{e\}_\Theta(x, \sigma) \simeq 0$ and $|\{e\}_\Theta(x, \sigma)|_\Theta < \nu$, then $|\{g_2(n)\}_\Theta(e, \sigma)|_\Theta < \mu$, and if for all x there is some $y \neq 0$ such that $\{e\}_\Theta(x, \sigma) \simeq y$ and $|\{e\}_\Theta(x, \sigma)| < \nu$, then $|\{g_2(n)\}_\Theta(e, \sigma)|_\Theta < \mu$, where $n = \mathrm{lh}(\sigma)$ and $g_2(n)$ is an index for $\mathbf{E}_\omega$.

We have similar clauses for other functions and functionals entering into the axiomatic description of Θ. Let $\rho(\nu)$ be the least μ satisfying the conditions above. ρ is seen to be Θ-computable.

We now start the proof of Lemma 5.4.16. Recall from 5.4.5 the notion $\mathrm{Ord}(f)$, defined whenever f is a total Θ-computable function.

Let f be total and Θ-computable. Let $\nu = \mathrm{Ord}(f)$ and set $\mu = \sup(\pi(\nu), \rho(\nu))$. Then

$$F_0(\langle f, n, 0\rangle) = \begin{cases} 0 & \text{if} \quad n = \langle e, \sigma, y\rangle, \{e\}_\Theta(\sigma) \simeq y, \quad |\{e\}_\Theta(\sigma)|_\Theta < \nu, \\ 1 & \text{otherwise.} \end{cases}$$

$$F_0(\langle f, n, 1\rangle) = \begin{cases} 0 & \text{if} \quad n = \langle e, \sigma, y\rangle, \{e\}_\Theta(\sigma) \simeq y, \quad |\{e\}_\Theta(\sigma)|_\Theta < \mu, \\ 1 & \text{otherwise.} \end{cases}$$

Let $F_0(g) = 1$ if g is not Θ-computable, or g is not of the forms $\langle f, n, 0\rangle$, $\langle f, n, 1\rangle$. F_0 is easily seen to be Θ-computable. Let F be the join of F_0 and E.

We note that if f is F-recursive and total, then f is also Θ-computable, hence $\text{Ord}(f)\downarrow$. Let

(a) $\lambda = \sup\{\text{Ord}(f) : f \text{ is total, } F\text{-recursive}\}$.

Let $f = \lambda x \cdot \{e\}_F$ be total and $\nu = \text{Ord}(f)$. We see that $\lambda n \cdot F_0(\langle f, n, 0\rangle)$ is the characteristic function of the set

$$B_e = \{\langle e', \sigma, y\rangle : \{e'\}_\Theta(\sigma) \simeq y \wedge |\{e'\}_\Theta(\sigma)|_\Theta < \nu\}.$$

In the same way $\lambda n \cdot F_0(\langle f, n, 1\rangle)$ is the characteristic function of

$$C_e = \{\langle e', \sigma, y\rangle : \{e'\}_\Theta(\sigma) \simeq y \wedge |\{e'\}_\Theta(\sigma)|_\Theta < \mu\},$$

where we remember that $\mu = \sup(\pi(\nu), \rho(\nu))$. Thus B_e, C_e are F-recursive, uniformly in e.

Also remember that if $\inf(|x|_\Theta, |y|_\Theta) < \nu$ then $|\{\hat{p}\}_\Theta(x, y)|_\Theta < \rho(\nu) \leqslant \mu$. The set

$$\{(x, y) : |x|_\Theta < \nu \text{ or } |y|_\Theta < \nu, \text{ and } \langle \hat{p}, x, y, 0\rangle \in C_e\},$$

is a prewellordering of length ν and is F-recursive, since B_e and C_e are F-recursive. Any F-recursive pwo has length $< \alpha_F$. Hence $\nu < \alpha_F$.

Since ν above is of the form $\text{Ord}(f)$, where f is total and F-recursive, it follows that the ordinal λ introduced in (a) satisfies $\lambda \leqslant \alpha_F$. If we had equality, $\lambda = \alpha_F$, Lemma 5.4.16 would immediately follow:

(1) To prove that α_F is a fixed-point for π, it suffices to prove that $\nu < \alpha_F$ implies $\pi(\nu) < \alpha_F$. So let $\nu < \alpha_F$ and choose an e such that $f = \lambda x\{e\}_F(x)$ is total and $\text{Ord}(f) = \nu' \geqslant \nu$. Such an e exists since $\lambda \geqslant \alpha_F$. Let $\mu = \sup(\pi(\nu'), \rho(\nu'))$. C_e is F-recursive and from it we can construct a total and F-recursive f' different from all f'' with $\text{Ord}(f'') \leqslant \mu$. Thus $\mu < \text{Ord}(f') < \lambda \leqslant \alpha_F$. Since $\nu \leqslant \nu'$ the desired inequalities follow.

(2) To prove that α_F is Θ-admissible it suffices to prove that if $|\{e\}_\Theta(x, \sigma)|_\Theta < \alpha_F$ for all x, then there is an ordinal $\mu < \alpha_F$ such that $|\{e\}_\Theta(x, \sigma)|_\Theta < \mu$ for all x. But up to length α_F enough information about Θ-computations is coded into F_0. In fact if $|\{e\}_\Theta(x, \sigma)| < \alpha_F$ for all x then it is possible to construct an index e' such that $\{e\}_\Theta(x, \sigma) \simeq \{e'\}_F(x, \sigma)$ and $|\{e\}_\Theta(x, \sigma)|_\Theta < |\{e'\}_F(x, \sigma)|_F$ for all x. Then let μ be the length of the computation $E(\lambda x \cdot \{e'\}_F(x, \sigma))$.

We know that $\lambda \leqslant \alpha_F$. It remains to prove equality. (And, note that equality was used in (1) above.)

We do this by constructing a function $\sigma: \lambda \to \alpha_F$ such that σ is F-recursive and cofinal in α_F. Admissibility, i.e. 5.4.11, then implies that $\lambda = \alpha_F$.

Replacing Θ by F we introduce a notion $\text{Ord}_F(f)$, for f total and F-recursive. A simple diagonal construction will tell us that

(b) $\alpha_F = \sup\{\text{Ord}_F(f) : f \text{ is total, } F\text{-recursive}\}$.

From (a) and (b) there is a short step to a suitable function σ, viz. for $\nu < \lambda$ set

(c) $\sigma(\nu) = \inf\{\text{Ord}_F(f) : \text{Ord}(f) > \nu\}$.

We must prove cofinality and F-recursiveness.

The F-recursiveness of σ is obtained by a painstaking analysis of the definition, writing out each part in its ultimate recursion-theoretic details (Moldestad [105], pp. 106–107). For cofinality, assume to the contrary that

$$\sup\{\sigma(\nu) : \nu < \lambda\} = \mu < \alpha_F,$$

which means that $\sup\{\text{Ord}(f) : \text{Ord}_F(f) < \mu\} = \lambda$. This, however, contradicts the following fact: If $\mu < \alpha_F$, then

(d) $\nu = \sup\{\text{Ord}(f) : \text{Ord}_F(f) < \mu\} < \lambda$.

For the proof of (d) let μ and ν be fixed. The set $D = \{e : |\{e_1\}_F(e)|_F < \mu\}$ is F-recursive, where e_1 is the index used in the definition of $\text{Ord}_F(f)$. Note that $\text{Ord}_F(f) < \mu$ iff $\exists e \in D \cdot [f = \lambda x \cdot \{e\}_F(x)]$. Let

$$B = \bigcup_{e \in D} B_e.$$

This set is F-recursive since B_e is F-recursive uniformly in e. And we observe that

$$\langle e', \sigma, y\rangle \in B \quad \text{iff} \quad \{e'\}_\Theta(\sigma) \simeq y \wedge |\{e'\}_\Theta(\sigma)|_\Theta < \nu,$$

where ν is the ordinal introduced in (d). Now, if f is a total Θ-computable function such that $\text{Ord}(f) \leqslant \nu$, then there is an index e such that $f = \lambda x \cdot \{e\}_\Theta(x)$ and $\forall x \exists y \langle e, x, y\rangle \in B$. Hence the set

$$E = \{e : \forall x \exists y \langle e, x, y\rangle \in B\},$$

which is F-recursive, contains Θ-indexes for all total Θ-computable functions f with $\text{Ord}(f) \leqslant \nu$, in particular, E contains Θ-indexes for all f such that $\text{Ord}_F(f) < \mu$. Once more, a diagonal construction will yield a function f' which is F-recursive and different from all total Θ-computable functions with Θ-index in E. Hence, $\nu < \text{Ord}(f') < \lambda$, which concludes the proof of (d).

5.4.17 Remark. We add a brief remark on the sources for the theory of Section 5.4 up to this point. The fact that Θ is equivalent to a theory PR[G] for a normal type-2 G iff the ordinal α_Θ of Θ is not Θ-Mahlo is due independently to S. Simpson and to L. Harrington and A. Kechris, see [58] and [56]. The notion of Θ-Mahlo (5.4.6) can be found in the seminar report of A. Kechris [74], where Theorem 5.4.7 for both one and two domains are proved. Implicit in this work are both the Fattening Lemma 5.4.4 and the fact that "Mahlo is Mahlo". We have followed

Moldestad [105] in our exposition. In particular the detailed and explicit constructions in Lemmas 5.4.15 and 5.4.16 are taken from his study.

We have now successfully characterized those Spector theories which are equivalent to prime computability in a normal type-2 functional over ω. Restricting ourselves to sections we can go further and show that for *every* Spector theory Θ on ω there is a normal type-2 functional F such that $\mathrm{sc}(\Theta) = \mathrm{sc}(F)$, but in general the envelopes will be different. This is the "plus-one" theorem of G. E. Sacks [142].

5.4.18 Definition. Let M be a non-empty transitive set. M is called an *abstract 1-section* if it is closed under pairing and union and satisfies the following axioms:

(1) *Local countability:* $\forall x$ [x is countable].
(2) Δ_0*-separation:* $\exists x \forall y [y \in x \leftrightarrow y \in a \wedge \varphi(y)]$.
(3) Δ_0*-dependent choice:* $\forall x \exists y \varphi(x, y) \rightarrow \exists h \forall n \varphi(h(n), h(n+1))$, where $\varphi(x)$ and $\varphi(x, y)$ are Δ_0-formulas (with parameters) and h is a function from ω to M.

The reader will note that if M is an abstract 1-section, then M is an admissible set and each element of M is hereditarily countable.

This leads to the topic of codings: Each set $x \in \mathrm{HC}$ (the hereditarily countable sets) can be encoded by an $\alpha \in \omega^\omega$. If α is a code, let $m(\alpha) \in \mathrm{HC}$ be the set encoded by α. By induction on the set theoretic rank of x we can define a relation:

(i) $\qquad \alpha$ is a code and $m(\alpha) = x$,

α is a code for the set $\{m(\alpha_n) : n \in \omega\}$, α_n being the usual projection of α. The set of codes is Π_1^1, i.e. semicomputable in ${}^2E$.

5.4.19. Proposition. *Let α be a code and $\varphi(x)$ a Δ_0-formula. The relation $P(\beta)$ iff $\exists n[\beta = \alpha_n \wedge \varphi(m(\beta))]$ is recursive in α, ${}^2E$.*

By now this is familiar: bounded quantification corresponds to number quantification over ω.

So we come to our main construction. Let Θ be a Spector theory on ω. Let $m(\mathrm{sc}(\Theta))$ be the set of all sets in HC with code in $\mathrm{sc}(\Theta)$.

5.4.20 Proposition. *Let Θ be a Spector theory on ω. Then $m(\mathrm{sc}(\Theta))$ is an abstract 1-section.*

We verify Δ_0-separation and Δ_0-dependent choice. For Δ_0-separation let $\varphi(y)$ be Δ_0 and $\alpha \in \mathrm{sc}(\Theta)$ be a code. We must find a code $\alpha_0 \in \mathrm{sc}(\Theta)$ such that

$$\forall y [y \in m(\alpha_0) \text{ iff } y \in m(\alpha) \wedge \varphi(y)].$$

By Proposition 5.4.19 the relation $P(\beta)$ is Θ-computable. Let α_0 be an enumeration of all β's that satisfy $P(\beta)$.

Turning to Δ_0-DC let $\varphi(x, y)$ be a Δ_0-formula such that

(ii) $\forall x \exists y \varphi(x, y)$,

is true in $m(\mathrm{sc}(\Theta))$.

Let $\{a\}_\Theta$ be a code in $\mathrm{sc}(\Theta)$. Define a set

$$Q_a = \{n \in \omega : \{n\}_\Theta \text{ is a code } \wedge \ \varphi(m(\{a\}_\Theta), m(\{n\}_\Theta))\}.$$

Q_a is Θ-semicomputable uniformly in a. We now use a selection operator $\nu(a)$ such that whenever $\{a\}_\Theta$ is a code then $\nu(a) \in Q_a$. We then define a function h by recursion

$$h(0) = n_0, \quad \{n_0\}_\Theta \text{ a code for } \varnothing.$$
$$h(n + 1) = \nu(h(n)).$$

Then, clearly, $\forall n \varphi(m(\{h(n)\}_\Theta), m(\{h(n + 1)\}_\Theta))$.

5.4.21 Remark. In the proof of 5.4.20 we asserted that the relation $P(\beta)$ is Θ-computable. Strictly speaking this makes no sense: Our Spector theories are the "light-faced" version of hyperarithmetic theory and $P(\beta)$ is a "bold-faced" relation. But using essentially Proposition 3.1.12, we can pass from the "light-faced" to the "bold-faced" version as in hyperarithmetic theory.

$m(\mathrm{sc}(\Theta))$ can be given a more precise description. It is in fact equal to $L_{\alpha_\Theta}[R_\Theta]$, where α_Θ and R_Θ have their usual meaning. And $\alpha_\Theta = m(\mathrm{sc}(\Theta)) \cap \mathrm{On}$. In this setting the imbedding theorem of Section 5.3 asserts that Θ-semicomputability corresponds to $\Sigma_1(\langle L_{\alpha_\Theta}[R_\Theta], \in, R_\Theta\rangle)$. We shall not prove this in detail, for it is not needed in the actual proof of Sack's "plus-one" theorem.

This theorem asks if it is possible to define a normal type-2 F such that $\mathrm{sc}(\Theta) = L_{\alpha_\Theta}[R_\Theta] \cap 2^\omega = \mathrm{sc}(F)$ for any Spector theory Θ on ω.

The answer is yes, but there are several stumbling-blocks in the proof. Theorems 5.4.7 and 5.4.14 should warn us that there is no trivial way of pulling R_Θ back to a functional over ω. One point is that the notion of code involves the notion of well-foundedness, and well-foundedness is not computable in every Spector theory. This is the first obstacle to get around.

A second obstacle comes from Proposition 5.4.12. It is not at all obvious (in fact, it may be false) that α_Θ is the least R_Θ-admissible ordinal, which it should be.

But here is a lead: use forcing. Given any abstract one-section M we can generically construct an R such that $M = L_\alpha[R]$, where $\alpha = M \cap \mathrm{On}$, and α is the least R-admissible ordinal. In particular, α will not be R-Mahlo. And this is a result we can apply. But let us first digress and make some historical remarks.

Forcing was rather soon applied to problems in arithmetic and recursion theory, some of the early and influential papers are Feferman [24], Gandy-Sacks [43], and Sacks [141]. In the context of admissible sets Jensen included a section on forcing in his lecture notes [70], proving, characteristically, some very deep theorems. Unfortunately Jensen's lecture notes have remained unpublished. Our next result is, in fact, a simple version of a result of Jensen. We follow the exposition in Normann [121], who saw how to apply this result to the "plus-one" theorem, avoiding the somewhat complicated hierarchy for recursion in higher types introduced by Sacks.

5.4.22 Proposition. *Let M be an abstract 1-section and $\alpha = M \cap \mathrm{On}$. There exists an $R \subseteq \alpha$ such that $M = L_\alpha[R]$ and α is the least R-admissible ordinal.*

Introduce the conditions $\mathbf{P}$ as follows:

(1) $p \in \mathbf{P}$ iff $p \subseteq \alpha$ and no ordinal $\leqslant \sup p$ is p-admissible.
(2) $p \leqslant q$ iff $q = p \cap \mathrm{rnk}(q)$.

The forcing relation will be defined directly for Δ_0-formulas and then extended to all formulas in the usual way.

(3) $p \Vdash \varphi(x_1, \ldots, x_n, p)$ iff $x_1, \ldots, x_n \in L_{\mathrm{rnk}(p)}[p]$ and $\langle L_{\mathrm{rnk}(p)}[p], \in, p\rangle \models \varphi(x_1, \ldots, x_n \dot{p})$.

Here $\varphi(x_1, \ldots, x_n, \dot{p})$ is a Δ_0-formula containing the symbol $\dot{p}$ to be interpreted by the set p. To be really careful (or pedantic?) we should also have distinguished between the variable x_i in $\varphi(x_1, \ldots, x_i, \ldots, x_n, \dot{p})$ and the set $x_i \in L_{\mathrm{rnk}(p)}[p]$. We also remind the reader that $\mathrm{rnk}(x)$ is the usual set-theoretic notion of rank.

From (3) we see immediately that $\Vdash^{\Delta_0}$ is Δ_1-definable. Let R be generic with respect to $\langle \mathbf{P}, \leqslant\rangle$. We want to show that

(4) $M = L_\alpha[R]$.

The part of (4) that requires some work follows from the following lemma.

5.4.23 Lemma. *For all conditions p and all $x \in M$ there is an extension $q \leqslant p$ such that $x \in L_{\mathrm{rnk}(q)}[q]$.*

Actually it is sufficient to prove the simpler result that there exists a $q \leqslant p$ such that the *code* for x belongs to $L_{\mathrm{rnk}(q)}[q]$, since a set belongs to an abstract 1-section iff its code, which is a subset of ω, belongs to the 1-section. And it is easy to extend p to a q encoding the code of x, viz. put $q = p \cup \{\mathrm{rnk}(p) + n;$ n belongs to the code of $x\}$.

The main thing to verify is $\Delta_0(R)$-collection. And as usual assume the contrary. Then there is some Δ_0-formula φ and a set $u \in M$ such that for some $p \subseteq R$

(5) $p \Vdash (\forall x)(\exists y)\varphi(x, y, \dot{R})$.
(6) $p \Vdash (\forall v)(\exists x \in u)(\forall y \in v)\neg\varphi(x, y, \dot{R})$.

We can rewrite (5) and (6) in the following forms

(7) $(\forall q \leqslant p)(\forall x)(\exists r \leqslant q)(\exists y) \cdot r \Vdash \varphi(x, y, \dot{R})$.

(8) $(\forall q \leqslant p)(\forall v)(\exists r \leqslant q) \cdot r \Vdash \exists x \in u\, \forall y \in v \neg \varphi(x, y, \dot{R})$.

From (7) we can derive

(9) $(\forall q \leqslant p)(\exists r \leqslant q)(\forall x \in L_{\mathrm{rnk}(q)}[q])(\exists y \in L_{\mathrm{rnk}(r)}[r]) \cdot r \Vdash \varphi(x, y, \dot{R})$.

This is obtained in the following way. Let $r_0 \leqslant q$ take a wellordering of type ω of $L_{\mathrm{rnk}(q)}[q]$ inside the model. By Σ_1-DC we may choose a sequence $\langle r_i \rangle_{i\in\omega}$ (here is a point where the definability of the forcing-relation enters) such that if $r = \bigcup_{i\in\omega} r_i$, then r_{i+1} is of minimal rank such that $r_{i+1} \leqslant r_i$ and if x_i is element number $i + 1$ in $L_{\mathrm{rnk}(q)}[q]$ then $r_{i+1} \Vdash \varphi(x_i, y_i, \dot{R})$, for some element y_i. rnk(r) is not r-admissible, hence $r \in \mathbf{P}$ and (9) is verified.

Wellorder $u = \{x_i : i \in \omega\}$ inside M and use (9) to get a sequence $\langle q_i, y_i \rangle$ such that

(10)
(i) $q_i \leqslant p$ and $u \in L_{\mathrm{rnk}(q_0)}[q_0]$.
(ii) $q_{i+1} \leqslant q_i$.
(iii) $(\forall x \in L_{\mathrm{rnk}(q_i)}[q_i])(\exists y \in L_{\mathrm{rnk}(q_{i+1})}[q_{i+1}]) \cdot q_{i+1} \Vdash \varphi(x, y, \dot{R})$.
(iv) $(\forall i \in \omega) \cdot q_{i+1} \Vdash \varphi(x_i, y_i, \dot{R})$.

Let $q = \bigcup_{i\in\omega} q_i$. We must first verify that q is a condition. First observe from (iii) that $L_{\mathrm{rnk}(q)}[q] \Vdash \forall x \exists y \varphi(x, y)$. Suppose that $\exists v \in L_{\mathrm{rnk}(q)}[q]$ such that for $(\forall i \in \omega)(\exists y \in v)\varphi(x_i, y)$. Since $q = \bigcup q_k$, there must be some q_k such that $v \in L_{\mathrm{rnk}(q_k)}[q_k]$. Then $q_k \Vdash (\forall x \in u)(\exists y \in v)\varphi(x, y, \dot{R})$ (recall the definition (3) of the forcing relation). But this contradicts (8), hence q is a condition.

The same type of argument applied once again will finish off the proof of $\Delta_0(R)$-collection. Let $s \leqslant q$ be such that $\langle y_i \rangle_{i\in\omega}$, $v = \{y_i : i \in \omega\} \in L_{\mathrm{rnk}(s)}[s]$. Then

$$L_{\mathrm{rnk}(s)}[s] \models (\forall x \in u)(\exists y \in v)\varphi(x, y, \dot{R}).$$

But this contradicts (8). Putting things together we now have a full proof of Proposition 5.4.22.

We are now ready for the main result supplementing the characterization Theorem 5.4.7.

5.4.24 Plus-One Theorem. *Let Θ be a Spector theory on ω. There exists a normal type-2 functional F on ω such that* sc(Θ) = sc(F).

Let Θ be given and construct its associated one-section $m(\mathrm{sc}(\Theta)) = L_\alpha[R]$, where $\alpha = \|\Theta\|$ and R obtained as in Proposition 5.4.22. α is the least R-admissible, hence not R-Mahlo. $\Sigma_1(\langle L_\alpha[R], \in, R \rangle)$ defines a Spector class, hence a Spector theory on ω, call the theory Θ^*. This theory is not Θ^*-Mahlo, hence by Theorem 5.4.7, it is equivalent to a Spector theory PR$[F]$, for some total, normal type-2 F over ω. Since it is easy to see that sc(Θ) is determined by $m(\mathrm{sc}(\Theta))$, it follows that

$\mathrm{sc}(\Theta) = \mathrm{sc}(F)$. But note that $\mathrm{en}(\Theta)$ may differ from $\mathrm{en}(\Theta^*)$. The latter corresponds to $\Sigma_1(\langle L_\alpha[R], \in, R\rangle)$ restricted to ω, but since R is obtained by a forcing argument, which does not preserve Δ_1-definability, this may differ from $\Sigma_1(\langle L_\alpha[R_\Theta], \in, R_\Theta\rangle)$.

5.4.25 Example. (This is a simple version of a result in J. Bergstra [15].) There exist two normal type-2 functionals F_1, F_2 such that

(1) $\mathrm{en}(F_1) \neq \mathrm{en}(F_2)$

(2) $\mathrm{sc}(F_1, \alpha) = \mathrm{sc}(F_2, \alpha)$, for all $\alpha \in \omega^\omega$.

The envelope cannot be reconstructed from its section.

We let $F_1 = {}^2E$ and F_2 the recursive join of ${}^2E$ and a functional F_{α_0}, where $\alpha_0 \notin 1\text{-sc}({}^3E)$ and

$$F_{\alpha_0}(\beta) = \begin{cases} 1 & \text{if } \alpha_0 = \beta \\ 0 & \text{otherwise.} \end{cases}$$

To verify (1) assume that $F_{\alpha_0} \leqslant {}^2E$. Then $1\text{-sc}(F_2, {}^3E) \subseteq 1\text{-sc}({}^2E, {}^3E) = 1\text{-sc}({}^3E)$. But this is a contradiction since $\alpha_0 \in 1\text{-sc}(F_2, {}^3E)$ but does not lie in $1\text{-sc}({}^3E)$.

Let α be given. If α_0 *is* $\Delta_1^{1,\alpha}$, then F_{α_0} is recursive in ${}^2E, \alpha$; hence $1\text{-sc}(F_2, \alpha) \subseteq 1\text{-sc}({}^2E, \alpha)$. If α_0 *is not* $\Delta_1^{1,\alpha}$, we will show that F_{α_0} has no effect on the 1-sections, and (2) will again follow.

So suppose that we are making a calculation

$$\{e\}(t) = F_{\alpha_0}(\lambda v \cdot \{e'\}(v, t)),$$

where for some t_0, $\{e\}(t_0) = 1$ (otherwise $\lambda t \cdot \{e\}(t)$ would just be the characteristic function of ω) and $\lambda t \cdot \{e\}(t)$ is total. And suppose that this is the "first" (in length of computations) where we are breaking out of $1\text{-sc}({}^2E, \alpha)$. This means that $\lambda v \cdot \{e'\}(v, t_0) \in 1\text{-sc}({}^2E, \alpha)$. But this is impossible since $\{e\}(t_0) = 1$ implies that

$$\alpha_0 = \lambda v \cdot \{e'\}(v, t_0),$$

and we had assumed that α_0 is not in $\Delta_1^{1,\alpha}$.

Chapter 6
Degree Structure

The study of degrees, in particular of r.e. degrees, is a characteristic and important part of recursion theory. And no account of general recursion theory can be claimed to be successful unless at least an introduction to notions of *reducibilities* and the associated *degree theory* is given. This is precisely what we will do in this chapter: to present an *introduction* to this topic within the general framework of infinite theories and to give an example of a non-trivial result in the extended framework.

But we should really like to do something more. In the spirit of an axiomatic analysis we want to determine the "true" domain for degree theory and priority arguments. This is the question we turn to in Section 6.3. Our discussion is fragmentary and we do not claim any complete solution. Indeed there may not be any well-defined "solution". But we hope that this section may give some clue as to how far recursion-theoretic regularities extend.

Our discussion is, in principle, self-contained, but some familiarity with the basic notions of α-recursion theory would be helpful: we recommend the introductory paper of R. A. Shore, *α-recursion theory* [152], in a precise sense we continue his discussion in this chapter.

6.1 Basic Notions

The setting is an infinite computation theory Θ on a prewellordered domain $(\mathfrak{A}, \preccurlyeq)$, see Definition 5.1.5. We shall need a suitable notion of enumeration and of parametrization of the Θ-semicomputable sets. But as usual we must preface our definitions by introducing some necessary notation.

Let f be a mapping which to every $x \in A$ gives us a canonical Θ-index for a Θ-finite set, i.e. $f(x)$ is an index for the function $\mathbf{E}_{W^x}$, where W^x is the Θ-finite set associated with x. It will be convenient to write the mapping as

$$f = \lambda x \cdot W^x,$$

but we should always remember that the value of f at x is a canonical Θ-index for the Θ-finite set W^x.

6.1.1 Definition. A $\preccurlyeq$-*enumeration* of a set W is a Θ-computable mapping $\lambda x \cdot W^x$ whose values are canonical Θ-indices for the Θ-finite sets W^x, such that

(i) $y \preccurlyeq x \Rightarrow W^y \subseteq W^x$,
(ii) $W = \bigcup \{W^x : x \in A\}$.

A $\preccurlyeq$-*parametrization* of the Θ-semicomputable sets is a Θ-computable mapping $\lambda ax.\ W_a^x$ such that

(iii) $y \preccurlyeq x \Rightarrow W_a^y \subseteq W_a^x$,
(iv) for each Θ-semicomputable set W there is an a such that $W = \bigcup \{W_a^x; x \in A\}$.

From axioms A and B (see 5.1.1 and 5.1.2) the reader will have no difficulty in constructing a $\preccurlyeq$-enumeration of the sets

$$\Theta_n = \{\langle a, \sigma, z\rangle : \{a\}_\Theta(\sigma) \simeq z \wedge \mathrm{lh}(\sigma) = n\},$$

from which he may derive a $\preccurlyeq$-parametrization of the Θ-semicomputable sets.

A number of arguments in α-recursion theory seem to require the use of the μ-operator. Let $R(\sigma, x)$ be a Θ-computable relation, we would like to introduce a function $\mu x R(\sigma, x)$ by the equivalence

$$\mu x R(\sigma, x) = z \quad \text{iff} \quad R(\sigma, z) \wedge (\forall y \prec z) \neg R(\sigma, y).$$

In α-recursion theory the domain, a segment of the ordinals, is well-ordered, so there is a unique z satisfying the equivalence. When the domain has a prewell-ordering, the μ-operator would in general have to be multiple-valued. But there is a way of getting around this obstacle.

6.1.2 Proposition. *Let $R(\sigma, x)$ be a Θ-semicomputable relation such that*

(i) *$R(\sigma, x) \Rightarrow x$ is a canonical Θ-index for some Θ-finite set K_x.*
(ii) *$R(\sigma, x) \wedge R(\sigma, y) \Rightarrow K_x = K_y$.*

Then there is a Θ-computable mapping $q(\sigma)$ obtained uniformly from an index r of R such that

$$K_x = K_{q(\sigma)},$$

for all x such that $R(\sigma, x)$

We concentrate on the key point of the proof. $q(\sigma)$ will be a canonical Θ-index for the set

$$N_\sigma = \bigcup \{K_x : R(\sigma, x)\}.$$

To see how to define $q(\sigma)$ we calculate:

$$\begin{aligned}\mathbf{E}_{N_\sigma}(\{\hat{f}\}) = 0 \quad &\text{iff} \quad \exists y \in N_\sigma \cdot \{\hat{f}\}(y) = 0\\ &\text{iff} \quad \exists y \exists w (R(\sigma, w) \wedge y \in K_w \wedge \{\hat{f}\}(y) = 0)\\ &\text{iff} \quad \exists w (R(\sigma, w) \wedge \{w\}(\hat{f}) = 0).\end{aligned}$$

In the same way we get

$$\begin{aligned}\mathbf{E}_{N_\sigma}(\{\hat{f}\}) = 1 \quad &\text{iff} \quad \forall y \in N_\sigma \cdot \{\hat{f}\}(y) = 1\\ &\text{iff} \quad \forall y \forall w (R(\sigma, w) \wedge y \in K_w \rightarrow \{\hat{f}\}(y) = 1)\\ &\text{iff} \quad \forall w (R(\sigma, w) \rightarrow \{w\}(\hat{f}) = 1)\\ &\text{iff} \quad \exists w (R(\sigma, w) \wedge \{w\}(\hat{f}) = 1).\end{aligned}$$

The last equivalence follows from assumption (ii). We now choose a code $q'(\sigma)$ such that $\{q'(\sigma)\}(\hat{f}, j) \simeq 0$ iff $\exists w(R(\sigma, w) \wedge \{w\}(\hat{f}) = j)$, and we get our function $q(\sigma)$ by using selection over the integers.

6.1.3 Theorem. *There is a* Θ*-computable function* $q(a, \sigma)$ *such that if* $B_\sigma = \{x : \{a\}(x, \sigma) \simeq 0\}$ *is a non-empty* Θ*-semicomputable set, then* $q(a, \sigma)$ *gives a canonical* Θ*-index for a non-empty* Θ*-finite subset* N *of* B_σ.

This is another variation of a familiar theme. We are not in general able to computably select a *unique* element of B_σ, but we can *effectively compute* an index of a finite subset of B_σ. For the proof let $\lambda z \cdot W^z$ be a $\preccurlyeq$-enumeration of the set $\{\langle a, x, \sigma, 0\rangle : \{a\}(x, \sigma) \simeq 0\}$. For each z we introduce the Θ-finite set $N_{\sigma,z} = \{y : y \prec z \wedge \langle a, y, \sigma, 0\rangle \in W^z\}$. Consider the following Θ-semicomputable relation

$$\begin{aligned}H(\sigma, z) \quad \text{iff} \quad &\exists y \prec z \cdot \langle a, y, \sigma, 0\rangle \in W^z\\ &\wedge (\forall w \prec z) \neg (\exists y \prec w) \cdot \langle a, y, \sigma, 0\rangle \in W^w.\end{aligned}$$

From this we see that if $H(\sigma, z_1)$ and $H(\sigma, z_2)$, then $z_1 \sim z_2$ and further $N_{\sigma,z_1} = N_{\sigma,z_2}$. Let

$$N = \bigcup \{N_{\sigma,z} : H(\sigma, z)\}.$$

Since the canonical Θ-indices involved are effectively computable from the given data, Proposition 6.1.2 allows us to compute $q(a, \sigma)$ as canonical Θ-index of N, and obviously $\varnothing \neq N \subseteq B_\sigma$.

6.1.4 Definition. A theory Θ is *projectible* into a subset W of its domain A if there is a Θ-computable function p such that $\text{dom}(p) \subseteq W$ and p maps onto all of A.

Here the set $p^{-1}(x)$ is a set of "notations" for $x \in A$, but, lacking a well-ordering, we have in general no unique notation for each $x \in A$. For some purposes it is important to know that $p^{-1}(x)$ is Θ-finite uniformly in x. We can always so arrange things by using 6.1.3. Usually we shall study projections into sets of the form

$$L^{\beta} = \{x \in A : x \prec \beta\}.$$

This is a slight abuse of notation; what we mean is that $|x|_{\preccurlyeq} < \beta$, where $|x|_{\preccurlyeq}$ is the ordinal of x in the prewellordering $\preccurlyeq$. In the same way $L^{x} = \{y \in A : y \prec x\}$. Note that we can always assume that $p^{-1}(L^{x})$ is Θ-finite. This is used e.g. in the proof of Theorem 6.1.18.

6.1.5 Definition. Let Θ be an infinite theory on a domain $(\mathfrak{A}, \preccurlyeq)$. The *projectum* of Θ, denoted by $|\preccurlyeq|^{*}$, is the least ordinal β such that Θ is projectible into L^{β}.

This means that we have a notation for each $x \in A$ below $|\preccurlyeq|^{*}$. And more importantly it follows that we have a $\preccurlyeq$-parametrization of the Θ-semicomputable sets below $|\preccurlyeq|^{*}$.

6.1.6 Lemma. (i) *Let $W = \{a : W_a \neq \varnothing\}$ for a given $\preccurlyeq$-parametrization of the Θ-semicomputable sets. Then Θ is projectible into W.*

(ii) *Let p be a projection. Then there is a $\preccurlyeq$-parametrization of the Θ-semicomputable sets such that $\{a : W_a \neq \varnothing\} \subseteq \operatorname{dom}(p)$.*

To clarify our notation, if $\lambda za \cdot W_a^z$ is a $\preccurlyeq$-parametrization, then $W_a = \bigcup \{W_a^z; z \in A\}$. We prove (ii): Let $\lambda az \cdot V_a^z$ be any $\preccurlyeq$-parametrization of the Θ-semicomputable sets. Let W be the domain of the projection p and $\lambda z \cdot W^z$ a $\preccurlyeq$-enumeration of W. Define a relation R by

$$R(a, x) \quad \text{iff} \quad p(a) = x.$$

By 6.1.3 let R_a be a Θ-finite subset of $\{x : R(a, x)\}$. Note that if $R(a, x)$, then $R_a = \{x\}$. Introduce sets $R_{a,z}$ by

$$R_{a,z} = \begin{cases} \varnothing & \text{if} \quad a \notin W^z \\ R_a & \text{if} \quad a \in W^z. \end{cases}$$

We have our $\preccurlyeq$-parametrization by setting

$$W_a^z = \{y \in L^z : (\exists x \in R_{a,z})[y \in V_x^z]\}.$$

So far things have extended. But now we come to a difficulty. An important technical lemma of α-recursion theory states that any α-semicomputable subset bounded below the projectum is α-finite. This may not be true for arbitrary infinite theories. It could also happen that the projectum $|\preccurlyeq|^{*}$ is not a limit ordinal. Both properties seem necessary for a decent degree theory. And since we cannot prove them in general we get around these difficulties by a definition. We shall return to this point in Section 6.3.

6.1.7 Definition. Let Θ be an infinite theory on a domain $(\mathfrak{A}, \preccurlyeq)$. The *r.e.-projectum* of Θ, denoted by $|\preccurlyeq|^+$, is the least ordinal β for which there is a Θ-semicomputable non Θ-finite set $W \subseteq L^\beta$.

It is always true that $|\preccurlyeq|^+ \leqslant |\preccurlyeq|^*$. The converse is a definition.

6.1.8 Definition. Let Θ be an infinite theory on a domain $(\mathfrak{A}, \preccurlyeq)$. Θ is called *adequate* if $|\preccurlyeq|^+ = |\preccurlyeq|^* =$ limit ordinal.

We shall in Section 6.3 discuss the "true" domain of degree theory and priority arguments. Here we just note that there are non-wellorderable adequate theories.

Let Θ be infinite and let $\lambda z \cdot K_z$ be an enumeration of the Θ-finite sets, i.e. the values of the function are canonical Θ-indices for Θ-finite sets. Every Θ-finite set K is K_z for some z. Sometimes we may require of the enumeration that $K_z \subseteq L^z$, i.e. every $x \in K_z$ satisfies $x \prec z$.

Given two subsets B, C of the domain A there is an immediate reducibility notion that comes to mind, viz. B is reducible to C if B is $\Theta[C]$-computable. But aside from fixing the proper version of $\Theta[C]$ there are difficulties. We want a notion of reducibility relative to a given theory Θ, i.e. we want to decide questions about B using Θ-finite information about C and its complement. But the notion of finiteness may change in passing from Θ to $\Theta[C]$. Thus we are led inevitably to the following notion of "Θ-computable in".

6.1.9 Definition. Let $B, C \subseteq A$, f a function, and $\lambda z \cdot K_z$ a fixed enumeration of Θ-finite sets.

(i) f is *weakly Θ-computable in C*, denoted $f \leqslant_w C$, if there is a Θ-semicomputable set W such that for all σ, y

$$f(\sigma) \simeq y \quad \text{iff} \quad \exists z, w(\langle \sigma, y, z, w\rangle \in W \wedge K_z \subseteq C \wedge K_w \cap C = \varnothing).$$

B is *weakly Θ-computable* in C, $B \leqslant_w C$, in case $c_B \leqslant_w C$.

(ii) B is *Θ-computable in* C, denoted $B \leqslant C$, if there is a Θ-semicomputable set W such that for all z_1, z_2

$$K_{z_1} \subseteq B \wedge K_{z_2} \cap B = \varnothing \quad \text{iff} \quad \exists w_1, w_2(\langle z_1, z_2, w_1, w_2\rangle \in W \wedge K_{w_1} \subseteq C \wedge K_{w_2} \cap C = \varnothing).$$

(iii) B is *weakly Θ-semicomputable in* C if there is a Θ-semicomputable set W such that for all x

$$x \in B \quad \text{iff} \quad \exists z, w(\langle x, z, w\rangle \in W \wedge K_z \subseteq C \wedge K_w \cap C = \varnothing).$$

(iv) B is *Θ-semicomputable* in C if there is a Θ-semicomputable set W such that for all z

$$K_z \subseteq B \quad \text{iff} \quad \exists w_1, w_2(\langle z, w_1, w_2\rangle \in W \wedge K_{w_1} \subseteq C \wedge K_{w_2} \cap C = \varnothing).$$

The definitions are independent of the particular enumeration of the Θ-finite sets. As usual we set $B \equiv C$ iff $B \leqslant C$ and $C \leqslant B$.

The reducibility notion $B \leqslant C$ will be a focus of our attention. It is the one among several possible generalizations of Turing reducibility in ORT which has led to the most interesting results in the general framework. However, the relation "B is $\Theta[C]$-computable" (i.e. c_B is $\Theta[C]$-computable) also merits some comment. We shall not pursue any philosophic discussions of notions of reducibilities here, the reader may want to consult Kriesel [90] and also the excellent and annotated bibliography of Shore [152]. We shall return to more general matters in Section 6.3.

Given a set $C \subseteq A$ we construct $\Theta[C]$ along the lines of the construction in Section 5.2, and arrange things such that a tuple (a, σ, z) is added at stage $\Theta^\beta[C]$ only if a, σ, z and $\langle a, \sigma, z\rangle$ are elements of L^β.

6.1.10 Definition. Let B, C be subsets of the domain of Θ and f a function.

(i) $f \leqslant_d C$ if f is $\Theta[C]$-computable.
(ii) $B \leqslant_d C$ if c_B is $\Theta[C]$-computable.

A simple argument shows that $\leqslant_d$ is transitive. The following lemma is also immediate.

6.1.11 Lemma. *Let f be an integer-valued function. Then $f \leqslant_w C$ implies $f \leqslant_d C$.*

$f \leqslant_w C$ means that for some Θ-semicomputable W

$$f(\sigma) \simeq x \quad \text{iff} \quad \exists z, w(\langle \sigma, x, z, w\rangle \in W \ \wedge \ K_z \subseteq C \ \wedge \ K_w \cap C = \varnothing).$$

Thus f has a $\Theta[C]$-semicomputable graph. Using selection over the integers, which is available in $\Theta[C]$, we define f as a $\Theta[C]$-computable function.

The lemma allows us to conclude that

$$B \leqslant C \Rightarrow B \leqslant_w C \Rightarrow B \leqslant_d C.$$

But none of these implications can be reversed. (See Driscoll [21] where an example is given that $\leqslant_w$ need not be transitive even on the Θ-semicomputable sets.)

But there is one case where $B \leqslant_d C$ implies $B \leqslant C$, viz. the *regular* and *hyper-regular* case. These notions are due to Sacks [140]. Before introducing the definition let us note that the sets weakly Θ-semicomputable in C are enumerated by setting

$$W_a^C = \{x : \exists z, w(\langle x, z, w\rangle \in W_a \ \wedge \ K_z \subseteq C \ \wedge \ K_w \cap C = \varnothing)\}.$$

As an approximation to W_a^C let

$${}^zW_a^C = \{x : \exists w_1, w_2(\langle x, w_1, w_2\rangle \in W_a^z \ \wedge \ K_{w_1} \subseteq C \ \wedge \ K_{w_2} \cap C = \varnothing)\}.$$

6.1.12 Definition. (i) A set B is *regular* if $B \cap K$ is Θ-finite whenever K is Θ-finite.

(ii) A set B is *hyperregular* if for all Θ-finite sets K and all indices a, $K \subseteq W_a^B$ implies that $K \subseteq {}^zW_a^B$, for some z.

The reader will notice the similarity of (i) to Δ_0-separation and of (ii) to Δ_0-collection. It is perhaps not too surprising that when B is regular and hyperregular, then $\Theta[B]$ will be an infinite theory and the notion of Θ-finite and $\Theta[B]$-finite will coincide. This is the substance of the following proposition.

6.1.13 Proposition. *Let B be regular. Then the following are equivalent:*

(i) *B is hyperregular.*
(ii) *$\Theta[B]$ is an infinite theory.*
(iii) *$f \leqslant_w B$ iff $f \leqslant_d B$, whenever f is integer-valued.*

In this setting the result is due to V. Stoltenberg-Hansen [163]. For the proof we need to be a bit more careful in how we construct $\Theta[B]$. Let B_1 and B_2 be disjoint sets and define a theory $\Theta[B_1, B_2]$ by the following modification of the construction of $\Theta[B]$. Let b be the index in $\Theta[B]$ which introduces the characteristic function of B. Then if $b, x, 0, \langle b, x, 0\rangle \in L^\beta$ and $x \in B_1$ we add $(b, x, 0)$ to $\Theta^\beta[B_1, B_2]$. And if $b, x, 1, \langle b, x, 1\rangle \in L^\beta$ and $x \in B_2$ we add $(b, x, 1)$ to $\Theta^\beta[B_1, B_2]$. Obviously $\Theta[B] = \Theta[B, A - B]$, where A is the domain of Θ. Now introduce

$$^mH_{z,w}^x = \{\langle a, \sigma, y\rangle : (a, \sigma, y) \in \Theta^{|x|}[K_z, K_w], \mathrm{lh}(\sigma) = m\}.$$

An analysis of the definitions will show that $^mH_{z,w}^x$ is Θ-finite uniformly in the parameters m, x, z, w. And we further note that if

$$\langle a, \sigma, y\rangle \in {}^mH_{z,w}^x \ \wedge\ K_z \subseteq B \ \wedge\ K_w \cap B = \varnothing,$$

then $(a, \sigma, y) \in \Theta^{|x|}[B]$.

We now return to the proof of Proposition 6.1.13.

(i) ⇒ (ii). It suffices to show that the inductive definition of $\Theta[B]$ closes off at the ordinal $|\preccurlyeq|$. This reduces to studying the case of bounded universal quantification. So assume that $(a, y, 1) \in \Theta^{<|\preccurlyeq|}[B]$ (i.e. has been added before stage $|\preccurlyeq|$) for each $y \prec x$. We must show that $(a_0, a, z, 1) \in \Theta^{<|\preccurlyeq|}[B]$, where a_0 is a code for $\mathbf{E}^{\preccurlyeq}$ in $\Theta[B]$.

By regularity of B there are for each $y \prec x$ some z, w_1, w_2 such that $\langle a, y, 1\rangle \in {}^1H_{w_1,w_2}^z$ where $K_{w_1} \subseteq B$ and $K_{w_2} \cap B = \varnothing$. But now we can play with our notation. Letting $W^z = \{\langle y, w_1, w_2\rangle \in L^z : \langle a, y, 1\rangle \in {}^1H_{w_1,w_2}^z\}$, we see that $L^x \subseteq W^B$ (where $\lambda z \cdot W^z$ is a $\preccurlyeq$-enumeration of W). By hyperregularity of B there is a z such that $L^x \subseteq {}^zW^B$. But then $(a, y, 1) \in \Theta^{|z|}[B]$ for all $y \prec x$, and hence $(a_0, a, x, 1) \in \Theta^{<|\preccurlyeq|}[B]$.

(ii) ⇒ (iii). Let f be $\Theta[B]$-computable with index a. We then see that

$$f(\sigma) \simeq y \quad \text{iff} \quad \exists\beta < |\preccurlyeq|((a, \sigma, y) \in \Theta^{\beta}[B])$$
$$\text{iff} \quad \exists z, w_1, w_2(\langle a, \sigma, y\rangle \in {}^{m}H^{z}_{w_1, w_2} \wedge K_{w_1} \subseteq B \wedge K_{w_2} \cap B = \varnothing).$$

It follows that $f \leqslant_w B$.

(iii) ⇒ (i). Let V be $\Theta[B]$-semicomputable, it follows from (iii) that V is of the form W^B for some Θ-semicomputable set W. Letting

$$V^z = \{x \prec z : \exists w_1 w_2 \prec z(\langle x, w_1, w_2\rangle \in W^z \wedge K_{w_1} \subseteq B \wedge K_{w_2} \cap B = \varnothing)\}$$

we see that $\lambda z \cdot V^z$ is a $\preccurlyeq$-enumeration of V in $\Theta[B]$. Since every $\Theta[B]$-semicomputable set has a $\preccurlyeq$-enumeration in $\Theta[B]$ it follows that the domain A is $\Theta[B]$-infinite.

Let now $K \subseteq W^B_a$ where K is Θ-finite. Introduce the relation $F(x, z)$ by defining z to be a minimal element such that $x \in {}^{z}W^B_a$. Let F_x be non-empty $\Theta[B]$-finite subset of $\{z : F(x, z)\}$ and let $M = \bigcup\{F_x : x \in K\}$. Then M is $\Theta[B]$-finite and hence bounded by some $w \in A$ (since the domain A is $\Theta[B]$-infinite). Then $K \subseteq {}^{w}W^B_a$, so B is hyperregular.

6.1.14 Remark. We now observe that when B is regular and hyperregular, then for any set C, $C \leqslant B$ iff $C \leqslant_d B$. Just let $f(z, w) \simeq 0$ iff $K_z \subseteq A \wedge K_w \cap A = \varnothing$ in 6.1.13 (iii).

Hyperregularity and the relation $\leqslant_d$ is a digression from the main line of development of this chapter, whereas regularity is not. The importance of regularity comes from the following observation. Let $\lambda z \cdot W^z$ be a $\preccurlyeq$-enumeration of the Θ-semicomputable set W. Let $V^z = W^z - \bigcup\{W^w : w \prec z\}$. Then $\lambda z \cdot V^z$ is a disjoint $\preccurlyeq$-enumeration of W. And W is regular iff $(\forall\beta < |\preccurlyeq|)(\exists z)(\forall w \succ z)\cdot(V^w \cap L^{\beta} = \varnothing)$. This means that in enumerating W, given any level β, there is a stage z after which we always enumerate beyond β.

The anomaly of non-regularity can be circumvented by the following theorem when studying Θ-semicomputable degrees for adequate theories.

6.1.15 Theorem. *Let Θ be an adequate theory. Then for every Θ-semicomputable set B there is a regular Θ-semicomputable set D such that $B \equiv D$. D can be taken to consist of levels, i.e. $\forall x(\forall y \sim x)(x \in D \rightarrow y \in D)$.*

This was proved in the context of α-recursion theory by Sacks [140]; his proof was simplified by Simpson [153]. For adequate theories the result is due to Stoltenberg-Hansen [164], who had to go back to the original and more complicated proof of Sacks due to the lack of a well-ordering of the domain.

We shall not prove the general version in this book. For many purposes a simpler result is sufficient. This we now develop.

6.1.16 Lemma. *Let Θ be an infinite theory, and B a Θ-semicomputable non Θ-finite* *set. Let $\lambda z \cdot B^z$ be a disjoint enumeration of B such that each B^z is non-empty and contained in one level of the* pwo $\preccurlyeq$. *Then the deficiency set of B,*

$$D = \{z : (\exists w \succ z)(B^w \prec B^z)\},$$

is a regular Θ-semicomputable set with unbounded complement, and further $\forall x(\forall y \sim x)(x \in D \rightarrow y \in D)$.

It is clear that D is Θ-semicomputable with unbounded complement. To prove regularity it suffices to show that $D \cap L^x$ is Θ-finite for each $x \in A$.

Fix x and let $z_0 = x$. Suppose that we have defined $z_0, z_1, \ldots, z_n$. Choose, if possible, z_{n+1} such that $z_{n+1} \succ z_n$ and $B^{z_{n+1}} \prec B^{z_n}$. By the well-foundedness of $\prec$ the sequence is finite. Let z_n be its last element. It is then easily seen that

$$D \cap L^x = \{z \prec x : (\exists w \preccurlyeq z_n)(w \succ z \;\wedge\; B^w \prec B^z)\},$$

which is Θ-finite.

We shall apply the construction of the lemma in two situations, both important for the theory in Section 6.2.

6.1.17 Corollary. *If the set B of Lemma* 6.1.16 *is regular, then $B \equiv D$.*

To show that $B \leqslant D$ we define a relation $Q(z, w)$ iff w is minimal such that $w \notin D \wedge K_z \subseteq L^{B^w}$. Observe then that $K_z \cap B = \varnothing$ iff $\exists w[Q(z, w) \;\wedge\; K_z \cap B^{\prec w} = \varnothing]$. (Note that when the sets involved are Θ-semicomputable we need only worry about the "negative" requirements $K \cap B = \varnothing$; the "positive" requirements $K \subseteq B$ take care of themselves.)

To prove that $D \leqslant B$ we first introduce a relation $F(z, w)$ iff w is minimal such that $(\forall w_1 \in K_z)(B^{\preccurlyeq w_1} \subseteq L^w)$. Then we define $N(z, w)$ iff w is of minimal level such that $\exists w_1[F(z, w_1) \;\wedge\; L^{w_1} - B^{\prec w} \subseteq \bar{B}]$. We see that $K_u \cap D = \varnothing$ iff $(\forall z \in K_u)(\exists w)[N(z, w) \;\wedge\; (\forall w_1 \preccurlyeq w)(w_1 \preccurlyeq z \;\vee\; B^z \preccurlyeq B^{w_1})$.

Where did we use the regularity of B? Simply to know that given z there is some w such that $Q(z, w)$, and similarly for N.

We shall now state our approximation to 6.1.15. The result in the multiple-valued setting is due to Stoltenberg-Hansen [162].

6.1.18 Theorem. *Let Θ be an adequate theory. Then for every Θ-semicomputable non Θ-computable set B there are regular Θ-semicomputable sets D_1 and D_2 such that D_1 is not Θ-computable and $D_1 \leqslant B \leqslant D_2$.*

From B we shall construct two sets B_1^* and B_2^* and then let D_1 and D_2 be the deficiency sets of B_1^* and B_2^*, respectively.

For the definition of B_2^* assume that B is not regular. Then by adequacy, $|\preccurlyeq|^* < |\preccurlyeq|$. Let p be a projection into $L^{|\preccurlyeq|^*}$ and set $B_2^* = \{z : p(z)\downarrow \;\wedge\; K_{p(z)} \cap B \neq \varnothing\}$. D_2 will be the deficiency set of B_2^*. We leave the proof of $B \leqslant D_2$ to the reader. (It is not entirely trivial, but see [162] for details.)

We now turn to the existence of D_1. Since B is not regular, there is an x such that $B \cap L^x$ is not regular. Let $B_1^* = p^{-1}(B \cap L^x)$. (At this point recall the discussion following Definition 6.1.4.) We observe that $K \subseteq \bar{B}^*$ iff $K \cap p^{-1}(L^x) \subseteq \overline{p^{-1}(B)}$ iff $p(K \cap p^{-1}(L^x)) \subseteq \bar{B}$, from which we conclude that $B_1^* \leqslant B$.

Toward defining D_1 we first note that B_1^* is not Θ-finite since $p(B_1^*) = B \cap L^x$ is not Θ-finite. Let $\lambda w \cdot B^w$ be an enumeration of B_1^* as described in Lemma 6.1.16 and let D_1 be the deficiency set of B_1^* with respect to this enumeration. Then one verifies that

$$K \subseteq \bar{D}_1 \quad \text{iff} \quad \bigcup_{z \in K} (L^{B^z} - B^{\prec z}) \subseteq \bar{B}_1^*.$$

Hence $D_1 \leqslant B_1^* \leqslant B$.

Finally, if D_1 were Θ-computable we see that

$$x \notin B_1^* \quad \text{iff} \quad x \geqslant |{\preccurlyeq}|^* \;\vee\; \exists z(x \prec B^z \;\wedge\; z \notin D \;\wedge\; x \notin B^{\prec z}).$$

This means that $\bar{B}_1^*$ would be Θ-semicomputable, hence B_1^* would be Θ-finite. But we argued above that it is not. (Note that the adequacy of Θ is used to ensure the existence of a suitable z for the last equivalence; $x < |{\preccurlyeq}|^*$, hence by adequacy there must be a z_0 such that $x \prec B^z$ for all $z \succcurlyeq z_0$. Since $\bar{D}$ is unbounded, there must be a z such that $x \prec B^z \wedge z \notin D$.)

We conclude this section by two definitions.

6.1.19 Definition. A set B is *many-one reducible* to a set C, $B \leqslant_m C$, if there is a Θ-computable mapping $\lambda z \cdot H_z$ where H_z is a non-empty Θ-finite set, such that

(i) $x \in B$ iff $H_x \subseteq C$
(ii) $x \notin B$ iff $H_x \cap C = \varnothing$.

6.1.20 Definition. The jump of a set B is the set

$$B' = \{a : \exists z, w(\langle z, w\rangle \in W_a \;\wedge\; K_z \subseteq B \;\wedge\; K_w \cap B = \varnothing\}.$$

Some basic facts now follow, e.g. a set D is weakly Θ-semicomputable in B iff $D \leqslant_m B'$.

6.2 The Splitting Theorem

We shall give one example of a non-trivial degree-theoretic result.

6.2.1 Theorem. *Let Θ be an adequate theory. Let C be a regular Θ-semicomputable set and let D be a Θ-semicomputable non Θ-computable set. Then there exist Θ-semicomputable sets A and B such that $C = A \cup B$, $A \cap B = \varnothing$, $A \leqslant C$, $B \leqslant C$ and*

(i) $\Theta[A]$ *and* $\Theta[B]$ *are adequate theories* (*so in particular* A *and* B *are hyperregular*)

(ii) $A' \equiv B' \equiv O'$

(iii) $D \not\leq_w A$ *and* $D \not\leq_w B$.

The splitting theorem in ORT is due to G. E. Sacks [138]. In the context of α-recursion theory it was proved by R. A. Shore [150]. S. Simpson could prove in the context of "thin" admissible sets that there are Θ-semicomputable sets A, B such that $A \not\leq_w B$ and $B \not\leq_w A$ [154]. The strong version above is due to V. Stoltenberg-Hansen [163]. We must, however, make one reservation; Stoltenberg-Hansen needs to assume for parts (i) and (ii) that the theory Θ has a *reasonable pairing function*. By this we mean that for each $\alpha < |\precsim|^*$ there is a $\beta < |\precsim|^*$ such that $L^\alpha \times L^\alpha = \{\langle x, y\rangle; x, y \in L^\alpha\} \subseteq L^\beta$. It is not known whether every adequate theory Θ admits a reasonable pairing function.

We shall in this section prove the following weak version of the splitting theorem.

6.2.2 Theorem. *Let* Θ *be an adequate theory. Let* C *be a regular* Θ*-semicomputable set and let* D *be a regular* Θ*-semicomputable non* Θ*-computable set. Then there are* Θ*-semicomputable sets* A *and* B *such that* $C = A \cup B$, $A \cap B = \varnothing$, $A \leqslant C$, $B \leqslant C$, $D \not\leq_w A$ *and* $D \not\leq_w B$.

6.2.3 Remarks. We have the usual corollaries. First note that if A and B are disjoint regular Θ-semicomputable sets, then the join of $\deg(A)$ and $\deg(B)$, $\deg(A) \vee \deg(B)$, is $\deg(A \cup B)$. If we let $\mathbf{a}, \mathbf{b}, \mathbf{c}$ range over Θ-semicomputable degrees, we can from 6.2.2 and 6.1.18 conclude that

$$(\forall \mathbf{c} > 0)(\exists \mathbf{a}, \mathbf{b})(\mathbf{a} \vee \mathbf{b} \leqslant \mathbf{c} \wedge \mathbf{a} < \mathbf{c} \wedge \mathbf{b} < \mathbf{c} \wedge \mathbf{a}|\mathbf{b}),$$

where as usual $\mathbf{a}|\mathbf{b}$ means that $\mathbf{a}$ and $\mathbf{b}$ are incomparable. Using the regular set Theorem 6.1.15 we may draw the stronger conclusion that $\mathbf{a} \vee \mathbf{b} = \mathbf{c}$.

Also note that the same results are true for d-degrees, i.e. degrees with respect to the relation $\leqslant_d$, see Definition 6.1.10; this is a consequence of 6.1.14 and 6.2.1.

But before we turn to a proof of Theorem 6.2.2 we have to develop a certain "blocking" technique due to Shore [150]. The reason for this is that when the domain of an infinite theory is not computably well-ordered one cannot consider a unique requirement at a given stage of a priority argument. But it will be possible to handle Θ-finite blocks at each stage.

The naive way to do this is to let one level of the pwo of the domain make up one block. And in his thesis [162] Stoltenberg-Hansen was able to obtain a weak positive solution to Post problem in this way.

But stronger results need more refined blocking techniques, even in the context of α-recursion theory. As noted above, this was developed by Shore [150] (see also his survey [152] for further motivation). Simpson [154] observed that this technique also worked for "thin" admissible sets. For adequate theories in general this was developed by Stoltenberg-Hansen [162, 163]. We present here a version for single-valued theories following closely the exposition in [163].

Remark. We have yielded to tradition and used A in the statement of the splitting theorem. For the rest of this section we will use U for the domain of Θ.

As always there are some technical preliminaries. A relation $F(\sigma, z)$ on the domain of Θ induces, in certain circumstances a function on the associated ordinal $|\preccurlyeq|$ of the domain. Let $\sigma \sim \sigma'$, where $\sigma = (x_1, \ldots, x_n)$ and $\sigma' = (x_1', \ldots, x_n')$, mean that $x_i \sim x_i'$, $i = 1, \ldots, n$. If F satisfies the requirement that

$$F(\sigma, z) \ \wedge \ F(\sigma', z') \ \wedge \ \sigma \sim \sigma' \ \Rightarrow \ z \sim z',$$

then F induces a function f on $|\preccurlyeq|$. We classify f in terms of the associated relation F. Thus f is called Θ-computable if the associated F is Θ-computable. It is called Σ_n if F is Σ_n, where we use the usual Σ_n, Π_n hierarchy starting with $\Sigma_0 = \Pi_0 =$ Θ-computable. For functions on $|\preccurlyeq|$ we use the standard notion of limit

$$\lim_{\alpha} f'(\alpha, \gamma) = \delta \quad \text{iff} \quad (\exists \beta)(\forall \alpha \geqslant \beta)[f'(\alpha, \gamma) = \delta].$$

With this bit of terminology we have the following standard approximation result.

6.2.4 Lemma. *Let Θ be adequate and f a total Σ_2 function on $|\preccurlyeq|$. Then there is a total Θ-computable f' on $|\preccurlyeq|$ such that* $\lim_\alpha f'(\alpha, \gamma) = f(\gamma)$.

The reader may first establish the following part of Post's theorem: If B is Σ_{n+1} then B is weakly Θ-semicomputable in a Π_n set. From this we may conclude that if B is Σ_2 then B is weakly Θ-semicomputable in a Θ-semicomputable set A, which by 6.1.18 can be taken to be regular.

Let G_f be the graph on the domain of Θ of the function f. By assumption G_f is Σ_2, hence by the remark above G_f is weakly Θ-semicomputable in a regular Θ-semicomputable set A *via* some Θ-semicomputable set W. Let $\lambda z \cdot A^z$ and $\lambda z \cdot W^z$ be $\preccurlyeq$-enumerations of A and W, respectively.

Let N_x^z be the Θ-finite set of all minimal $w \prec z$ such that

$$(\exists y \prec z)(\exists x' \sim x)[\langle x', y, w\rangle \in W^z \ \wedge \ K_w \cap A^z = \varnothing].$$

We define a relation F' by the following requirements: If $N_x^z = \varnothing$, then $\langle z, x, z\rangle \in F'$. If $N_x^z \neq \varnothing$, then

$$\langle z, x, y\rangle \in F' \quad \text{iff} \quad y \text{ is a minimal element such that } (\exists w \in N_x^z)(\exists x' \sim x)[\langle x', y, w\rangle \in W^z].$$

F' is Θ-computable and induces a total function f' on $|\preccurlyeq|$. We must prove that f' converges to f.

Suppose $f(\alpha) = \beta$. Choose elements x, y such that $|x| = \alpha$, $|y| = \beta$ and

$\langle x, y\rangle \in G_f$. Since G_f is weakly Θ-semicomputable in A *via* W, choose some z such that

$$\langle x, y, z\rangle \in W \;\wedge\; K_z \cap A = \varnothing.$$

By the regularity of A we can choose w so large that $y \prec w$, $\langle x, y, z\rangle \in W^z$ and $(U - A) \cap L^z = (U - A^w) \cap L^z$. We want to show that for $\gamma \geqslant |w|, f'(\gamma, \alpha) = \beta$, i.e. $\lim_\gamma f'(\gamma, \alpha) = f(\alpha)$.

Let $w' \succcurlyeq w$. Then $N_x^{w'} \neq \varnothing$ since (possibly except for minimality) z is a candidate for membership. (Note that the enumeration for finite sets is such that $K_z \subseteq L^z$.) Let $u \in N_x^{w'}$. There are elements $x' \sim x$ and y' such that

$$\langle x', y', u\rangle \in W^{w'} \;\wedge\; K_u \cap A^{w'} = \varnothing.$$

Since $u \preccurlyeq z$ and $K_u \subseteq L^u$, $K_u \cap A = \varnothing$. But then $\langle x', y', u\rangle$ is a correct computation of f, i.e. $f(|x'|) = |y'|$, since $|x'| = \alpha$, $|y'| = \beta$. The definition of F' shows that $\langle w', x', y'\rangle \in F'$, i.e. letting $\gamma = |w'|$ we get $f'(\gamma, \alpha) = \beta$, and convergence is proved.

6.2.5 Definition. The Σ_2-cf(α) is the least ordinal β for which there is a Σ_2 function f with domain β and range unbounded in α.

6.2.6 Lemma. *Let Θ be adequate. Then* Σ_2-cf$(|\preccurlyeq|) = \Sigma_2$-cf$(|\preccurlyeq|^*)$.

Let k be a total Θ-computable function on $|\preccurlyeq|$ with range in $|\preccurlyeq|^*$ such that $\{\beta; k(\beta) < \alpha\}$ is bounded for each $\alpha < |\preccurlyeq|^*$. k can be defined from a $\preccurlyeq$-enumeration of a Θ-semicomputable non Θ-computable set $W \subseteq L^{|\preccurlyeq|^*}$.

Let f be Σ_2 with domain β and unbounded in $|\preccurlyeq|$. Then $g(\alpha) = k(f(\alpha))$ is Σ_2 and unbounded in $|\preccurlyeq|^*$. This proves Σ_2-cf$(|\preccurlyeq|^*) \leqslant \Sigma_2$-cf$(|\preccurlyeq|)$.

For the converse, let f be Σ_2 with domain β and unbounded in $|\preccurlyeq|^*$. Let $g(\alpha) = \mu\gamma[(\forall \xi \geqslant \gamma)(f(\alpha) < k(\xi))]$. g is unbounded in $|\preccurlyeq|$. Use 6.2.4 to write f as a limit. Then a simple quantifier analysis shows that g is Σ_2.

By a *$\preccurlyeq$-sequence* of Θ-semicomputable sets we mean a Θ-computable mapping r such that $x \sim y \Rightarrow W_{r(x)} = W_{r(y)}$.

6.2.7 Lemma. *Suppose $\alpha < \Sigma_2$-cf$(|\preccurlyeq|)$ and $\langle I_x : x \prec \alpha\rangle$ is a $\preccurlyeq$-sequence of Θ-semicomputable sets such that I_x is Θ-finite for each $x \prec \alpha$. Then $\bigcup_{x\prec\alpha} I_x$ is Θ-finite.*

This simple lemma is crucial for the later priority construction. The proof is by contradiction. Let α be the least ordinal for which we have a sequence $\langle I_x : x \prec \alpha\rangle$ whose union is not Θ-finite. Let $W = \bigcup_{x\prec\alpha} I_x$ and let $\lambda z \cdot W^z$ be a $\preccurlyeq$-enumeration of W. Define a function g on the associated ordinal $|\preccurlyeq|$ by

$$g(\gamma) = \mu\eta[I_\gamma \subseteq W^\eta].$$

Of course, we should have defined a relation G and let g be the induced function.

In any case, g is Σ_2. g is defined for every $\beta \prec \alpha$, but $g(\alpha)$ is unbounded in $|\preccurlyeq|$ since W is not Θ-finite, hence $\Sigma_2\text{-cf}(|\preccurlyeq|) \leqslant \alpha$.

6.2.8 Blocking Procedure. *Suppose that Θ is an adequate theory. The projectum $L^{|\preccurlyeq|^*}$ can be divided into $\Sigma_2\text{-cf}(|\preccurlyeq|)$ many Θ-finite blocks M_α, each bounded strictly below $|\preccurlyeq|^*$. Each block M_α can be approximated by Θ-finite sets M_α^z. The approximation is uniform in α and z, and is "tame" in the sense that*

$$(\forall\alpha < \Sigma_2\text{-cf}(|\preccurlyeq|))(\exists z)(\forall w \succcurlyeq z)(\forall\beta < \alpha)[M_\beta^w = M_\beta].$$

We know that $\Sigma_2\text{-cf}(|\preccurlyeq|) \leqslant |\preccurlyeq|^*$. If we have equality, we simply set $M_\alpha = M_\alpha^z = \{x : x \sim \alpha\}$. More care is needed when $\Sigma_2\text{-cf}(|\preccurlyeq|) < |\preccurlyeq|^*$.

Let g be a Σ_2 function from the $\Sigma_2\text{-cf}(|\preccurlyeq|)$ to $|\preccurlyeq|^*$ unbounded in $|\preccurlyeq|^*$, and let g' be Θ-computable such that $\lim_\sigma g'(\sigma, \alpha) = g(\alpha)$. These functions exist by 6.2.4 and 6.2.6.

Define an ordinal function

$$h(\sigma, \alpha) = \mu\gamma[(\forall\beta < \alpha)(g'(\sigma, \beta) < \gamma)].$$

Since $\alpha < \Sigma_2\text{-cf}(|\preccurlyeq|)$ there always exists a $\gamma < |\preccurlyeq|^*$ satisfying the requirements inside the brackets [...]. Now put

$$M_\alpha^z = \{\varepsilon : h(|z|, \alpha) \leqslant |\varepsilon| < h(|z|, \alpha + 1)\}.$$

Each M_α^z is obviously bounded strictly below $|\preccurlyeq|^*$. But we also need to know that a canonical Θ-index can be obtained for M_α^z uniformly in α and z.

To this purpose, let $H(z, a, x)$ be a Θ-computable relation which induces the function h. By the selection principle 6.1.3 we have Θ-finite set $H_{z,a}$ uniformly in z, a such that $x \in H_{z,a}$ implies $H(z, a, x)$. Next observe that given any a in the domain of the pwo $\preccurlyeq$ there is a Θ-finite set S_a (uniform in a) of "successor" notations, i.e. if $a' \in S_a$ then $|a'| = |a| + 1$. Let $H_{z,a}^* = \bigcup\{H_{z,b} : b \in S_a\}$. Then $H_{z,a}^*$ is Θ-finite, and if $y \in H_{z,a}^*$, then $|y| = h(|z|, |a| + 1)$. From $H_{a,z}$ and $H_{a,z}^*$ we now define M_a^z.

To show that the approximation is tame, let

$$I_\beta = \{w : (\exists w' \succ w)(g'(|w'|, \beta) \neq g'(|w|, \beta))\}.$$

Fix $\alpha < \Sigma_2\text{-cf}(|\preccurlyeq|)$. Then $\langle I_\beta : \beta < \alpha + 1\rangle$ is a $\preccurlyeq$-sequence of Θ-semicomputable sets such that each I_β is Θ-finite. By Lemma 6.2.7 the union is Θ-finite. Hence there is some z such that for all $\beta \leqslant \alpha$ and all $w \succcurlyeq z$ we must have $g'(|w|, \beta) = g'(|z|, \beta)$, i.e. tameness.

Let $M_\beta = M_\beta^z$ for sufficiently large z. It remains to verify that

$$L^{|\preccurlyeq|^*} = \bigcup\{M_\beta : \beta < \Sigma_2\text{-cf}(|\preccurlyeq|)\}.$$

Fix $\varepsilon \prec |\preccurlyeq|^*$ and choose the least α for which $\varepsilon \prec h(\sigma, \alpha)$ where σ is fixed and sufficiently large. Such α exists since g is unbounded in $|\preccurlyeq|^*$. By the definition of h there exists a $\beta < \alpha$ such that $\varepsilon \preccurlyeq g'(\sigma, \beta)$. But then $\varepsilon \prec h(\sigma, \beta + 1)$, which by the choice of α means that $\alpha = \beta + 1$. By the minimality of α we also get $h(\sigma, \beta) \preccurlyeq \varepsilon$, hence $\varepsilon \in M_\beta^\sigma$.

We can now return to the proof of Theorem 6.2.2. Let the sets C and D be given. By 6.1.16 and 6.1.17 we may assume that D satisfies in addition the requirement $\forall x(\forall y \sim x)(x \in D \Rightarrow y \in D)$. Let $\lambda z \cdot D^z$ be a $\preccurlyeq$-enumeration of D and $\lambda z \cdot C^z$ a disjoint $\preccurlyeq$-enumeration of C. The sets A and B will be defined via $\preccurlyeq$-enumerations $\lambda z \cdot A^z$ and $\lambda z \cdot B^z$, inductively on the pwo $\preccurlyeq$. If $z \sim w$ then the construction at stages z and w will be identical, but indices may differ. At stage z, C^z will be added to precisely one of $A^{\prec z}$ and $B^{\prec z}$, where $A^{\prec z}$ means $\bigcup \{A^w : w \prec z\}$.

This will ensure that A and B will be Θ-semicomputable, $C = A \cup B$, and $A \cap B = \varnothing$. It is also easy to see that $A \leqslant C$ and $B \leqslant C$. We simply have

$$K_z \cap A = \varnothing \quad \text{iff} \quad \exists w[(K_z - C^{\prec w}) \cap C = \varnothing \ \wedge \ K_z \cap A^w = \varnothing].$$

By regularity of C there is always an element w such that $(K_z - C^{\prec w}) \cap C = \varnothing$.

It remains to ensure that $D \not\leqslant_w A$ and $D \not\leqslant_w B$. We restrict attention to A, the case for B being similar. It suffices to show that $U - D$ is not weakly Θ-semicomputable in A, i.e. for no index ε is $(U - D) = W_\varepsilon^A$. (For notation see the paragraph immediately preceding 6.1.12.)

To violate the equality $(U - D) = W_\varepsilon^A$ we follow the original procedure of Sacks [138]. The idea is to try to preserve computations $x \in W_\varepsilon^A$ for minimal x not in D. In case $(U - D) = W_\varepsilon^A$ for some ε we would eventually preserve a correct computation for each $x \in W_\varepsilon^A$. But then W_ε^A would be Θ-semicomputable and this would contradict the assumption that D is non Θ-computable. Hence computations $x \in W_\varepsilon^A$ will eventually stop being preserved, and we will eventually violate the equality $U - D = W_\varepsilon^A$.

But there are obstacles to overcome. We need e.g. to have Θ-finite blocks of requirements to settle down by some stage of the computation; to this end we can use the blocking procedure developed in 6.2.8, letting each block play the role of a single requirement in trying to preserve a computation $x \in W_\varepsilon^A$ for $x \notin D$ and some ε in the block considered. And we use the fact that D has the property $\forall x(\forall y \sim x)(x \in D \Rightarrow y \in D)$ to avoid the problem of never finishing creating requirements with arguments from a fixed level of the pwo $\preccurlyeq$; there is a need to create a requirement preserving a computation $x \in W_\varepsilon^A$ only if no other computation $y \in W_\varepsilon^A$ for $y \sim x$ is being preserved.

We turn to the details of the construction. Let M_α^z and M_α for $\alpha < \Sigma_2\text{-cf}(|\preccurlyeq|)$ be the Θ-finite blocks described in 6.2.8. Sets R_A and R_B of requirements will be created, R_A to ensure that $D \not\leqslant_w A$. S_A will denote the set of A-requirements, i.e. requirements in R_A, injured during the construction. R_A^z and S_A^z are the Θ-finite parts of R_A and S_A obtained by stage z. Each requirement will be of the form $\langle \varepsilon, x, F \rangle$ where F is (a canonical Θ-index for) a Θ-finite set. Such a requirement in

R_A is called an ε-A requirement or an α-A requirement (at z) in case $\varepsilon \in M_\alpha$ ($\varepsilon \in M_\alpha^z$): it is said to have argument x. In case $F \cap A^z = \varnothing$ it is said to be *active* at z, else it is *inactive*. $\varepsilon \in M_\alpha^z$ is an *inactive* α-A *reduction procedure at* z in case there is an active ε-A requirement in R_A^z preserving a computation $x \in W_\varepsilon^A$ for some $x \in D^z$, i.e. there is $\langle \varepsilon, x, F \rangle \in R_A^{\prec z} - S_A^{\prec z}$ such that

$$(\exists w \prec z)[\langle x, w \rangle \in W_\varepsilon^z \ \wedge \ K_w \subseteq F \ \wedge \ x \in D^z].$$

If no such requirement exists, then ε is an *active* α-A *reduction procedure at* z.

Let $r\colon |\preccurlyeq| \to \Sigma_2\text{-cf}(|\preccurlyeq|)$ be a Θ-computable function such that

$$(\forall \alpha < \Sigma_2\text{-cf}(|\preccurlyeq|))(\forall \beta)(\exists \gamma > \beta)(r(\gamma) = \alpha), \text{ where } \alpha, \beta, \gamma \text{ vary over } |\preccurlyeq|.$$

r indicates which part of the construction to concern ourselves with at a given stage.

The construction at stage z: Let $r(z) = \alpha$. As remarked above, we only treat the A-requirements, the construction of B-requirements being similar.

We recall the motivation above. Every block M_α plays the role of a single A-requirement. This means that if there is an $\varepsilon \in M_\alpha$ which is an active α-A reduction procedure and if there is an active ε-A requirement with argument x, then there is no need to create a new α-A requirement with argument $x' \sim x$. Otherwise we shall contemplate creating new requirements.

To this end let H^z be the Θ-finite set of minimal x such that for each $x' \sim x$, $x' \notin D^z$ and $\neg(\exists \langle \varepsilon, x', F \rangle \in R_A^{\prec z} - S^{\prec z})$ ("ε is an active α-A reduction procedure at z").

Next let

$$N^z = \{\langle \varepsilon, x \rangle \in M_\alpha^z \times H^z : \text{"}\varepsilon \text{ is an active } \alpha\text{-}A \text{ reduction procedure at } z\text{"} \\ \wedge \ (\exists w \prec z)[\langle x, w \rangle \in W_\varepsilon^z \ \wedge \ K_w \cap A^{\prec z} = \varnothing]\}.$$

and

$$F_\varepsilon^z = \bigcup \{K_w : (\exists x \in H^z)[\langle x, w \rangle \in W_\varepsilon^z \ \wedge \ K_w \cap A^{\prec z} = \varnothing]\}.$$

Then put

$$R_A^z = R_A^{\prec z} \cup \{\langle \varepsilon, x, F_\varepsilon^z \rangle : \langle \varepsilon, x \rangle \in N^z\}.$$

We must now decide whether to add C^z to $A^{\prec z}$ or to $B^{\prec z}$. Let

$$J_A^z = \{\langle \varepsilon, x, F \rangle \in R_A^z - S_A^{\prec z} : F \cap C^z \neq \varnothing\},$$

i.e. J_A^z is the set of active A requirements which would be injured in case C^z were added to A. Define

$$f_A(z) = \mu\beta[(\exists \langle \varepsilon, x, F \rangle \in J_A^z)(\varepsilon \in M_\beta^z)],$$

if such an ordinal β exists, otherwise let $f_A(z) = |\preccurlyeq|$. In a similar way introduce a function $f_B(z)$. An analysis of the definition tells us that we can Θ-computably decide whether $f_A(z) \leqslant f_B(z)$ or $f_B(z) < f_A(z)$.

If $f_A(z) \leqslant f_B(z)$, let $B^z = B^{\prec z} \cup C^z$ and $A^z = A^{\prec z}$. If $f_B(z) < f_A(z)$, let $A^z = A^{\prec z} \cup C^z$ and $B^z = B^{\prec z}$.

To complete the construction let

$$S_A^z = \{\langle \varepsilon, x, F \rangle \in R_A^z : F \cap A^z \neq \varnothing\}.$$

6.2.9 Lemma. *For each* $\alpha < \Sigma_2\text{-cf}(|\preccurlyeq|)$ *the set of* α-A *and* α-B *requirements is* Θ-*finite.*

This is proved by induction on $\alpha < \Sigma_2\text{-cf}(|\preccurlyeq|)$. Fix α and assume that the set of β-A and β-B requirements is Θ-finite for each $\beta < \alpha$. By the tameness of the blocking there is a stage z_0 by which all blocks M_β^z for $\beta \leqslant \alpha$ have settled down. Let

$$I_\beta = \{z \succ z_0 : (\exists \langle \varepsilon, x, F \rangle \in R_A^z \cup R_B^z - R_A^{\prec z} \cup R_B^{\prec z})(\varepsilon \in M_\beta^z)\}.$$

Then I_β is Θ-finite for each $\beta < \alpha$ by the induction hypothesis, thus $\bigcup_{\beta < \alpha} I_\beta$ is Θ-finite by Lemma 6.2.7. C is assumed regular so there exists a stage $z_1 \succcurlyeq z_0$ such that all β-requirements for $\beta < \alpha$ have been created by z_1 and no such β-requirement will meet C^w for any $w \succcurlyeq z_1$. It follows that $f_A(w) \geqslant \alpha$ and $f_B(w) \geqslant \alpha$ for $w \succcurlyeq z_1$. Hence, by the way we have arranged the priorities, no α-A requirement will be injured beyond z_1.

This means that an α-A reduction procedure inactive at some $w \succcurlyeq z_1$ will remain inactive forever. We see that the set of α-A reduction procedures which becomes inactive beyond z_1 is Θ-semirecursive and hence Θ-finite. Thus there is a $z_2 \succcurlyeq z_1$ beyond which no α-A reduction procedure is made inactive.

Suppose $z_2 \preccurlyeq z \prec w$ and $r(z) = r(w) = \alpha$. From the choice of z_2 we see that $H^z \preccurlyeq H^w$, i.e. $x \in H^z$ and $y \in H^w$ implies that $x \preccurlyeq y$. Moreover, if an α-A requirement is created at z then $H^z \prec H^w$. From this we may conclude that either the set of α-A requirements it Θ-finite, or for each $x \notin D$ there is a permanent α-A requirement $\langle \varepsilon, x', F \rangle$ where $x' \sim x$ and ε is a reduction procedure active beyond z_2.

But the latter cannot be the case since D then would be Θ-computable, in fact

$$x \notin D \quad \text{iff} \quad (\exists w \succcurlyeq z_2)(\exists x' \sim x)(\exists \langle \varepsilon, x', F \rangle \in R_A^w - S_A^w)$$

("ε is an active α-A reduction procedure at w").

This completes the proof that the set of α-A requirements is Θ-finite. Using the regularity of C choose $z_3 \succcurlyeq z_2$ so large that all α-A requirements have been created and such that no C^w will meet an α-A requirement for $w \succcurlyeq z_3$. No α-B requirement is injured beyond z_3 since $f_A(w) > \alpha$ whenever $w \succcurlyeq z_3$. Now repeat the argument above with B in place of A and starting with z_3 in place of z_1 and conclude that the set of α-B requirements is also Θ-finite.

This ends the proof of Lemma 6.2.9.

It remains to show that $U - D$ is not weakly Θ-semicomputable in A. We argue by contradiction, so suppose that $(U - D) = W_\varepsilon^A$. Choose α and z_0 such that $\varepsilon \in M_\alpha$, all α-A requirements have settled down by stage z_0, and no $\delta \in M_\alpha$ becomes an inactive α-A reduction procedure beyond z_0. Note that ε is an active α-A reduction procedure at z_0, otherwise an erroneous computation would be preserved. Choose a minimal $x \notin D$ such that there is no $x' \sim x$ for which $\langle \delta, x', F \rangle \in R_A^{z_0} - S_A^{z_0}$ where δ is an active α-A reduction procedure at z_0. By the regularity of D there is a stage $z_1 \succcurlyeq z_0$ such that $L^x \cap D = L^x \cap D^{z_1}$. Let $w \succcurlyeq z_1$ be such that $x \in {}^wW_\varepsilon^A$ and $r(w) = \alpha$. Then $H^w = \{x' : x' \sim x\}$ and $\langle \varepsilon, x \rangle \in N^w$. It follows that an ε-requirement with argument x will be created at w, contradicting the fact that $w \succcurlyeq z_0$.

The proof of Theorem 6.2.2 is now complete.

6.3 The Theory Extended

In the last section we gave a non-trivial finite injury argument in a non-wellordered setting: the Sacks' splitting theorem for an arbitrary adequate theory.

But one example is no general proof that the class of adequate theories is the "correct" setting of an axiomatic degree theory. Beyond the finite lies the domain of infinite injuries.

The splitting theorem was our paradigm for the finite injury arguments. The density theorem could be a test case for infinite injury case.

6.3.1 Density Theorem. *Let* **a** *and* **b** *be two Θ-semicomputable degrees such that* $\mathbf{a} < \mathbf{b}$. *There exists a Θ-semicomputable degree* **c** *such that* $\mathbf{a} < \mathbf{c} < \mathbf{b}$.

In the context of ORT over ω this theorem was proved by G. Sacks [139]. R. A. Shore was able to further refine the techniques he used in [150] (such as the blocking technique 6.2.8) to produce a proof of the density theorem in the setting of α-recursion theory (see [151]). Beyond this the question is open: Give an adequate axiomatic analysis of the density theorem! Which is an injunction to study infinite injury arguments in the abstract.

6.3.2 Remark. There exists at least one example in the context of adequate theories. Stoltenberg-Hansen has proved (unpublished) by an infinite injury argument that if $\Theta[\mathbf{O}']$ is adequate, then there exists a Θ-semicomputable set $A < \mathbf{O}'$ such that $A' \equiv \mathbf{O}''$.

Leaving infinite injuries aside we may still ask whether adequate theories is a "good" category for degree theory. In a certain sense it is *reasonable.* We started with a class of infinite theories corresponding to admissible pwo's, i.e. resolvable admissible structures (possibly with urelements), and we imposed some necessary properties on the projectum: Keep in mind that the Θ-semicomputable sets can be indexed below the projectum; our requirement of adequacy stated that the projectum is a limit ordinal and every Θ-semicomputable set bounded below the projectum is in fact Θ-finite. This is a key combinatorial property, see e.g. Lemma 6.2.9.

The "key combinatorial property" is provable in α-recursion theory, but not in the general setting of infinite theories (in the precise sense of Definition 5.1.5.) We discuss two examples.

6.3.3 Determinacy and Degrees: a Counterexample to Post's Problem. We present a result due to S. Simpson [155]. First a definition: Let M be an admissible set. $B \subseteq M$ is *complete-*$\Sigma(M)$ if B is $\Sigma(M)$ and for each $\Sigma(M)$ set $A \subseteq M$ there is a $\Sigma(M)$ relation C such that:

(a) $\forall x \exists y C(x, y)$

(b) $\forall x \forall y [C(x, y) \to (x \in A \leftrightarrow y \in B)]$,

i.e. A is $\leqslant_m$-reducible to B.

Simpson proved the following

Theorem. *Assume the Axiom of Determinateness. Let* $M = \mathbb{R}^+$, *the next admissible set after the continuum. Then every* $\Sigma(M)$ *set is either* $\Delta(M)$ *or complete and every regular* $\Sigma(M)$ *set is* $\Delta(M)$.

Here we have a total breakdown of the theory of semicomputable degrees. Note that M is an admissible pwo, hence supports an infinite theory. The only blemish of the counterexample is the use of AD.

Hence the obvious question: Can one get rid of determinacy? There is an unpublished example due to L. Harrington which answers this in the affirmative, but his admissible set is not resolvable.

Thus some restriction, perhaps adequacy, seems necessary. However, there is a rich degree theory in certain non-adequate contexts.

6.3.4 A Non-adequate Theory with a "Good" Degree Structure. This example is due to D. Normann and V. Stoltenberg-Hansen [130]. The setting is Barwise's theory of admissible sets with urelements and the structure is $L(\omega_1)_{\mathfrak{M}}$, where $\mathfrak{M} = V^{\omega}(\mathbb{Q})$ is a countable-dimensional vectorspace over $\mathbb{Q}$.

$L(\omega_1)_{\mathfrak{M}}$ has a degree structure isomorphic to $L(\omega_1)$. This comes from the fact that the structure $\mathfrak{M} = V^{\omega}(\mathbb{Q})$ has a *natural representative* in $L(\omega_1)$. ($\mathfrak{M}' \in L(\omega_1)$ is a natural representative of $\mathfrak{M}$ if (i) $\mathfrak{M}'$ and $\mathfrak{M}$ are isomorphic, and (ii) the set of finite (in the good old sense!) $\tau: \mathfrak{M} \to \mathfrak{M}'$ which can be extended to an isomorphism from $\mathfrak{M}$ to $\mathfrak{M}'$ is $L(\omega_1)_{\mathfrak{M}}$-computable.) A definability analysis shows that $V^{\omega}(\mathbb{Q})$ has indeed a natural representative in $L(\omega_1)$. And thus the degree theory of $L(\omega_1)_{\mathfrak{M}}$ is reduced to the *adequate* theory $L(\omega_1)$.

But, as shown by Normann and Stoltenberg-Hansen, $L(\omega_1)_{\mathfrak{M}}$ being obviously admissible and resolvable, is *not* adequate. It does, however, satisfy the "regular set" theorem.

We sketch the proof that the structure is not adequate: Let $<$ be a Σ_1 pwo of the structure $L(\omega_1)_{\mathfrak{M}}$, and let $\pi(a)$ be a projection, i.e. a set of "notations" for a below the projectum. An analysis of the construction shows that π and $<$ are definable in a finite set of parameters $p_1, \ldots, p_k$ from the underlying structure. Thus there is a finite-dimensional subspace $\mathfrak{M}_0$ of $\mathfrak{M}$ such that $p_1, \ldots, p_k$ belongs to $\mathfrak{M}_0$. We have the following *basic lemma*: For any two elements $r_1, r_2 \in \mathfrak{M}\backslash\mathfrak{M}_0$ if there is an automorphism τ of $\mathfrak{M}$ leaving $\mathfrak{M}_0$ fixed such that $\tau(r_1) = r_2$, then the sets $\pi(r_1)$ and $\pi(r_2)$ will be at the same level below the projectum of the pwo.

Choose an $r \in \mathfrak{M}\backslash\mathfrak{M}_0$. Let π_1 be a projectum of ω_1 into ω. Define π_2 for $\alpha < \omega_1$ by

$$\pi_2(\alpha) = \pi((1 + \pi_1(\alpha)\cdot r)).$$

It is easily seen that the image of the projection is semirecursive. By the basic lemma the image will be bounded below the projectum; we have a set of non-zero elements $1 + \pi_1(\alpha)\cdot r$ in $\mathfrak{M}\backslash\mathfrak{M}_0$, for any two such elements an automorphism of the required kind exists, hence their projections will lie at the same level.

The projection is not $L(\omega_1)_{\mathfrak{M}}$-finite; for π_2 is essentially a projection of ω_1, the ordinal of the structure.

What is the moral of example 6.3.4? Perhaps this, that adequacy always lurks in the background?

So far *admissibility* has been a dogmatic assumption of the general theory. But irreverent questions began to be asked: Is Σ_1-admissibility a crude global hypothesis which obscures the finer points of recursion theory? (Sacks [144]). And irreverence is always an instrument for progress.

6.3.5 Inadmissible Extensions of the Theory. The basic ideas were announced in S. Friedman and G. Sacks [37] and have been extensively developed in papers of S. Friedman and W. Maass, see [35] and [101] for a preliminary guide to the field.

The Friedman-Sacks' setting for β-recursion theory is the S-hierarchy for L introduced by R. Jensen. An introduction to β-recursion theory from this point of view is given in Friedman [35]. But for our purposes we can equally well, as is done in Maass [101], stick to the more familiar L-hierarchy.

So replace the structure $\langle L_\alpha, \in\rangle$, α admissible, by $\langle L_\beta, \in\rangle$, where β is any limit ordinal. We must choose the "correct" notions of *semicomputable* and *finite* for the structures.

L_β admits a natural hierarchy; this motivates the following definitions: Let $A \subseteq L_\beta$

A is *β-r.e.* iff A is Σ_1-definable over $\langle L_\beta, \in\rangle$
A is *β-recursive* iff $A, \bar{A} = L_\beta - A$ are both β-r.e.

From this we derive, more or less canonically, the notion of a β-recursive function: A partial function f from L_β to L_β is called (partial) *β-recursive* iff its graph is β-r.e.

In the Friedman-Sacks' approach the notion of β-finite is the same as in the admissible case: Let $A \subseteq L_\beta$

A is *β-finite* iff $A \in L_\beta$.

One can now introduce the *reducibilities* of β-recursion theory: $\leqslant_{w\beta}$ and $\leqslant_\beta$, exactly as in Definition 6.1.9.

There is a threefold split in the recursion theory on the ordinals. Let $\kappa = \Sigma_1\text{-cf}(\beta)$, i.e. the least ordinal κ such that some Σ_1-function with domain κ has range unbounded in β. Let β^* be the usual (Σ_1-) projectum of β. Given any limit β,

(i) β is *admissible* iff $\kappa = \beta$.
(ii) β is called *weakly inadmissible* if $\beta^* \leqslant \kappa < \beta$.
(iii) β is called *strongly inadmissible* if $\kappa < \beta^*$.

The basic patterns of admissibility theory extend to the weakly inadmissible case. Below κ everything looks very admissible, and since $\beta^* \leqslant \kappa$ we can work below κ. W. Maass developed in [100] a technique which gave a precise meaning to this remark. He associated to each weakly inadmissible β an admissible structure $\mathfrak{A}_\beta$, the *admissible collapse*, such that $\mathfrak{A}_\beta$-r.e. degrees embed into the β-r.e. degrees. Thus one can transfer results from the admissible case to the weakly inadmissible one. A generalization of Maass' construction is given in V. Stoltenberg-Hansen [165].

There are, however, certain complications. "Strange" things start to happen in Friedman-Sacks' β-recursion theory. For example, it is not always possible to effectively enumerate the β-finite subsets of an β-r.e. set. The following definition is not trivial: A set A is said to be *tamely r.e.* (t.r.e.) if the set $\{a \in L_\beta : a \subseteq A\}$ is β-r.e. A β-recursive enumeration $\lambda\sigma \cdot A^\sigma$ of A is called *tame* if $a \subset A$ implies $\exists\sigma(a \subseteq A^\sigma)$. Of course, in the admissible case every enumeration of a β-r.e. set is tame; not so in the inadmissible case.

One can now give a quick summary of the results of Maass and Stoltenberg-Hansen: the structure of the regular t.r.e. β-degrees is non-trivial (and even rich) iff β is admissible or weakly inadmissible.

We will not go into details at this point since we shall outline an alternate approach to β-recursion theory in a moment. But a few results must be mentioned.

For any inadmissible β the following is true: Let W be a universal β-r.e. set. There is a β-recursive set A such that $\mathbf{0} <_\beta A <_\beta W$, and every β-recursive or tamely r.e. set is β-reducible to A, see Friedman [35].

Letting $\mathbf{0}^{1/2}$ denote the degree of A we always have at least three β-r.e. degrees, $\mathbf{0}$, $\mathbf{0}^{1/2}$, and $\mathbf{0}'$. In the weakly inadmissible case there are always more, we have incomparable β-r.e. degrees below $\mathbf{0}'$, even below $\mathbf{0}^{1/2}$.

The situation is at present more "confusing" in the strongly inadmissible case, i.e. when the $\Sigma_1\text{-cf}(\beta)$ is strictly smaller than β^*. Both results and methods become different.

For a large class of ordinals one may answer in the affirmative the following version of Post's problem:

(*) There are β-r.e. sets A, B such that $A \not\leqslant_{w\beta} B$, $B \,{}_{w\beta}\!\not\leqslant A$?

But there are also ordinals where (*) fails. In [36], S. Friedman shows that (*)

is not true when β is the ω-th primitive-recursively closed ordinal greater than $\aleph_{\omega_1}^L$. But what of the following version?

(**) There are β-r.e. sets A, B such that $A \not\leq_\beta B$, $B \not\leq_\beta A$?

Obviously (**) is true in the admissible or weakly inadmissible case. It is unknown if there is a strongly inadmissible β where (**) fails.

Let us return to basics. The definition of β-finite was lifted verbatim from the admissible case. And one may argue that it is natural from a set-theoretic point of view. But the necessity of the notion of *tameness* indicates that the concept of β-finiteness is awkward for a recursion-theoretic analysis.

W. Maass has recently reanalyzed the foundation of β-recursion theory. He accepts that Σ_1-definable over $\langle L_\beta, \in \rangle$ is the "correct" notion of β-r.e. This gives a unique choice for the notion of a partial β-recursive function.

In his analysis of finiteness Maass was guided by the time-honored principle of seeking a notion invariant under a suitable group of symmetries, in this case the group of β-recursive permutations of the domain.

One cannot start in thin air. Two elementary properties of finiteness seem almost unavoidable: (i) a "finite" set must be β-recursive; (ii) a "finite" set must be bounded. Granted this much Maass [102] proved the following result.

6.3.6 Proposition. *There is a largest class I of subsets of L_β satisfying*

(i) *every $K \in I$ is β-recursive,*
(ii) *every $K \in I$ is bounded, i.e. $K \subseteq L_\gamma$ for some $\gamma < \beta$,*
(iii) *if $K \in I$ and f is a β-recursive permutation of L_β, then $f''K \in I$.*

Moreover, I is explicitly given as

$$I = \{K \subseteq L_\beta : K \in L_\beta \wedge \beta\text{-card}(K) < \Sigma_1\text{-cf}(\beta)\}.$$

Note that if β is admissible we have the usual notion of finiteness of admissibility theory. We also note that the assumption on β is not known to be necessary.

We shall modify Maass' analysis a little, replacing (i) and the invariance property (iii) by

(i′) every $K \in I$ is β-r.e.

(iii′) if K is "finite", f a partial β-recursive function, and $K \subseteq \operatorname{dom} f$, then $f''K$ is "finite".

Our exposition is based on an unpublished note by V. Stoltenberg-Hansen.

One may argue what is more basic. Here we just note that the modified approach expresses the idea that the "finite" sets should be precisely those β-r.e. sets for which *every* enumeration is short compared with those of the universal β-r.e. set.

6.3.7 Lemma. *The class*

$$I = \{K \subseteq L_\beta : K \in L_\beta \wedge \beta\text{-card}(K) < \Sigma_1\text{-cf}(\beta)\}$$

is the largest class satisfying (i′), (ii), *and* (iii′).

It is easy to verify that I satisfies (i′), (ii), and (iii′). Suppose $\mathbf{C}$ is any class satisfying (i′), (ii), and (iii′) and not contained in I. Let $M \in \mathbf{C} - I$. We argue first that $M \in L_\beta$: Let $\lambda\sigma \cdot M^\sigma$ be a β-recursive enumeration of M which exists by (i′). Define for $x \in M$ a function

$$h(x) = \text{some } \sigma[x \in M^\sigma].$$

h is a partial β-recursive function and by (iii′) $h''M$ belongs to $\mathbf{C}$. Thus $h''M$ is bounded using (ii); let σ_0 be a bound. But then $M = M^{\sigma_0} \in L_\beta$.

Since $M \in L_\beta - I$ there is an $f\colon \kappa \xrightarrow{1\text{-}1} M$, $f \in L_\beta$. But then $f''\kappa \in L_\beta$ since L_β is rudimentarily closed. Define a function $g\colon M \to \kappa$ by

$$g(x) = \begin{cases} f^{-1}(x) & \text{if } x \in f''\kappa \\ 0 & \text{otherwise.} \end{cases}$$

Let $h\colon \kappa \to \beta$ be β-recursive and unbounded. By property (iii′) $h \circ g''M \in \mathbf{C}$. But $g''M = \kappa$, therefore $h \circ g''M$ is unbounded, contradicting (ii).

6.3.8 Definition. A set $K \subseteq L_\beta$ is called *invariantly finite*, *i*-finite, if $K \in I$.

This is the notion which will replace the notion of β-finite as used by Friedman-Sacks.

6.3.9 Proposition. *A set $M \subseteq L_\beta$ is i-finite iff for every partial β-recursive function $\varphi(x, y)$ there is a partial β-recursive function $\psi(x)$ such that for all $x \in L_\beta$*

$$\psi(x) \simeq \begin{cases} 0 & \textit{if } \exists y \in M \cdot \varphi(x, y) \simeq 0 \\ 1 & \textit{if } \forall y \in M \cdot \varphi(x, y) \simeq 1. \end{cases}$$

Thus *i*-finiteness is nothing but the usual axiomatic notion of finiteness, viz. the computability of the functional $\mathbf{E}_M$, $M \in I$. The proof is not difficult. One can argue as in the following 6.3.12 that $\mathbf{E}_M$ is computable if $M \in I$. Conversely, the set $\{M \subseteq L_\beta : \mathbf{E}_M \text{ is computable}\}$ will satisfy (i′), (ii), and (iii′), thus by maximality of I be included in I.

We have the following useful fact:

6.3.10 Proposition. *I is β-recursive.*

Obviously I is β-r.e. As in the proof of 6.3.7 $x \in L_\beta - I$ iff $\exists f \in L_\beta[f\colon \kappa \xrightarrow{1\text{-}1} x]$.
We come to the main construction. Let $h\colon \kappa \to \beta$ be β-recursive and unbounded,

and let $\exists y \cdot \varphi(e, x, y)$ be a universal $\Sigma_1(\langle L_\beta, \in\rangle)$ relation where φ is Δ_0. Define the following β-recursive relations:

(i) $U(x)$ iff $x \in L_\beta - I$
(ii) $\psi(e, x, \delta)$ iff $\delta < \kappa \wedge \exists y \in L_{h(\delta)} \cdot \varphi(e, x, y)$.

6.3.11 Definition. The structure

$$\mathfrak{A}_\beta = \langle L_\beta; U, I, \in \restriction L_\beta \times I, \psi \rangle$$

is called the *admissible collapse* of L_β.

Here U is the class of urelements and I the class of sets. When working in $\mathfrak{A}_\beta$ we write $\in$ for the more correct but hopelessly pedantic $\in \restriction L_\beta \times I$.

The construction of $\mathfrak{A}_\beta$ was suggested by Maass [102]. It is the "i-finite" version of the original construction [100].

6.3.12 Lemma. *Every $\Delta_0(\mathfrak{A}_\beta)$ relation is $\Delta_1(L_\beta)$.*

The following case is representative: Consider $\forall x \in a \cdot \varphi(x, q)$ where $a \in I$ and φ is $\Delta_0(\mathfrak{A}_\beta)$. By the induction hypothesis $\varphi(x, q)$ is $\Delta_1(L_\beta)$. Suppose that

$$\mathfrak{A}_\beta \models \varphi(x, q) \quad \text{iff} \quad L_\beta \models \exists z \theta(z, x, q),$$

where θ is $\Delta_0(L_\beta)$. Suppose further that $\mathfrak{A}_\beta \models \forall x \in a \cdot \varphi(x, q)$. Define f on a by $f(x) =$ some x such that $\theta(z, x, q)$. Since $a \in I$, $f''a$ is i-finite. Letting $b = f''a$ it follows that $L_\beta \models \forall x \in a \cdot \exists z \in b \cdot \theta(z, x, q)$. We conclude that

$$\mathfrak{A}_\beta \models \forall x \in a \cdot \varphi(x, q) \quad \text{iff} \quad L_\beta \models \exists b \cdot \forall x \in a \cdot \exists z \in b \cdot \theta(z, x, q),$$

i.e. $\forall x \in a \cdot \varphi(x, q)$ is $\Delta_1(L_\beta)$.

6.3.13 Theorem. *$\mathfrak{A}_\beta$ is an admissible set with urelements. Furthermore, a set $W \subseteq L_\beta$ is β-r.e. iff it is $\mathfrak{A}_\beta$-r.e.*

The proof that $\mathfrak{A}_\beta$ is admissible is straightforward. For example the union axiom follows since an i-finite union of i-finite sets is i-finite. Δ_0-separation follows since a β-recursive subset of an i-finite set is i-finite. And Δ_0-collection follows as in Lemma 6.3.12.

Now let $W \subseteq L_\beta$. If W is β-r.e. then for some index e, $x \in W$ iff $L_\beta \models \exists \delta < \kappa \cdot \psi(e, x, \delta)$ iff $\mathfrak{A}_\beta \models \exists \delta \psi(e, x, \delta)$, so W is $\mathfrak{A}_\beta$-r.e. Conversely if W is $\mathfrak{A}_\beta$-r.e. then $x \in W$ iff $\mathfrak{A}_\beta \models \exists z \varphi(x, z)$ where φ is $\Delta_0(\mathfrak{A}_\beta)$. By Lemma 6.3.12 φ is a $\Delta_1(L_\beta)$ relation; we conclude that W is β-r.e.

Over L_β we can now define the notion of an *i-degree* as in Definition 6.1.9. We shall use $\leqslant_{i\beta}$ for the associated reducibility notion. Obviously $A \leqslant_{i\beta} B$ iff

$A \leqslant_{\mathfrak{A}_\beta} B$. Thus the study of i-degrees over L_β is reduced to the study of degrees over the admissible structure $\mathfrak{A}_\beta$.

6.3.14 Theorem. *$\mathfrak{A}_\beta$ is resolvable iff β is admissible or weakly inadmissible.*

The admissible case is trivial, $\mathfrak{A}_\beta$ is then equal to L_β itself. Suppose β is weakly inadmissible. Then there is a β-recursive bijection $q: \kappa \leftrightarrow L_\beta$. For each $\gamma < \kappa$, $q''\gamma$ is i-finite. Thus q induces a β-recursive wellordering on L_β whose initial segments are i-finite, i.e. $\mathfrak{A}_\beta$ is resolvable.

For the converse assume that $\mathfrak{A}_\beta$ is resolvable. Let $\preccurlyeq$ be the induced prewellordering of L_β whose initial segments are $\mathfrak{A}_\beta$-finite. Let $<_\beta$ be the standard β-recursive wellordering of L_β. Define

$$x < y \quad \text{iff} \quad x \prec y \ \vee \ (x \sim y \ \wedge \ x <_\beta y).$$

Then $<$ is an $\mathfrak{A}_\beta$-recursive wellordering of L_β whose initial segments are $\mathfrak{A}_\beta$-finite. $\mathfrak{A}_\beta$ is therefore adequate and hence the deficiency set D (see 6.1.16) of a complete regular $\mathfrak{A}_\beta$-r.e. set is a β-r.e. non-β-recursive subset of κ. In particular, $D \notin L_\beta$; but then $\beta^* \leqslant \kappa$.

We have arrived at the following conclusion: *L_β with the notion of "finite" being i-finite in the sense of Definition* 6.3.8 *is an infinite and adequate computation theory iff β is either admissible or weakly inadmissible.*

Thus there is a natural computation-theoretic analysis of the weakly inadmissible case. But even in the strongly inadmissible case we have the beginnings of a computation-theoretic analysis. Theorem 6.3.13 still holds, but the associated computation theory is no longer resolvable. It is, however, s-normal and p-normal in the sense of the general theory. It is a topic for further research as to how far into the inadmissible one can extend the coherence and unity of concepts and methods that we find in the axiomatic analysis.

Part D

Higher Types

Chapter 7

Computations Over Two Types

In Chapter 4 we started our study of computation theories on domains of two types $\mathfrak{A} = \langle A, S, \mathbf{S}\rangle$ where $A = S \cup \mathrm{Tp}(S)$ and $\mathbf{S}$ is a coding scheme for S. Given a *normal* list $\mathbf{L}$ on $\mathfrak{A}$ we defined a recursion theory $\mathrm{PR}(\mathbf{L})$ generalizing Kleene recursion in higher types, and all of Chapter 4 was aimed at proving the following result (Theorem 4.4.1):

(a) $\mathrm{PR}(\mathbf{L})$ is p-normal, hence admits a selection operator over N,

(b) A is weakly but not strongly $\mathbf{L}$-finite, i.e. the $\mathbf{L}$-semicomputable relations are not closed under $\exists x \in \mathrm{Tp}(S)$.

(c) S is strongly $\mathbf{L}$-finite, i.e. the $\mathbf{L}$-semicomputable relations are closed under $\exists s \in S$.

The ultimate goal of this chapter is to see how far properties (a)–(c) determine normal recursion in higher types. In this study we meet a new and characteristic feature of computations on two types which was entirely absent in the general study of finite theories on one type (Chapter 3), viz. *reflection*.

7.1 Computations and Reflection

The setting is a computation theory Θ on $\mathfrak{A} = \langle A, S, \mathbf{S}\rangle$ satisfying properties (a)–(c), such theories are called *normal*. Since we are on two types, Θ also allows functional evaluation

$$f(x, y, \sigma) = x(y),$$

if $x \in \mathrm{Tp}(S)$ and $y \in S$. We also have extended our functional substitution

$$f(\sigma) = h(\lambda s \cdot g(s, \sigma), \sigma).$$

The code set C is assumed to be equal to N.

The reader should at this point recall Definitions 4.1.1, 4.1.2, 4.1.4 as well as Remark 4.1.5. Also recall the need of including the equality relation on S in the general case.

7.1.1 Definition. A computation theory Θ on $\mathfrak{A} = \langle A, S, \mathbf{S}\rangle$ is called *normal* if

(i) the equality relation on S is Θ-computable,
(ii) A is weakly and S is strongly Θ-finite,
(iii) Θ is p-normal.

As usual in the "finite" case we work with the norm or length function $|\cdot|_\Theta$ rather than with the subcomputation relation. Our computation theories are single-valued so we shall abbreviate $|a, \sigma, z|_\Theta$ to $|a, \sigma|_\Theta$ or in some cases to $|\{a\}_\Theta(\sigma)|_\Theta$.

Let $N \subseteq X \subseteq A$, and $\mathrm{Ord}(X) = \{|e, \sigma|_\Theta : \{e\}_\Theta(\sigma)\downarrow$ and σ a list from $X\}$. We introduce

$$\kappa^X = \sup \mathrm{Ord}(X)$$
$$\lambda^X = \text{order-type of } \mathrm{Ord}(X).$$

We shall be interested in the special cases $X = N,\ N \cup \{x_1, \ldots, x_m\},\ S,\ S \cup \{x_1, \ldots, x_m\},\ A$. Letting σ be the list $x_1, \ldots, x_m$, κ^X will be denoted by κ^0, κ^σ, κ^S, $\kappa^{S,\sigma}$, κ_Θ, respectively. And similarly for λ^X.

7.1.2 Remark. There is a well-known connection between prewellorderings and the ordinals κ^X. In particular, if $N \subseteq X \subseteq A$, then κ^X is the supremum of the lengths of the pwo's with domain $\subseteq A$ which are Θ-computable in elements from X. The supremum is not attained.

And if X admits a pairing scheme $\langle M_X, K_X, L_X\rangle$ Θ-computable in elements from X, then λ^X is the supremum of the lengths of the pwo's with domain $\subseteq X$ which are Θ-computable in elements from X. Again the supremum is not attained.

From this we may conclude that if $X = N,\ N \cup \{x_1, \ldots, x_m\},\ S,\ S \cup \{x_1, \ldots, x_m\}$, then

$$\lambda^X < \kappa^X < \kappa_\Theta.$$

If $\sigma = (x_1, \ldots, x_m)$, then $\lambda^\sigma \leqslant \lambda^{S,\sigma} < \kappa^\sigma$.

Detailed proofs of these facts are given in lemmas 20 and 21 of Moldestad [105]. To give the flavor of such proofs we include a brief hint for the first case. Let P be a pwo which is Θ-computable in $\sigma = (x_1, \ldots, x_n)$, where $x_1, \ldots, x_m \in X$ and $\mathrm{dom}\, P \subseteq A$. First find an index e_1 such that $\{e_1\}_\Theta(x, \sigma) \simeq 0$ if $x \in \mathrm{dom}\, P$, and if $x \in \mathrm{dom}\, P$ then $|e_1, y, \sigma|_\Theta < |e_1, x, \sigma|_\Theta$ for all y below x in the pwo P. Next, let e_2 be an index such that $\{e_2\}_\Theta(\sigma)\downarrow$ and $|e_1, x, \sigma|_\Theta < |e_2, \sigma|_\Theta$ for all $x \in \mathrm{dom}\, P$, e.g. let $\{e_2\}_\Theta(\sigma) \simeq E(f)$, where $f(x) \simeq \{e_1\}_\Theta(x, \sigma)$ if $x \in \mathrm{dom}\, P$, $f(x) \simeq 1$, otherwise. Obviously, $\kappa^X > |e_2, \sigma|_\Theta \geqslant |P|$. Conversely, given any $\nu < \kappa^X$, there is a computation in a list σ from X, $\{e\}_\Theta(\sigma)$, such that $|e, \sigma|_\Theta > \nu$. The subcomputation tree below $\{e\}_\Theta(\sigma)$ gives a pwo P such that $|P| = |e, \sigma|_\Theta > \nu$.

We now turn to a brief study of *reflection phenomena in higher types*. This notion was introduced in recursion theory by G. E. Sacks and further developed by L. Harrington in his thesis [53]. A Kechris in a set of unpublished notes from MIT [74] developed the general theory, see also his account in [76]. For the use of reflection in forcing arguments in higher types, see G. Sacks [143]. As we saw in Section 3.3, similar reflection properties are also of great importance for the general theory of inductive definability.

A computation-theoretic approach was developed by J. Moldestad [105]; we shall follow his account. As an introduction we present the following simple result.

7.1.3 Simple Reflection. *For all e, σ: If $\exists x \cdot |e, x, \sigma|_\Theta < \kappa^{S,\sigma}$, then $\exists x \cdot |e, x, \sigma| < \kappa^\sigma$.*

The premiss simply says:

$$\exists e' \in N \quad \exists s \in S[\{e'\}(s, \sigma)\downarrow \ \wedge \ \exists x \cdot |e, x, \sigma|_\Theta < |e', s, \sigma|_\Theta].$$

This is a Θ-semicomputable relation of σ: "$\exists x$" can be expressed by the E functional since A is weakly Θ-finite. "$\exists s \in S$" can be handled since S is strongly Θ-finite, and "$\exists e' \in N$" is no problem since Θ is p-normal, and we have selection over N.

7.1.4 Definition. Let μ be an ordinal $\leqslant \kappa$ ($= \kappa_\Theta$). μ is called *σ-reflecting* if for all e:

$$\text{If } \exists x \cdot |e, x, \sigma| < \mu, \quad \text{then} \quad \exists x \cdot |e, x, \sigma| < \kappa^\sigma.$$

σ is here a list of elements from A. Note that the σ-reflecting ordinals are an initial segment. And κ is not σ-reflecting for all σ if A is not strongly Θ-finite.

7.1.5 Remark. As remarked above we shall not develop the general theory of σ-reflecting ordinals, but only that part of the theory which is needed for the characterization results of Section 7.2. But we cannot resist mentioning the following *characterization of strong Θ-finiteness*:

Let $B \subseteq A$ be Θ-computable. Then B is strongly Θ-finite iff for all e, σ: if $\exists x \in B \cdot \{e\}_\Theta(x, \sigma)\downarrow$, then $\exists x \in B \cdot |e, x, \sigma|_\Theta < \kappa^\sigma$. (For a proof and further refinements, see Lemma 25 in Moldestad [105].)

7.1.6 Further Reflection. *$\kappa^{S,P,\sigma}$ is σ-reflecting.*

Here $P = \{\langle e, \tau\rangle : \{e\}_\Theta(\tau, \sigma)\downarrow, \tau$ is a list from $S\}$. Note that P has a natural pwo of length $\lambda^{S,\sigma}$.

The proof of 7.1.6 will be a consequence of the following four propositions. We shall compare $\lambda^{S,\sigma}$ with the ordinal $|x|$ for any pwo x on S. Let $\text{pwo}_S(x)$ mean that x is a pwo with domain $\subseteq S$.

Proposition 1. "$\text{pwo}_S(x) \ \wedge \ |x| < \lambda^{S,\sigma}$" *is Θ-semicomputable as a relation of x, σ.*

The statement is equivalent to

$$\text{pwo}_S(x) \;\wedge\; \exists y \in S[y \in P \;\wedge\; \forall z \in \text{dom}(x)[|z|_x < |y|_P]],$$

where the notations $||_x$ and $||_P$ are self-explanatory. Note that the relation $\text{pwo}_S(x)$ is Θ-computable.

Proposition 2. *If* $\text{pwo}_S(x)$ *and* $|x| \geqslant \lambda^{S,\sigma}$, *then* P *is* Θ*-computable in* x, σ *and parameters from* S.

First compute an x' such that $|x'| > \lambda^{S,\sigma}$. Then there exists $r \in \text{dom}(x')$ such that $|r|_{x'} = \lambda^{S,\sigma}$. From x, σ, r we can decide P.

Proposition 3. *If* P *is* Θ*-computable in* σ, x *and parameters from* S, *then* $\kappa^{S,P,\sigma} \leqslant \kappa^{S,x,\sigma}$.

Proposition 4. "$\exists y \cdot |e, y, \sigma|_\Theta < \kappa^{S,x,\sigma}$" *is* Θ*-semicomputable as a relation of* e, σ, x. *There is an index* $\hat{e}$ *for this relation such that* $|\hat{e}, e, \sigma, x|_\Theta > \inf\{|e, y, \sigma|_\Theta : y \in A\}$.

The statement is equivalent to

$$\exists e' \in N \,\exists r \in S[\{e'\}_\Theta(r, x, \sigma)\downarrow \;\wedge\; \exists y \cdot |e, y, \sigma|_\Theta \leqslant |e', r, x, \sigma|_\Theta].$$

From these four propositions further reflection easily follows. So assume that $\exists y \cdot |e, y, \sigma| < \kappa^{S,P,\sigma}$. Define the following relation

$$R(\sigma, x) \quad \text{iff} \quad \text{pwo}_S(x) \;\wedge\; [|x| < \lambda^{S,\sigma} \;\cdot\vee\cdot\; \exists y \cdot |e, y, \sigma| < \kappa^{S,x,\sigma}].$$

(i) By Propositions 1 and 4 $R(\sigma, x)$ is Θ-semicomputable. There is an index f for R such that if $R(\sigma, x)$ and $|x| \geqslant \lambda^{S,\sigma}$, then

$$|f, \sigma, x|_\Theta > \inf\{|e, y, \sigma|_\Theta : y \in A\}.$$

(ii) From Propositions 1–3 and the assumption we conclude that if $\text{pwo}_S(x)$, then $R(\sigma, x)$ is true. Hence we can assume that $\lambda x \cdot \{f\}_\Theta(\sigma, x)$ is total. Let $\{g\}(\sigma) \simeq E(\lambda x \cdot \{f\}(\sigma, x))$, then $|g, \sigma|_\Theta \downarrow$ and $|g, \sigma|_\Theta > |f, \sigma, x|_\Theta$ for all x.

(iii) There is some x such that $\text{pwo}_S(x)$ and $|x| \geqslant \lambda^{S,\sigma}$. Combining (i) and (ii) we get

$$\kappa^\sigma > |g, \sigma|_\Theta > |f, \sigma, x|_\Theta > \inf\{|e, y, \sigma|_\Theta : y \in A\}.$$

Hence we conclude $\exists y \cdot |e, y, \sigma|_\Theta < \kappa^\sigma$, which completes the proof of 7.1.6.

Further reflection has the following compactness property as a corollary.

7.1.7 Compactness. *Assume that* B *is a set of subsets of* S *and that* B *is* Θ*-semicomputable in some list* σ. *Assume that* B *has as element a non-empty subset* α_0 *of*

S such that α_0 is Θ-semicomputable in σ. Then B contains a subset of S which is non-empty and Θ-computable in σ.

We indicate the proof. If $\alpha_0 \in B$ let α_μ be the approximation to α_0 up to length less than μ. Let g be an index such that if $|y| = \mu$, then $\alpha_\mu \in B$ iff $\{g\}(y, \sigma)\downarrow$.

If we can show that $\exists y \cdot |g, y, \sigma|_\Theta < \kappa^{P,\sigma}$, then by further reflection $\exists y \cdot |g, y, \sigma|_\Theta < \kappa^\sigma$. And from this we get a subset $\alpha_\tau \subseteq \alpha_0$ such that $\alpha_\tau \in B$ and α_τ is Θ-computable in σ.

But since $\kappa^{P,\sigma} > \kappa^{S,\sigma}$, let y be a computation $\{e\}(P, \sigma)$ such that $\kappa^{P,\sigma} > |y| > \kappa^{S,\sigma}$. Then $\alpha_{|y|} = \alpha_0$ because y is convergent and $|y| > \kappa^{S,\sigma}$, and $|g, y, \sigma| < \kappa^{P,\sigma}$ because $y = \langle e, P, \sigma\rangle$.

7.2 The General Plus-2 and Plus-1 Theorem

We start by fixing some notations. Let Θ be a normal theory on $\mathfrak{A} = \langle A, S, \mathrm{S}\rangle$:

$$\begin{aligned} \mathrm{sc}(\Theta) &= \{X \subseteq A : X \text{ is } \Theta\text{-computable}\},\\ \mathrm{sc}(\Theta, \sigma) &= \{X \subseteq A : X \text{ is } \Theta\text{-computable in } \sigma\},\\ \mathrm{en}(\Theta) &= \{X \subseteq A : X \text{ is } \Theta\text{-semicomputable}\},\\ S\text{-en}(\Theta) &= \{X \subseteq S : X \text{ is } \Theta\text{-semicomputable}\}.\end{aligned}$$

7.2.1 Theorem. *Let Θ be a normal theory on $\mathfrak{A}$. Then there exists a normal list* **L** *such that* S-en(Θ) = S-en(**L**) *and* sc(Θ, r) = sc(**L**, r) *for all $r \in S$.*

This is an abstract version of the plus-2 theorem of Harrington [53]. Harrington's original version was a reduction result: Starting out with a normal functional G of type $>n + 2$ he constructed a functional F of type $n + 2$ such that ${}_n\mathrm{en}(G) = {}_n\mathrm{en}(F)$. The proof used the fact that Tp(n) is *strongly* finite in G. Theorem 7.2.1 is an improvement, here we assume that A, which in the concrete setting of higher types corresponds to Tp(n), is *weakly* Θ-finite. And in general we should not assume more since Tp(n) is *not* strongly PR(${}^{n+2}E$)-finite. Thus Theorem 7.2.1 gives a kind of characterization result which we will supplement in Section 7.3.

We follow the detailed proof in Moldestad [105]. Moldestad's proof is patterned on the original Harrington proof in [53]. However, one refinement is necessary in order to go from strong to weak finiteness in the assumption; this refinement is the joint effort of Harrington and Moldestad.

We have also labelled the theorem a "plus-1" result. A plus-1 result was first proved by G. Sacks [143]. He generalized the notion of an abstract 1-section to the appropriate notion of abstract $k + 1$-section, and constructed by a forcing-type argument a functional F giving a concrete representation of the abstract $k + 1$-section. And Further Reflection 7.1.6 was an essential ingredient in his proof. In the present setting the starting point is different, viz. a normal theory Θ; thus we have both the section and the envelope. The section result is, as we shall see, a consequence of the result about envelopes, and the proof is an exercise in the use of reflection principles. We leave it to the reader to decide on how different the forcing construction of Sacks' is from the present construction. Note in this connection that the notion of abstract $k + 1$-section is not entirely "pure", there is something semicomputable involved.

We divide the proof of Theorem 7.2.1 into several parts:

7.2.2 Some Preliminary Material. Let $P = \{\langle e, \sigma\rangle : \{e\}_{\Theta}(\sigma)\downarrow$ and σ is a list from $S\}$. The ordinals $|e, \sigma|_{\Theta}$, where $\langle e, \sigma\rangle \in P$, are called Θ-*subconstructive*. Their order-type is $\lambda = \lambda^S$; let $\langle \eta_\nu : \nu < \lambda\rangle$ be an enumeration of the Θ-subconstructive ordinals.

We will construct a normal list $\mathbf{L}$ consisting of the equality relation on S, the quantifier functional E, and a functional G which shall code up information about S-en(Θ). G will be constructed in stages G_τ, each G_τ being a partial approximation to G containing sufficient information to generate the set $H_\tau = \{\langle e, \sigma\rangle : |e, \sigma|_{\mathbf{L}} < \tau, \sigma$ a list from $A\}$. If $\tau < \tau'$, then $G_{\tau'}$ will be an extension of G_τ.

G will be defined by

$$G(f) = \begin{cases} \left(\bigcup_\tau G_\tau\right)(f) & \text{if } f \in \bigcup_\tau \operatorname{dom} G_\tau \\ 0 & \text{otherwise.} \end{cases}$$

Finally, let μ_ν be an enumeration of the $\mathbf{L}$-*subconstructive* ordinals.

7.2.3 Further Preliminary Material. Let G_τ and H_τ be given. We note that

$$H_{\tau+1} = \{\langle e, \sigma\rangle : \text{all immediate subcomputations of } \{e\}_{\mathbf{L}}(\sigma) \text{ are in } H_\tau\}.$$

We observe that if $\lambda x\cdot\{e\}_{\mathbf{L}}(x, \sigma)$ is total and $\langle e, x, \sigma\rangle \in H_\tau$ for all x, then we must define G on this function at stage $\tau + 1$ if we have not previously done so. Thus we let $f = \lambda x\cdot\{e\}_{\mathbf{L}}(x, \sigma) \in \operatorname{dom} G_{\tau+1}$ and set $G_{\tau+1}(f) = 0$ or 1. $G_{\tau+1}$ is called a *trivial extension* of G_τ if $\operatorname{dom} G_{\tau+1} = \operatorname{dom} G_\tau \cup \{f : f$ as above$\}$ and $G_{\tau+1}(f) = 0$ if $f \in \operatorname{dom} G_{\tau+1} - G_\tau$.

7.2.4 On How G Shall Contain Information About the S-en(Θ). Not every extension should be trivial. Let τ be $\mathbf{L}$-subconstructive, in fact, let $\tau = \mu_\nu$ for some $\nu < \lambda$. Information about the S-en(Θ) will be coded into G_τ at this stage.

If $x \in \Theta$ and $|x|_{\Theta} \leqslant \eta_\nu$, we shall take some function f_{xy} where $y \in \mathrm{PR}[\mathbf{L}]$ and $|y|_{\mathbf{L}} = \mu_\nu$ and such that $f_{xy} \notin \operatorname{dom} G_\mu$, $\mu < \tau$. We then let $f_{xy} \in \operatorname{dom} G_\tau$ and set $G_\tau(f_{xy}) = 1$.

From f_{xy} and G_τ we shall recover information about x inside $\mathbf{L}$ in order to get S-en(Θ) $\subseteq$ S-en($\mathbf{L}$). And f_{xy} should be $\mathbf{L}$-computable in x, y when y is an $\mathbf{L}$-computation of length τ.

Such functions f_{xy} exist:

Proposition 1. *Let* $\mathbf{L}$ *be any normal list. Let* y *be an* $\mathbf{L}$-*computation of length* τ.

For each $x \in A$ there is a total function f_{xy} such that f_{xy} is **L**-*computable in* x, y *and if* $x \neq x'$, *then* $f_{xy} \neq f_{x'y}$. *If* $f_{xy} = \lambda t \cdot \{e\}_{\mathbf{L}}(t, \sigma)$ *for some* e, σ *then* $\tau \leqslant |e, t, \sigma|_{\mathbf{L}}$ *for some* $t \in A$.

We indicate briefly the proof. Let τ^+ be the least limit ordinal $\geqslant \tau$. The set of **L**-computations with length $< \tau^+$ is **L**-computable in y. Let f_y be defined by the following instructions: (i) $f_y(u) = 0$, if u is not of the form $\langle e, \sigma \rangle$. If $u = \langle e, \sigma \rangle$, then ask if $|e, t, \sigma|_{\mathbf{L}} < \tau^+$ for all t; (ii) If the answer is no, let $f_y(u) = 0$. (iii) If the answer is yes, let $f_y(u)$ be different from $\{e\}_{\mathbf{L}}(\langle e, \sigma \rangle, \sigma)$.

f_y is recursive in **L**, y. And if $f = \lambda t \cdot \{e\}_{\mathbf{L}}(t, \sigma)$ is a total function such that $|e, t, \sigma|_{\mathbf{L}} < \tau^+$ for all t, then f and f_y differs at $t = \langle e, \sigma \rangle$.

We can now define f_{xy}:

$$f_{xy}(t) = \begin{cases} \langle f_y(t), x, 0 \rangle & \text{if } \ x \in S \\ \langle f_y(t), x(t), 1 \rangle & \text{if } \ x \in \mathrm{Tp}(S),\ t \in S \\ f_y(t) & \text{if } \ x, t \in \mathrm{Tp}(S). \end{cases}$$

Then f_{xy} is **L**-computable uniformly in x, y. And $f_{xy} \neq f_{x'y}$ if $x \neq x'$. Let $f_{xy} = \lambda t \cdot \{e\}_{\mathbf{L}}(t, \sigma)$ for some e, σ. It is not difficult to see that we can obtain f_y from f_{xy}. There is, in fact, an index e' such that $f_y = \lambda t \cdot \{e'\}_{\mathbf{L}}(t, \sigma)$ and $|e', t, \sigma|_{\mathbf{L}} < |e, t, \sigma|_{\mathbf{L}} + \omega$, for all t. If $|e, t, \sigma|_{\mathbf{L}} < \tau$ for all t, then $|e', t, \sigma|_{\mathbf{L}} < \tau^+$ for all t. This means that $f = \lambda t \cdot \{e'\}_{\mathbf{L}}(t, \sigma)$ would have to differ from f_y at $\langle e', \sigma \rangle$, but $f = f_y$.

7.2.5 Toward the Definition of G_τ. Actually, we have almost arrived. Suppose G_μ is defined for $\mu < \tau$. Then H_μ is constructed for $\mu < \tau$. Note that we can decide from $\bigcup_{\mu < \tau} H_\mu$ whether τ is **L**-subconstructive.

In the subconstructive case let $y \in \mathbf{PR}[\mathbf{L}]$ such that $|y|_{\mathbf{L}} = \tau$. If $\tau = \mu_\nu$, $\nu < \lambda$, let $x \in \Theta$ and $|x|_{\Theta} \leqslant \eta_\nu$. Define G^0 as the "zero-extension" of $\bigcup_{\mu < \tau} G_\mu$:

$$G^0(f) = \begin{cases} \left(\bigcup_{\mu < \tau} G_\mu\right)(f) & \text{if } \ f \in \bigcup_{\mu < \tau} \operatorname{dom} G_\mu \\ 0 & \text{otherwise.} \end{cases}$$

Then $\mathbf{L}^0 = E, G^0, =_S$ is a normal list. Choose f_{xy} according to Proposition 1. Then f_{xy} is $\mathbf{L}^0$-computable in x, y, hence **L**-computable in x, y since G^0 is **L**-computable in y, and $f_{xy} \notin \operatorname{dom} G_\mu$ for any $\mu < \tau$.

7.2.6 Defining G_τ. Suppose G_μ and H_μ are defined for all $\mu < \tau$. There are two cases in the definition of G_τ.

Case 1. There exists an ordinal $\nu < \lambda$ such that ν is the order-type of all ordinals $< \tau$ which are **L**-subconstructive, i.e. $\{\mu_\rho : \mu_\rho < \tau\} = \{\mu_\rho : \rho < \nu < \lambda\}$.

I. τ *is* **L**-*subconstructive*, i.e. $\tau = \mu_\nu$. Let

$$G_\tau(f_{xy}) = 1$$

for all x, y such that $x \in \Theta$, $|x|_{\Theta} \leqslant \eta_\nu$ and $y \in \mathrm{PR}[\mathbf{L}]$ and $|y|_{\mathbf{L}} = \tau$, where f_{xy} is chosen according to 7.2.5 above. Note that we always include trivial extensions whenever relevant.

II. τ *is not* **L**-*subconstructive*, i.e. $\tau < \mu_\nu$. Let

$$\varepsilon = \eta_\nu - \sup\{\eta_\rho : \rho < \nu\}.$$

We ask the following *question*: Does there exist an ordinal π such that $\tau < \pi \leqslant \tau + \varepsilon$ and π is $\mathbf{L}^0$-subconstructive?

If the answer is *yes*: Let $G_\tau = \bigcup_{\mu < \tau} G_\mu$ if τ is a limit ordinal, and let G_τ be the trivial extension of G_μ if $\tau = \mu + 1$.

If the answer is *no*: Let G_τ be defined as in subcase I.

Case 2. Otherwise: Let $G_\tau = \bigcup_{\mu < \tau} G_\mu$ if τ is a limit ordinal, and let G_τ be the trivial extension of G_μ if $\tau = \mu + 1$.

This completes the definition of G_τ, and from the sequence G_τ we define G as in Section 7.2.2 above.

7.2.7 Proposition 2. *The order-type of the* **L**-*subconstructive ordinals is* $\geqslant \lambda$.

This should not come as a surprise. In fact, this is the way we have arranged the construction of G. Nevertheless, assume that ν = order-type of the **L**-subconstructives $< \lambda$.

Let $\tau = \sup\{\mu_\rho : \rho < \nu\}$; τ is not **L**-subconstructive. Back to the construction of G_τ: We must be in subcase II of case 1, and the answer to the question is no!

Let $x \in \Theta$, $|x|_{\Theta} = \eta_\nu$, and $x \in S$. Then $G(f_{xy}) = 1$ for all y such that $|y|_{\mathbf{L}} = \tau$. And

$$\tau = \text{least ordinal } \tau \text{ such that } \exists y[|y|_{\mathbf{L}} = \tau \wedge G(f_{xy}) = 1].$$

We should not forget to point out that there exists such $y \in \mathrm{PR}[\mathbf{L}]$.

We have now set the stage for an application of further reflection. Let Q code all **L**-computations with arguments from S. By Theorem 7.1.6 $\kappa^{Q,x}$ is x-reflecting and $\kappa^{Q,x} > \kappa^{S,x} = \kappa^S = \tau$, where the last equality follows from our assumption, and $\kappa^{S,x} = \kappa^S$ since $x \in S$.

Let m be an index such that $\{m\}_{\mathbf{L}}(Q, x)\downarrow$ and $|m, Q, x|_{\mathbf{L}} > \kappa^{S,x} = \tau$. Then

$$\exists y[|y|_{\mathbf{L}} < |m, Q, x|_{\mathbf{L}} \wedge G(f_{xy}) = 1].$$

By reflection, omitting a few pedantic details, we conclude

$$\exists y[|y|_{\mathbf{L}} < \tau \wedge G(f_{xy}) = 1].$$

But this is impossible by the definition of τ.

7.2.8 Proposition 3. S-en$(\Theta) \subseteq S$-en$(\mathbf{L})$ *and for all* $r \in S$, $\mathrm{sc}(\Theta, r) \subseteq \mathrm{sc}(\mathbf{L}, r)$.

Let $X \in S\text{-en}(\Theta)$; then $r \in X$ iff $\langle e, r\rangle \in P$ for some index e. Since we have enough subconstructives on the **L**-side we are in the "normal" case 1, I and conclude

$$r \in X \quad \text{iff} \quad \langle e, r\rangle \in P \quad \text{iff} \quad \exists y \in S[y \in \mathrm{PR}[\mathbf{L}] \ \wedge \ G(f_{\langle e,r\rangle y}) = 1].$$

Hence $X \in S\text{-en}(\mathbf{L})$.

For the section part, note that if $X \in \mathrm{sc}(\Theta, r)$ then we have indices e_1, e_2 such that

$$\begin{aligned} x \in X \quad &\text{iff} \quad \{e_1\}_\Theta(x, r)\downarrow. \\ x \notin X \quad &\text{iff} \quad \{e_2\}_\Theta(x, r)\downarrow. \end{aligned}$$

From this construct an index e such that $\lambda x \cdot \{e\}_\Theta(x, r)$ is total and $|e, x, r|_\Theta \geqslant \inf\{|e_1, x, r|_\Theta, |e_2, x, r|_\Theta\}$. Then we compute $E(\lambda x \cdot \{e\}_\Theta(x, r))$ to get a Θ-subconstructive level larger than the ordinals associated to e_1 and e_2 when they are defined. This can be matched by an **L**-subconstructive level. As above, this allows us to conclude that both X and $A - X$ are **L**-semicomputable in r.

7.2.9 Toward the Second Half of the Theorem. For the converse we need to analyze the construction of G and hence of $\mathrm{PR}[\mathbf{L}]$ inside Θ. We fix some notation

$$\begin{aligned} \eta &= \sup\{\eta_\nu : \nu < \lambda\} = \kappa_\Theta^S \\ \mu &= \sup\{\mu_\nu : \nu < \lambda\} \leqslant \kappa_{\mathbf{L}}^S. \end{aligned}$$

Proposition 4. (a) *There exists a total Θ-computable function f and a partial Θ-computable p such that:*

(i) *If $|e, \sigma|_{\mathbf{L}} < \mu$, then $\{e\}_{\mathbf{L}}(\sigma) \simeq \{f(e)\}_\Theta(\sigma)$.*

(ii) *If $|x|_{\mathbf{L}} < \mu$ or $|y|_{\mathbf{L}} < \mu$, then $p(x, y)\downarrow$, and $x \in \mathrm{PR}[\mathbf{L}] \wedge |y|_{\mathbf{L}} < \mu \wedge |x|_{\mathbf{L}} \leqslant |y|_{\mathbf{L}} \Rightarrow p(x, y) \simeq 0$. $|y|_{\mathbf{L}} < \mu \wedge |x|_{\mathbf{L}} > |y|_{\mathbf{L}} \Rightarrow p(x, y) \simeq 1$.*

(b) *There exists a total Θ-computable function f' and a partial Θ-computable function p' such that:*

(i) $\{e\}_{\mathbf{L}}(\sigma) \simeq \{f'(e)\}_\Theta(\sigma, P)$,

(*where P is the set defined in* 7.2.2).

(ii) *If $|x|_{\mathbf{L}} < \kappa_{\mathbf{L}}$ or $|y|_{\mathbf{L}} < \kappa_{\mathbf{L}}$, then $p'(x, y)\downarrow$, and $x \in \mathrm{PR}[\mathbf{L}] \wedge |x|_{\mathbf{L}} \leqslant |y|_{\mathbf{L}} \Rightarrow p'(x, y) \simeq 0$. $|x|_{\mathbf{L}} > |y|_{\mathbf{L}} \Rightarrow p'(x, y) \simeq 1$.*

The proof being an exercise in the use of the second recursion theorem is long and very computational. The overall strategy is as follows (we restrict ourselves to part (a) for the moment). Let $\rho < \mu$ and suppose that $\{e\}_{\mathbf{L}}(\sigma) \simeq \{f(e)\}_\Theta(\sigma)$ for all e, σ such that $|e, \sigma|_{\mathbf{L}} < \rho$ and that $p(x, y)$ is defined and has the right value when $\inf(|x|_{\mathbf{L}}, |y|_{\mathbf{L}}) < \rho$. When $|e, \sigma|_{\mathbf{L}} = \rho$ we shall describe $\{f(e)\}_\Theta(\sigma)$ in terms of $\{f(e')\}_\Theta(\sigma')$ and $p(x', y')$, where $\{e'\}_{\mathbf{L}}(\sigma')$ is an immediate subcomputation of $\{e\}_{\mathbf{L}}(\sigma)$ and $\inf(|x'|_{\mathbf{L}}, |y'|_{\mathbf{L}}) < \rho$. When $\inf(|x|_{\mathbf{L}}, |y|_{\mathbf{L}}) = \rho$ we shall describe $p(x, y)$ in terms of $p(x', y')$, $\{f(e')\}_\Theta(\sigma')$, where $\inf(|x'|_{\mathbf{L}}, |y'|_{\mathbf{L}}) < \rho$ and $|e', \sigma'|_{\mathbf{L}} < \rho$.

In the construction of f the case to worry about is an application of G, so suppose $|e, \sigma|_{\mathbf{L}} = \rho$ and

$$\{e\}_{\mathbf{L}}(\sigma) \simeq G(\lambda u \cdot \{e'\}_{\mathbf{L}}(u, \sigma)).$$

By the induction hypothesis $\{e'\}_{\mathbf{L}}(u, \sigma) \simeq \{f(e')\}_{\Theta}(u, \sigma)$ for all u. We must now be able to decide inside Θ if $\lambda u \cdot \{e'\}_{\mathbf{L}}(u, \sigma) = f_{xy}$ for some x, y, and if this is true calculate $G(f_{xy})$. We ask five questions (and note that by construction $\lambda u \cdot \{e'\}_{\mathbf{L}}(u, \sigma) \neq f_{xy}$ if $|y|_{\mathbf{L}} \geqslant \rho$).

Question 1: Are there x, y such that $|y|_{\mathbf{L}} < \rho$ and $\lambda u\{e'\}_{\mathbf{L}}(u, \sigma) = f_{xy}$?

NO: Set $G(\lambda u \cdot \{e'\}_{\mathbf{L}}(u, \sigma)) = 0$
YES: Go to question 2.

Question 2: Let $\tau < \rho$ be the ordinal such that for some x and y, $\tau = |y|_{\mathbf{L}}$ and $f_{xy} = \lambda u \cdot \{e'\}_{\mathbf{L}}(u, \sigma)$. Is there an ordinal $\nu < \lambda$ such that $\mu_\xi < \tau$ when $\xi < \nu$ and $\mu_\nu \geqslant \tau$?

NO: Set $G(\lambda u \cdot \{e'\}_{\mathbf{L}}(u, \sigma)) = 0$
YES: Go to question 3.

Question 3: Let ν, τ be as above. Is there an x such that $|x|_{\Theta} \leqslant \eta_\nu$ and $\lambda u \cdot \{e'\}_{\mathbf{L}}(u, \sigma) = f_{xy}$, where $|y|_{\mathbf{L}} = \tau$?

NO: Set $G(\lambda u \cdot \{e'\}_{\mathbf{L}}(u, \sigma)) = 0$
YES: Go to question 4.

Question 4: Is τ **L**-subconstructive?

YES: Set $G(\lambda u \cdot \{e'\}_{\mathbf{L}}(u, \sigma)) = 1$
NO: Go to question 5.

Question 5: Let $\varepsilon = \eta_\nu - \sup\{\eta_\xi : \xi < x\}$. Is there an ordinal π such that $\tau < \pi \leqslant \tau + \varepsilon$ and π is $\mathbf{L}^0$-subconstructive?

YES: Set $G(\lambda u \cdot \{e'\}_{\mathbf{L}}(u, \sigma)) = 0$
NO: Set $G(\lambda u \cdot \{e'\}_{\mathbf{L}}(u, \sigma)) = 1$.

Each question must now be analyzed inside Θ. As an example we make some comments on the first question. We see that

$$\begin{aligned} |y|_{\mathbf{L}} < \rho \quad &\text{iff} \quad \exists u(|y|_{\mathbf{L}} \leqslant |e', u, \sigma|_{\mathbf{L}}) \\ &\text{iff} \quad \exists u \cdot p(y, \langle e', u, \sigma \rangle) \simeq 0. \end{aligned}$$

Observe that $\lambda u \cdot p(y, \langle e', u, \sigma \rangle)$ is total, hence $\exists u$ can be expressed by the E-functional, which means that the relation $|y|_{\mathbf{L}} < \rho$ is Θ-computable, uniformly in e, σ.

To describe f_{xy}, information about **L**-computations of length $< |y|_{\mathbf{L}}$ is needed. By the induction hypothesis this can be obtained from $\lambda e\sigma \cdot \{f(e)\}_{\Theta}(\sigma)$ and p when $|y|_{\mathbf{L}} < \rho$. In this case there is an index e_1 such that $f_{xy} = \lambda u\{e_1\}_{\Theta}(u, x, y, f(e), \sigma)$. We then have to decide the question

$$\exists x \exists y(|y|_{\mathbf{L}} < \rho \ \wedge \ f_{xy} = \lambda u\{f(e')\}_{\Theta}(u, \sigma)).$$

And this we argued that we can do, using E to express the quantifiers $\exists x \exists y$.

Question 2 is trivial in this case since $\rho < \mu$. It is when we come to part (b) of the proposition that we have to ask questions about P, hence the need to include P as argument. Also note that by bounding the search in question 5 we need not assume strong finiteness of the total domain A. With these hints we wish the reader the best of luck with the remaining details of the proof.

Remark. We need some more notation and a simple computational result (which actually is used in the proof of part (b) of the proposition).

We recall that P is the complete Θ-semicomputable set over S. Let Q be the corresponding **L**-set over S.

If $r \in Q$, then $|r|_{\mathbf{L}} = \mu_\nu$ for some ν, let $|r|_Q = \nu$, and set $|r|_Q =$ order-type of the **L**-subconstructives if $r \notin Q$.

If $r \in P$, then $|r|_{\Theta} = \eta_\nu$ for some ν, let $|r|_P = \nu$, and set $|r|_P =$ order-type of the Θ-subconstructives if $r \notin P$.

A simple computation from Proposition 4 shows that

$$\text{“} r \in P \ \wedge \ s \in Q \ \wedge \ |s|_{\mathbf{L}} < \mu \ \wedge \ |r|_P = |s|_Q \text{”},$$

is a Θ-semicomputable relation.

7.2.10 Proposition 5. *The order-type of the* **L**-*subconstructive ordinals is* λ.

We shall assume *not* and reflect down to a contradiction. So let there be an $s_0 \in Q$ such that $|s_0|_{\mathbf{L}} = \mu_\lambda$.

Some technical preliminaries are needed in order to reflect. Let $\lambda' \leqslant \lambda$ and set $P' = P_{\lambda'} = \{r \in P : |r|_P < \lambda'\}$ and $\eta' = \sup\{\eta_\rho + 1 : \rho < \lambda'\}$. Using P' in place of P we construct a functional G' and a list $\mathbf{L}'$. Note that G' and G agree up to μ', where $\mu' = \sup\{\mu_\rho + 1 : \rho < \lambda'\}$. And the functions f' and p' of Proposition 4 (b) act the same way with respect to P' as they do with respect to P.

We can now set the stage for the reflecting statement. Let e_1 be a Θ-index for the following statement which says that λ' is the order-type of the $\mathbf{L}'$-subconstructives below $|s_0|_{\mathbf{L}'}$.

(i) s_0 is an $\mathbf{L}'$-computation $\wedge \ \forall r \in P' \exists s(|r|_P = |s|_Q \ \wedge \ |s|_{\mathbf{L}'} < |s_0|_{\mathbf{L}'}) \ \wedge \ \forall s(|s|_{\mathbf{L}'} < |s_0|_{\mathbf{L}'} \rightarrow \exists r \in P'(|r|_{\mathbf{L}} = |s|_Q))$.

There further exist indices e_2, e_3, and e_4 such that

(ii) $\{e_2\}(P', s_0)\downarrow$ iff s_0 is an $\mathbf{L}'$-computation, in which case $|s_0|_{\mathbf{L}'} < |e_2, P', s_0|_{\Theta}$.

(iii) $\{e_3\}_{\Theta}(P')\downarrow$ and $|e_3, P'|_{\Theta} \geqslant \eta'$.

(iv) $\{e_4\}_{\Theta}(P', s_0)\downarrow$ iff s_0 is an $\mathbf{L}'$-computation, in which case $|e_4, P', s_0|_{\Theta} > \eta' + |s_0|_{\mathbf{L}'}$.

e_4 is constructed from e_2 and e_3.

The Reflecting Statement: The statement has three parts: There exists an $\lambda' \leqslant \lambda$ such that:

(a) $x = P_{\lambda'}$;
(b) s_0 is a convergent $\mathbf{L}'$-computation and λ' is the order-type of the $\mathbf{L}'$-subconstructives below $|s_0|_{\mathbf{L}'}$;
(c) if $r \notin x$, then $\eta' + |s_0|_{\mathbf{L}'} < |r|_\Theta$.

Part (a) of the statement can simply be expressed as "$x \subseteq P$ and $\forall r, r'(r \in x \wedge |r'|_\Theta \leqslant |r|_\Theta \rightarrow r' \in x)$". Part (b) is statement (i) above. Part (c) can be replaced by $\forall r(r \notin x \rightarrow |r|_\Theta > |e_4, x, s_0|_\Theta)$, which implies (c) when $x = P'$. Let e_5 be an Θ-index for these statements and define

$$\mathbb{B} = \{x : \{e_5\}_\Theta(x, s_0)\downarrow\}.$$

Then $P \in \mathbb{B}$ and by the compactness result 7.1.7 there exists a *proper* initial segment $P' = P_{\lambda'}$ of P, i.e. $\lambda' < \lambda$, such that $P' \in \mathbb{B}$.

Thus s_0 is a convergent $\mathbf{L}'$-computation, λ' is the order-type of the $\mathbf{L}'$-subconstructives below $|s_0|_{\mathbf{L}'}$ and if $r \notin P'$, then $\eta' + |s_0|_{\mathbf{L}'} < |r|_\Theta$.

We have reflected in order to show that s_0 is secured at an earlier stage than μ_λ. To do this we must go back to the construction of G and G'. We remarked above that G_ρ and G'_ρ are equal for all $\rho < \mu' = \sup\{\mu_\rho + 1 : \rho < \lambda'\}$. If we could extend this up to all $\rho < |s_0|_{\mathbf{L}'}$, then we would get $|s_0|_{\mathbf{L}} = |s_0|_{\mathbf{L}'} = \mu_{\lambda'} < \mu_\lambda = |s_0|_{\mathbf{L}}$. This will then be the desired contradiction which shows that the order-type of the $\mathbf{L}'$-subconstructives is at most, and hence equal to, λ.

It remains to fill in some of the details of this sketch.

Let $\tau = \mu'$. Note that the $\mathbf{L}$- and $\mathbf{L}'$-subconstructives agree below τ and have the same order-type λ'. And τ is $\mathbf{L}$-subconstructive iff it is $\mathbf{L}'$-subconstructive (see 7.2.5). And by part (b) of the reflecting statement $|s_0|_{\mathbf{L}'}$ is the λ'-th $\mathbf{L}'$-subconstructive.

Claim. $G_\rho = G'_\rho$ *for all* $\rho < |s_0|_{\mathbf{L}'}$.

If $\tau = |s_0|_{\mathbf{L}'}$, which is the case if τ is $\mathbf{L}$-subconstructive, then the claim follows from our preliminary remarks. So suppose $\tau < |s_0|_{\mathbf{L}'}$. τ is then *not* $\mathbf{L}$-subconstructive. We shall prove that $G_\tau = G'_\tau$.

In the construction of G'_τ we are in case 2 for the first time. In the construction of G_τ we are in case 1 since the order-type of the $\mathbf{L}$-subconstructive ordinals $< \tau$ is λ' and $\lambda' < \lambda$. Further, we are in subcase II because τ is not $\mathbf{L}$-subconstructive. The functional G^0 in the list $\mathbf{L}^0$ is the same as G', hence s_0 is a convergent $\mathbf{L}^0$-computation, $|s_0|_{\mathbf{L}'} = |s_0|_{\mathbf{L}^0}$, and $|s_0|_{\mathbf{L}^0}$ is the first ordinal $> \tau$ which is $\mathbf{L}^0$-subconstructive.

Let $\varepsilon = \eta_{\lambda'} - \sup\{\eta_\rho : \rho < \lambda'\}$. By (c) in the reflecting statement $\eta' + |s_0|_{\mathbf{L}^0} < |r|_\Theta$ if $r \notin P'$, i.e. if $|r|_\Theta \geqslant \eta_{\lambda'}$. Hence $\eta' + |s_0|_{\mathbf{L}^0} < \eta_{\lambda'}$. By definition $\eta' = \sup\{\eta_\rho + 1 : \rho < \lambda'\}$. Hence $\sup\{\eta_\rho : \rho < \lambda'\} + |s_0|_{\mathbf{L}^0} < \eta_{\lambda'}$. From this we conclude that

$$|s_0|_{\mathbf{L}^0} - \tau < \eta_{\lambda'} - \sup\{\eta_\rho : \rho < \lambda'\},$$

or

$$\tau < |s_0|_{\mathbf{L}^0} < \tau + \varepsilon.$$

But this means, since $|s_0|_{\mathbf{L}^0}$ is $\mathbf{L}^0$-subconstructive, that the answer to the question in II of case 1 is yes. We conclude that $G_\tau = G'_\tau$, and, in fact, that $G_\rho = G'_\rho$ for all ρ less than the next $\mathbf{L}^0$-subconstructive, which is $|s_0|_{\mathbf{L}'}$.

This proves the claim.

This means that the $\mathbf{L}'$-computations of length $< |s_0|_{\mathbf{L}'}$ are identical to the $\mathbf{L}$-computations of length $< |s_0|_{\mathbf{L}'}$, which means that $|s_0|_{\mathbf{L}'} = |s_0|_{\mathbf{L}}$. And this is impossible as explained above. The proof of Proposition 5 is complete.

7.2.11 Proposition 6. S-en(L) $\subseteq$ S-en(Θ) *and for all* $r \in S$, sc(L, r) $\subseteq$ sc(Θ, r).

Let $H = \{\langle e, \sigma\rangle : \{e\}_{\mathbf{L}}(\sigma)\downarrow \;\wedge\; |e, \sigma|_{\mathbf{L}} < \mu\}$. By Proposition 5 we know that $\mu = \sup\{\tau : \tau \text{ is } \mathbf{L}\text{-subconstructive}\} = \sup\{\mu_\nu : \nu < \lambda\}$. If $X \in S$-en(L), then there is an index e such that

$$r \in X \quad \text{iff} \quad \langle e, r\rangle \in H.$$

H, which is not necessarily a subset of S, is easily reduced to en(Θ), viz.

$$x \in H \quad \text{iff} \quad \exists r \in S\, \exists s \in S(r \in P \;\wedge\; s \in Q \;\wedge\; |s|_{\mathbf{L}} < \mu \;\wedge\; |r|_P = |s|_Q \;\wedge\; |x|_{\mathbf{L}} < |s|_{\mathbf{L}}).$$

By the remark under 7.2.9 and part (a) of Proposition 4 we see that H is Θ-semicomputable.

For the final part observe that if $X \in$ sc(L, r), then there are indices e_1, e_2 such that

$$x \in X \quad \text{iff} \quad \langle e_1, x, r\rangle \in H$$
$$x \notin X \quad \text{iff} \quad \langle e_2, x, r\rangle \in H,$$

because the lengths can be dominated by an L-subconstructive level, see the similar proof of Proposition 3 in 7.2.8.

And, if the reader has not yet noticed, Propositions 3 and 6 prove Theorem 7.2.1!

7.3 Characterization in Higher Types

Theorem 7.2.1 is in a well-defined sense a lifting of the plus-1 Theorem 5.4.24 from one to higher types. Of course, something more is added. In one type, specifically over ω, there are no subindividuals S, hence there is no S-envelope part to Theorem 5.4.24. And proofs are quite different.

On the other hand, with minor changes the proof of Theorem 5.4.7 which characterizes Spector theories over ω, generalizes to higher types.

7.3.1 Theorem. *Let Θ be a normal theory on $\mathfrak{A}$. Then Θ is equivalent to* PR[**L**] *for a normal list* **L** *iff Θ is not Θ-Mahlo.*

Only minor changes are needed in the previous proof. First we need to modify Definition 5.4.1.

7.3.2 Definition. Let Θ and Ψ be normal computation theories on $\mathfrak{A}$. We define

$$\Psi <_1 \Theta \quad \text{iff} \quad \text{en}(\Psi) \subseteq \text{en}(\Theta) \wedge \exists x[\kappa_\Psi^x < \kappa_\Theta^x].$$

Over ω the ordinal κ_Ψ^x, $x \in \omega$, would just be the ordinal of the Spector theory.

Corresponding to the Fattening Lemma 5.4.4 we have the following result.

7.3.3 Fattening Lemma. *Let Θ be normal and* **L** *a Θ-computable list such that $\forall x[\kappa_{\mathbf{L}}^x = \kappa_\Theta^x]$. Then there is a normal list* **L**$'$ *such that $\Theta \sim$* PR[**L**$'$].

The proof is the same, we just have to relativize to a parameter from the domain A at certain places. For example, Definition 5.4.5 of Ord(f) by the following. Let f be a total unary function from $A \to S$. Then Ord(f) is the least ordinal τ such that for some $e, y, f = \lambda x\{e\}_\Theta(x, y)$ and $\tau = |e_1, e, y|_\Theta$. (Here e_1 is an index such that $\{e_1\}_\Theta(e, y)\downarrow$ iff $\lambda x\cdot\{e\}(x, y)$ is total, in which case $|e, x, y|_\Theta < |e_1, e, y|_\Theta$ for all x.)

We now construct G_0 exactly as in the proof of 5.4.4 and let **L**$' =$ **L**, $G, E, =_S$. We immediately conclude that en(**L**$'$) $\subseteq$ en(Θ).

To prove the converse we introduce a relativized version of κ, viz. for $y \in A$, let

$$\kappa_y = \sup\{\text{Ord}(f); f \text{ is } \mathbf{L}'\text{-computable in } y\}.$$

And corresponding to the claim we now have: $\kappa_y = \kappa_\Theta^y$ for all $y \in A$.

From this claim the fattening lemma immediately follows: Suppose $\{e\}_\Theta(\sigma)\downarrow$. By the claim there is an index m such that $f = \lambda t\cdot\{m\}_{\mathbf{L}'}(t, \langle e, \sigma\rangle)$ is total and $|e, \sigma|_\Theta \leqslant$ Ord(f).

By construction of G_0 we see that

$$G_0(\langle\lambda t\cdot\{m\}_{\mathbf{L}'}(t, \langle e, \sigma\rangle), e, \sigma\rangle) \simeq \{e\}_\Theta(\sigma) + 1.$$

Using selection over N, we pick an m as a function of e, σ, which gives us, exactly as before, the converse inclusion, viz. en(Θ) $\subseteq$ en(**L**$'$).

7.3.4 Definition. A normal theory Θ on $\mathfrak{A}$ is called Θ-*Mahlo* if for all normal and Θ-computable lists **L**, PR[**L**] $<_1 \Theta$.

Theorem 7.3.1 now follows as before. If $\Theta \sim \mathrm{PR}[\mathbf{L}]$ for a normal list $\mathbf{L}$, then $\mathbf{L}$ is Θ-computable, and Θ, obviously, cannot be Θ-Mahlo, using $\mathbf{L}$ as a counter-example. Conversely, suppose that Θ is not Θ-Mahlo. Then there exists a Θ-computable list $\mathbf{L}$ such that $\mathrm{PR}[\mathbf{L}]$ is not $<_1$ than Θ. Since by the Θ-computability of $\mathbf{L}$, $\mathrm{en}(\mathbf{L}) \subseteq \mathrm{en}(\Theta)$, this means that $\forall x[\kappa_{\mathbf{L}}^x = \kappa_{\Theta}^x]$. The fattening lemma then gives us a normal list $\mathbf{L}'$ such that $\Theta \sim \mathrm{PR}[\mathbf{L}']$.

7.3.5 Remark. We have been discussing computation theories on *one* and *two* types. In the one-type case we have a computation domain A with no extra structure assumed, in the two-type case A has the structure $A = S \cup \mathrm{Tp}(S)$, where $\mathrm{Tp}(S) = \omega^S$.

Various finiteness assumptions have been placed on the domains, the crucial distinction being *strong* versus *weak* finiteness. In the Spector theory case we have a theory on one type A which is assumed to be strongly finite in the sense of the theory. In the normal type-2 case we imposed the requirement that A is weakly finite, but S is strongly finite.

There are a number of comments to make. First recall that weak and strong finiteness coincides over ω.

In two types we could drop the requirement that S is strongly finite. In this case reflection phenomena disappear, and we are essentially back to the case of one weakly finite domain. This case is not without interest: The characterization Theorem 7.3.1 would still be true, and there is a non-trivial result in the theory of inductive definability here, see our discussion in Theorem 3.3.15 on $\mathrm{IND}(\Sigma_2^0)$ versus $\mathrm{IND}(\Pi_1^0)$.

But we could in the type-2 case make a move in the opposite direction, strengthening the axioms to the strong finiteness of A. This case has been studied by Hinman and Moschovakis [62] under the name of hyperprojective theory. Here again we can have some effect of the type structure, but the fascinating interplay of strong and weak finiteness in the normal case, is lost.

Chapter 8
Set Recursion and Higher Types

Primitive recursive set functions and rudimentary set functions have been around for some time, see Jensen-Karp [72], Gandy [40], and also Devlin [19] for their basic properties and further references. The step to a general recursion theory on sets came rather late. There is, perhaps, a reason why: Primitive recursive and rudimentary set functions were introduced to elucidate the rather restricted recursion-theoretic nature of the constructible hierarchy, and in the hands of Jensen [71] have become an important tool in the fine structure theory of L.

Full set recursion was introduced by D. Normann [124], and later rediscovered by Y. Moschovakis, as a tool for developing a companion theory for Kleene-recursion in higher types. The theory has, however, a wider scope and we shall present a general version in the first part of this chapter. In this we follow the exposition in Normann [124]. The approach of Moschovakis uses inductive definability, but the end result is substantially the same.

In Section 3 we work out the detailed connection with Kleene-recursion in higher types. Some of this work has its origin in the theses of Harrington [53], MacQueen [98], and Normann [122]. We believe that the general set-theoretic approach adds both simplicity and insight.

As a testing ground for this belief we turn in Section 4 to the degree theory in higher types. We present a fairly simple priority argument involving ${}^3E$, allowing the reader to explore the full intricacies of the general theory for him- or herself. We just want to make the point that set recursion is a very natural computation theory to use in the study of degrees of functionals.

8.1 Basic Definitions

Set recursion in a relation R on the universe of sets V is generated by the schemes for the functions rudimentary in R augmented with the diagonalization scheme.

8.1.1 Definition. Let $R \subseteq V$ be a relation. The class of partial functions *set-recursive relative to* R is inductively defined by the following clauses

(i) $f(x_1, \ldots, x_n) = x_i$ $\qquad e = \langle 1, n, i \rangle$

(ii) $f(x_1, \ldots, x_n) = x_i - x_j$ $\qquad e = \langle 2, n, i, j \rangle$

(iii) $f(x_1, \ldots, x_n) = \{x_i, x_j\}$ $\quad e = \langle 3, n, i, j\rangle$

(iv) $f(x_1, \ldots, x_n) \simeq \bigcup_{y \in x_1} h(y, x, \ldots, x_n)$

$e = \langle 4, n, e'\rangle$ where e' is an index for h.

(v) $f(x_1, \ldots, x_n) \simeq h(g_1(x_1, \ldots, x_n), \ldots, g_m(x_1, \ldots, x_n))$

$e = \langle 5, n, m, e', e_1, \ldots, e_m\rangle$ where e' is an index for h and $e_1, \ldots, e_m$ are indices for $g_1, \ldots, g_m$, respectively.

(vi) $f(x_1, \ldots, x_n) \simeq x_i \cap R$ $\quad e = \langle 6, n, j\rangle$

(vii) $f(e_1, x_1, \ldots, x_n, y_1, \ldots, y_m) \simeq \{e_1\}_R(x_1, \ldots, x_n)$

$e = \langle 7, n, m\rangle$.

There are a number of comments to make. The functions generated by clauses (i) to (vi) are the functions rudimentary in R. These functions are all total. In the general case we get partial functions and we then assume in scheme (iv) that the computation is defined only if $h(y, x_2, \ldots, x_n)$ is defined for all $y \in x_1$.

The index e is always a natural number. This is a limiting feature of this version of set-recursion which squares with a characteristic feature of Kleene-recursion in higher types, but makes the theory differ from general admissibility theory. Note that since for each $n \in \omega$ the constant function with value n is rudimentary, these functions will be set-recursive.

The basic theory of the rudimentary set functions is developed in Devlin [19]. We shall use only their most elementary properties: For the more delicate parts of the theory see Jensen [71] (or the exposition in Devlin [19]); the reader will also profit from a study of Simpson's lectures [156].

For emphasis we repeat: Let $R \subseteq V$. A function f on V is called set-recursive in R if there is an index e such that for all $x_1, \ldots, x_n$

$$f(x_1, \ldots, x_n) \simeq \{e\}_R(x_1, \ldots, x_n),$$

where the notation $\{e\}_R$ obtains its meaning through Definition 8.1.1.

The set $\Theta_R = \{(e, \sigma, z); \{e\}_R(\sigma) \simeq z\}$ is a computation theory in the sense of part A of this book. From the schemes in Definition 8.1.1 we derive in a canonical way the notions of

a. *length of a computation*,
b. *subcomputation*,
c. *computation tree*.

These concepts should by now be thoroughly familiar to the reader and we need not repeat the detailed constructions. (A complete exposition exists in Gurrik [52], to which any reader who wishes to learn Norwegian can turn.)

We let $\|a, \sigma, z\|$, for $(a, \sigma, z) \in \Theta_R$, denote the length of the computation $\{e\}_R(\sigma) \simeq z$.

Set-recursion in R is p-normal. This is a consequence of the following simple lemma which combines an application of the schemes (iv) and (vii).

8.1.2 Lemma. *There is an index e such that for arbitrary* R, x, e_1, σ

$$\{e\}_R(x, e_1, \sigma) \simeq \begin{cases} 0 & \text{if } \forall y \in x \cdot \{e_1\}_R(y, \sigma) \simeq 0 \\ 1 & \text{if } \forall y \in x \cdot \{e_1\}_R(y, \sigma)\downarrow \quad \text{and} \quad \exists y \in x \cdot \{e_1\}_R(y, \sigma) \neq 0. \end{cases}$$

The proof is simple. By elementary properties of rudimentary functions we may assume that $\{e_1\}_R$ takes values 0 ($= \varnothing$) and 1 ($= \{\varnothing\}$) only. Let

$$\{e\}_R(x, e_1, \sigma) \simeq \bigcup_{y \in x} \{e_1\}_R(y, \sigma).$$

This is the proof. The lemma shows that we have the crucial property of normality from recursion in higher types built into set-recursion through the scheme (iv) of "bounded" union. If $x = \omega$, bounded union means quantification over ω, i.e. the computability of ${}^2E$. This explains the name "E-recursion" rather than "set-recursion" in Normann [124].

With the usual proof, Lemma 8.1.2 implies p-normality, i.e. stage comparison, and the existence of a selection operator over ω. (See Chapter 3 for details.)

8.1.3 Proposition. *There is a set-recursive function p such that if* $\sigma = (e, \sigma_1, z) \in \Theta_R$ *or* $\sigma' = (e', \sigma_1', z) \in \Theta_R$ *then* $p(\sigma, \sigma')\downarrow$, *and*

$$\sigma \in \Theta_R \wedge \|\sigma\| \leqslant \|\sigma'\| \Rightarrow p(\sigma, \sigma') \simeq 0$$
$$\|\sigma\| > \|\sigma'\| \Rightarrow p(\sigma, \sigma') \simeq 1.$$

8.1.4 Proposition. *There is an index e such that for arbitrary* R, e_1, σ

$$\{e\}_R(e_1, \sigma)\downarrow \quad \textit{iff} \quad \exists n \in \omega \{e_1\}_R(n, \sigma)\downarrow.$$

And if $\exists n \in \omega\{e_1\}_R(n, \sigma)\downarrow$ *then* $\{e_1\}_R(\{e\}_R(e_1, \sigma), \sigma)\downarrow$, *and*

$$\|e, e_1, \sigma\| > \inf\{\|e, n, \sigma\| : \{e\}_R(n, \sigma)\downarrow\}.$$

As our computations are single-valued we abbreviate in the usual fashion, i.e. if $\{e\}_R(\sigma) \simeq z$, we let $\|e, \sigma\|$ denote the length of the computation.

8.1.5 Remark. It is also possible to develop a version of the selection principle of Theorem 4.3.1 in the context of set-recursion. But proving "Grilliot-selection" once is sufficient for us! The reader should, however, consult a forthcoming paper, *A note on reflection*, in Math. Scandinavica by Dag Normann.

8.2 Companion Theory

We aim at a construction generalizing the "next admissible", see the introductory discussion in Section 5.3 as well as the construction of the abstract 1-section corresponding to a Spector theory Θ on ω in Proposition 5.4.20 of the same chapter.

As always there are some preliminary definitions. For completeness we put down the somewhat unexciting

8.2.1 Definition. Let $R \subseteq V$, $\tau \in V^m$. Let φ be a partial function from V^n to V. We say that φ is *set-recursive in τ relative to R* if there is an index e such that for all $\sigma \in V^n$

$$\varphi(\sigma) \simeq \{e\}_R(\sigma, \tau).$$

From this we have the obvious notions of sets recursive and semirecursive in σ relative to R.

The following definition is essential.

8.2.2 Definition. Let $A \subseteq V$, $R \subseteq V$. The *set-recursive closure of A relative to R* is the set

$$M(A; R) = \{\{e\}_R(\sigma) : e \in \omega, \sigma \in A^n, n \in \omega\}.$$

If A is set-recursively closed relative to R, we may split up A as follows

$$\langle M(B; R)\rangle_{B \in {}^f A},$$

where ${}^f A$ is the set of finite subsets of A. This splitting will be of crucial importance in the following theory. As a first result we shall characterize semicomputability relative to R in terms of a special kind of Σ_1-definability over the splitting $\langle M(B; R)\rangle_{B \in {}^f A}$.

But first a linguistic convention. We write *R-recursive in τ* and *R-semirecursive in τ* instead of "set-recursive in τ relative to R" and "set-semirecursive in τ relative to R". Similarly we shall use the phrase *R-recursive closure of A* for the notion introduced in Definition 8.2.2.

8.2.3 Definition. Assume that A is R-recursively closed and that B is a finite subset of A. A set $C \subseteq A$ is called *$\Sigma_B^*(R)$-definable* if for some Δ_0-formula φ with parameters from $M(B; R)$

$$x \in C \quad \text{iff} \quad \exists y \in M(B \cup \{x\}; R) \cdot \varphi(x, y, R).$$

$C \subseteq A$ is called $\Delta_B^*(R)$ if both C and $A - C$ are $\Sigma_B^*(R)$-definable.

If the correct R is clear from the context we write for simplicity Σ_B^*, Δ_B^*, and even $M(B)$.

8.2.4 Proposition. *Let A be R-recursively closed and transitive. Then $\langle M(B; R)\rangle_{B \in {}^f A}$ satisfies Σ^*-collection: Let φ be a Δ_0-formula with parameters from $M(B; R)$ and let $u \in M(B)$. Assume*

$$\forall x \in u\, \exists y \in M(B \cup \{x\}; R) \cdot \varphi(x, y, R),$$

then

$$\exists v \in M(B) \forall x \in u\, \exists y \in v \cdot \varphi(x, y, R).$$

The proof is exactly the same as the proof of Proposition 5.4.20. Let σ_B be a listing of the finite set B. By assumption

$$\forall x \in u\, \exists e \in \omega \cdot \varphi(x, \{e\}_R(\sigma_B, x), R).$$

By Gandy selection, i.e. Proposition 8.1.4, choose one e to each x and use the union scheme (iv) to find the set v. Formally let $\nu(x)$ be the index corresponding to x. Then we can set

$$v = \bigcup_{x \in u} \{\{\nu(x)\}_R(\sigma_B, x, R)\},$$

which is easily seen to belong to $M(B)$.

8.2.5 Proposition. *Well-foundedness is Σ^*-definable.*

We indicate the proof. By the recursion theorem find an index e such that if y is a well-founded relation on x, then $\{e\}_R(y, x)\downarrow$ and $\{e\}_R(y, x)$ is the rank function of y. So, for any x, y, y is a well-founded relation on x iff $\exists f \in M(\{x, y\})$ such that f is a rank function for y. (Of course, this is independent of, hence uniform in, R.)

8.2.6 Theorem. *Let A be R-recursively closed and B a finite subset of A. A set $C \subseteq A$ is R-semirecursive in σ_B iff C is $\Sigma_B^*(R)$-definable.*

This is the promised definability characterization of R-semirecursive sets. We shall return to this result in connection with Kleene-recursion in higher types in the next section.

First, assume that C is R-semirecursive in σ_B, i.e. for some index e, $x \in C$ iff $\{e\}_R(x, \sigma_B)\downarrow$. By a somewhat lengthy analysis using the recursion theorem we may prove that if $\{e\}_R(x, \sigma_B)\downarrow$ then the associated computation tree will be in $M(B \cup \{x\}; R)$. Therefore,

$$x \in C \quad \text{iff} \quad \exists T \in M(B \cup \{x\}; R)[T \text{ is well-founded and } T \text{ is a computation-tree for } \{e\}_R(x, \sigma_B)].$$

By Proposition 8.2.5, this is seen to be a $\Sigma_B^*(R)$-definition of C.

For the converse, assume that C is $\Sigma_B^*(R)$-definable. We may use the same trick as in the proof of Proposition 8.2.4 to find for each $x \in C$ an index $e = \nu(x)$ for the y in the $\Sigma_B^*(R)$-definition of C. $\nu(x)$ will diverge if $x \notin C$, hence $C = \{x : \nu(x)\downarrow\}$, i.e. C is R-semirecursive in σ_B.

We conclude this section by a brief discussion of R-admissibility.

8.2.7 Definition. A family $\langle M_B \rangle_{B \in {}^f A}$ is called *R-admissible* if it satisfies the following three requirements:

(i) each M_B is rudimentarily closed in R,
(ii) for $B, C \in {}^f A$, $M_B \subseteq M_C$ iff $B \subseteq M_C$,
(iii) the family satisfies $\Sigma^*(R)$-collection.

Note that this is an "arbitrary" splitting, i.e. we have a map from ${}^f A$ into the universe of sets satisfying (i) to (iii) above. We have the following closure result:

8.2.8 Proposition. *Let $\langle M_B \rangle_{B \in {}^f A}$ be R-admissible. Then each M_B is closed under R-recursion.*

The proof proceeds in the following steps. First by induction on the height of a well-founded relation we prove by Σ^*-collection that if y is a well-founded relation on x, then the rank-function is in $M_{\{x,y\}}$. Thus well-foundedness is Σ^*-definable over $\langle M_B \rangle_{B \in {}^f A}$. Next, and by the same method, we prove that if $\{e\}_R(\sigma_B)\downarrow$, then the computation tree is in M_B. We then finish the proof by observing that the value of a computation is rudimentary in the computation tree. We remark that this proof also shows that the relation $\{e\}_R(\sigma_B) \simeq z$ is $\Sigma^*(R)$-definable over the family $\langle M_B \rangle_{B \in {}^f A}$.

Putting Propositions 8.2.4 and 8.2.8 together we see that if A is transitive and R-recursively closed, then $\langle M(B; R) \rangle_{B \in {}^f A}$ is the *finest splitting* of A into an R-admissible family.

8.3 Set Recursion and Kleene-Recursion in Higher Types

We shall now explain how set-recursion generalizes Kleene-recursion in higher types. Let $I = \mathrm{Tp}(k)$, i.e. I is the set of all total functions of type k. Tp(0) is then the set of natural numbers. We note that I has a natural pairing function, hence we may identify finite subsets of I with elements of I.

In order to get the effect of normality, i.e. the computability of the functional ${}^{k+2}E$ over I, we see from the proof of Lemma 8.1.2 that I must not only be a domain for the computation theory, we must also have recourse to I as an input to computations. This motivates the following definition.

8.3.1 Definition. Let $I = \mathrm{Tp}(k)$ and let $R \subseteq V$ be a relation. By the *spectrum of R over I* is understood the family

$$\mathrm{Spec}(R; I) = \langle M(\{a, I\}; R) \rangle_{a \in I}.$$

For simplicity we often write $M_a(I; R)$ for $M(\{a, I\}; R)$, and occasionally drop the I or the R or both, if their presence is clear from the context.

We let

$$M(I; R) = \bigcup_{a \in I} M_a(I; R).$$

As in Proposition 8.2.4 we see that Spec$(R; I)$ will satisfy $\Sigma^*(R)$-*collection over* I, i.e. if φ is a Δ_0-formula with parameters from $M_a(I; R)$, and if

$$\forall b \in I \, \exists y \in M_{\langle a,b \rangle}(I; R) \cdot \varphi(b, y, R),$$

then

$$\exists v \in M_a(I; R) \forall b \in I \, \exists y \in v \cdot \varphi(b, y, R).$$

We further note that each $M_a(I; R)$ will be rudimentarily closed relative to R. Looking back to Definition 8.2.7 we are led to the following notion of R-admissibility over I.

8.3.2 Definition. A family $\langle M_a \rangle_{a \in I}$ is called *R-admissible over I* if it satisfies the following requirements:

(i) each M_a is rudimentarily closed relative to R,
(ii) $I \in M_a$ for all $a \in I$, and for all $a, b \in I$
$a \in M_b$ iff $M_a \subseteq M_b$,
(iii) the family satisfies $\Sigma^*(R)$-collection over I.

Obviously the family Spec$(R; I)$ is R-admissible over I. We shall prove that it is the minimal family that is R-admissible over I. To do this we need to discuss how to code elements of the spectrum.

8.3.3 Definition. (i) Let $A \subseteq I \times I$ be a transitive, reflexive relation. For $a, b \in I$ let $a \simeq b$ iff $A(a, b)$ and $A(b, a)$. We say that A is a *code for a set* x if $A/\simeq$ is isomorphic to $\langle \mathrm{TC}(\{x\}), \in \rangle$, where TC as usual denotes the transitive closure.

(ii) Let $\langle M_a \rangle_{a \in I}$ be a family over I. $\langle M_a \rangle_{a \in I}$ is called *locally of type I* if for any set x and $a \in I$

$$x \in M_a \quad \text{iff} \quad x \text{ has a code in } M_a.$$

8.3.4 Lemma. *Spec$(R; I)$ is locally of type I.*

The proof is an exercise in the use of the recursion theorem. One first establishes that there is an index e_1 such that if A is a code for x, then $\{e_1\}_R(A, I) = x$. From this, once more by the recursion theorem, one constructs an index e_2 such that given any $e \in \omega$, and any sequence of codes $A_1, \ldots, A_n$ for sets $y_1, \ldots, y_n$, if $\{e\}_R(y_1, \ldots, y_n) \simeq x$, then $\{e_2\}_R(e, A_1, \ldots, A_n)$ is a code for x.

8.3.5 Remark. As the reader will understand from the above hint-of-a-proof we do not wish to go into details of proofs involving codes for sets. Complete expositions can be extracted from D. Normann [124], and also from Gurrik [52].

We can now prove the minimality of the spectrum.

8.3.6 Theorem. Spec(R; I) *is the minimal family R-admissible over I.*

It remains to verify that Spec(R; I) is included in any family $\langle M_a \rangle_{a \in I}$ which is R-admissible over I. The proof is very similar to the proof of Proposition 8.2.8. We prove by induction on the length of the computation that for any $x_1, \ldots, x_n$, if $x_1, \ldots, x_n$ have codes in M_a and $\{e\}_R(x_1, \ldots, x_n)\downarrow$, then both the computation tree and the value will be in M_a. The use of Σ^*-collection over a set x in 8.2.8 is now replaced by the use of a code for x and Σ^*-collection over I.

We come now to the main characterization theorem.

8.3.7 Theorem. *Let F be a functional of type $k + 2$, let $I =$* Tp(k) *and $C \subseteq$* Tp($k + 1$). *The following three statements are equivalent:*

(i) *C is Kleene-semirecursive in ^{k+2}E, F.*
(ii) *C is F-semirecursive in I.*
(iii) *C is $\Sigma_I^*(F)$-definable.*

We recall the notion of $\Sigma_I^*(F)$-definable. C is $\Sigma_I^*(F)$-definable if

$$x \in C \quad \text{iff} \quad \exists y \in M_{\{x\}}(I; F)\varphi(x, y, F).$$

We note that F is here a relation, I is an element. The equivalence of (ii) and (iii) follows from Theorem 8.2.6.

The proof that (i) implies (ii) consists in constructing an index e in F-recursion such that

$$\{e\}_F(e_1, \sigma) \simeq \{e_1\}_{\text{Kleene}}(\sigma, {}^{k+2}E, F).$$

This is tedious but straightforward. For example the scheme

$$\{e_1\}_K(f, \sigma) \simeq f(\lambda t \cdot \{e_2\}_K(f, \sigma, t)),$$

is handled by rewriting in the following form

$$\{e_1\}_K(f, \sigma) \simeq z \quad \text{iff} \quad \exists x \in I\, \forall y \in \text{Tp}(k-1)[x(y) = \{e_2\}_K(f, \sigma, y) \wedge f(x) = z].$$

A combination of schemes (iv) and (vii) from Definition 8.1.1 will take care of this case, everything else is rudimentary.

For the converse assume that C is F-semirecursive in I, i.e. for some index e

$$f \in C \quad \text{iff} \quad \{e\}_F(f, I)\downarrow.$$

The method is now to simulate the computation $\{e\}_F(f, I)$ as a Kleene-computation in ${}^{k+2}E, F$ on codes. Again the details are not particularly exciting. The reader who for some reason wants to reconstruct the details, will need the following facts.

In Kleene-recursion there is an index e such that if f and g are characteristic functions for codes for x and y, respectively, then

$$\{e_1\}_K({}^{k+2}E, f, g) \simeq \begin{cases} 0 & \text{if } \; x = y \\ 1 & \text{if } \; x \neq y. \end{cases}$$

(Use the recursion theorem and induction on min(rank(x), rank(y)).)

From e_1 we construct a Kleene-index e_2 such that if $f_1, \ldots, f_k$ are characteristic functions of codes for $x_1, \ldots, x_k$ and $\{e\}_F(x_1, \ldots, x_n) \simeq y$, then

$$\lambda a \in I \cdot \{e_2\}_K(e, f_1, \ldots, f_k, a, {}^{k+2}E, F),$$

is the characteristic function of a code for y.

As we remarked in the introduction to this chapter, the sources of this theorem, as well as for Theorems 8.2.6 and 8.3.6 are the theses of Harrington [53], MacQueen [98], and Normann [122]. A first version of 8.2.6 can be found in Harrington's thesis, and the first version of Theorem 8.3.6 stems from MacQueen [98]. Further developments can be found in Normann [122] as well as in the (unpublished) lecture notes of Kechris [74].

8.3.8 Examples. We shall separately consider the case $k = 0$ and the case $k > 0$. And we shall first supplement Theorem 8.3.7 with the following observation: *Let F be of type $k + 2$, let $a \in I$ and $A \subseteq I$, then $A \in k + 1 -$ sc$({}^{k+2}E, F, a)$ iff $A \in M_a(I; F)$.* (The proof is implicit in 8.2.6 and 8.3.7.)

First to the case $k = 0$. Then I is the set of natural numbers and we are essentially studying Kleene-recursion in ${}^2E, F$, where F is a total type-2 functional over ω. In this case Spec(F; ω) reduces to one set, viz. the "next admissible" set corresponding to the Spector theory PR[${}^2E, F$]. We have recovered the theory of Section 5.3, and Theorem 8.2.6 is the genuine Gandy-Spector theorem.

In the case $k > 0$, Spec(F; I) $= \langle M_a(I; F)\rangle_{a \in I}$ is non-trivial (since not every $a \in I =$ Tp(k) is recursive in ${}^{k+2}E, F$). Since the spectrum is locally of type I, we see that $M_a(I; F)$ consists of exactly those sets which have a code in $k + 1$-sc$({}^{k+2}E, F, a)$. (Note, that this was exactly the way we constructed the abstract 1-section of a Spector theory over ω, see Proposition 5.4.20.)

Each set $M_a(I; F)$ is countable, rudimentarily closed in F and satisfies suitable versions of Δ_0-separation and Δ_0-dependent choices. These sets thus have all "good" properties of admissible sets and abstract 1-sections, but they do contain "gaps", i.e. the sets are not transitive. The gaps are necessarily present to reflect gap phenomena in computations in higher types.

The sets in the family Spec(F; I) interact via the $\Sigma^*(F)$-collection principle which leads to the characterization in Theorem 8.3.7. In this case the equivalence of (ii) and (iii) is the best possible version of the original Gandy-Spector theorem, the domain of the existential quantifier cannot be chosen independently of the element we are testing for membership in the given semicomputable set.

Theorem 8.3.7 shows that Kleene-recursion in a normal functional is a special case of set-recursion relative to some relation. We shall now prove a converse.

The following recursive approximation of the spectrum will prove useful.

8.3.9 Definition. Let α be an ordinal and A a set. The *α-approximation* to $M(A; R)$ is the set

$$M^{\alpha}(A; R) = \{\{e\}_R(\sigma_B) : B \in^f A,\ e \in \omega \text{ and } \|e, \sigma_B\| < \alpha\}.$$

From this we derive the correct definitions of $\langle M^{\alpha}(B; R)\rangle_{B \in^f A}$ and $\langle M^{\alpha}(I; R)\rangle_{a \in I}$.

We shall also need the following notion of weak Σ^*-definability.

8.3.10 Definition. Let $\langle M_a\rangle_{a\in I}$ be an R-admissible family over I and let $C \subseteq M$, where $M = \bigcup_{a\in I} M_a$. We say that C is *weakly Σ^*-definable in a relative to R*, in symbols, C is w-$\Sigma^*_a(R)$, if for some Δ_0-formula φ with parameters from M_a,

$$x \in C \quad \text{iff} \quad \forall b \in I(x \in M_{\langle a,b\rangle} \Rightarrow \exists y \in M_{\langle a,b\rangle} \cdot \varphi(x, y, R)).$$

C is w-$\Delta^*_a(R)$ if both C and $M - C$ are w-$\Sigma^*_a(R)$.

We shall comment below on why we need the weak notion. Here we first remark that if $C \subseteq I$, then w-$\Sigma^*_a(R)$ and $\Sigma^*_a(R)$ coincide since $M_{\langle a,x\rangle}$ is the least $M_{\langle a,b\rangle}$ such that $x \in M_{\langle a,b\rangle}$. We also have the following lemma.

8.3.11 Lemma. *Let R be a relation and $\langle M_a\rangle_{a\in I} =$ Spec$(R; I)$. If $C \subseteq M$ is $\Sigma^*_a(R)$-definable, then C is w-$\Sigma^*_a(R)$-definable.*

Since C is $\Sigma^*_a(R)$-definable, we have

$$x \in C \quad \text{iff} \quad \exists y \in M(\{x, a\}; R)\varphi(x, y, R).$$

Let $x \in M_{\langle a,b\rangle}$. Then

$$x \in C \quad \text{iff} \quad \exists \alpha \in M_{\langle a,b\rangle} \exists y \in M^{\alpha}(\{x, a\}; R)\varphi(x, y, R).$$

It is sufficient to show that the relation $z = M_{\{x,a\}}(R)$ is w-$\Sigma^*(R)$ and that if $\alpha \in M_{\langle a,b\rangle}$ and $x \in M_{\langle a,b\rangle}$, then $M^{\alpha}_{\{x,a\}} \in M_{\langle a,b\rangle}$. This we do by a careful analysis of the inductive definition of $M^{\alpha}_{\{x,a\}}$.

We need one more lemma.

8.3.12 Lemma. *Let R_1 and R_2 be relations. If R_1 is w-$\Delta^*(R_2)$, then* Spec$(R_1; I) \subseteq$ Spec$(R_2; I)$ *and if $C \subseteq I$ is $\Sigma^*(R_1)$-definable, then C is also $\Sigma^*(R_2)$-definable.*

For the proof we note that if R_1 is w-$\Delta^*(R_2)$ then $\text{Spec}(R_2; I)$ will be R_1-admissible over I; by Theorem 8.3.6 we conclude that $\text{Spec}(R_1; I) \subseteq \text{Spec}(R_2; I)$.

In connection with Definition 8.3.10 we observe that over I Σ^* and w-Σ^* are the same. So it suffices to prove that $w\text{-}\Sigma^*(R_1) \subseteq w\text{-}\Sigma^*(R_2)$. But this will follow if we can prove that w-$\Delta^*(R_2)$ is closed under bounded quantification.

Since we are not in the admissible case we will give the proof of this closure property. Or rather, we prove this for $\Delta^*(R)$ definability: the w-Δ^* case is similar.

Let θ be $\Delta^*(R)$ definable, i.e. we have formulas ψ_1 and ψ_2 such that

$$\begin{aligned}\theta(z, x) &\text{ iff } \exists w \in M_{\{x,z\}}\psi_1(x, z, w)\\ \neg\theta(z, x) &\text{ iff } \exists w \in M_{\{x,z\}}\psi_2(x, z, w).\end{aligned}$$

We want to show that

$$\psi(x) \text{ iff } \forall z \in y\theta(z, x)$$

is $\Delta^*(R)$ definable. But this is an immediate consequence of the following two equivalences

$$\begin{aligned}\psi(x) &\text{ iff } \exists f \in M_{\{x\}}\forall z \in y\psi_1(x, z, f(z))\\ \psi(x) &\text{ iff } \forall f \in M_{\{x\}}(\forall z \in y[\psi_1(x, z, f(z) \vee \psi_2(x, z, f(z))]\\ &\qquad \to \forall z \in y\psi_1(x, z, f(z))).\end{aligned}$$

In connection with this argument the reader should appreciate the fact that Σ^* is *not* closed under bounded quantification.

We are now in a position to prove that Kleene-recursion in a normal type $k + 2$ functional over $I = \text{Tp}(k)$ is the same as set-recursion over I relative to some relation.

8.3.13 Theorem. *Let $I = \text{Tp}(k)$ and let R be a relation. Then there is a total type $k + 2$ functional F such that*

(i) $\text{Spec}(R; I) = \text{Spec}(F; I)$.

(ii) *Over I, $\Sigma^*(R)$ and $\Sigma^*(F)$ are the same.*

At this point the reader would do well to recall our discussion of abstract 1-sections in Section 5.4; note that $k = 0$ is permissible in the above theorem.

We proceed to the proof. Let R be given, by Lemma 8.3.12 it is sufficient to construct an F such that F and R are w-Δ^* in each other.

We define approximations F_α to F by induction on α. If $F_\alpha(f)$ is undefined for all ordinals α, we complete the definition of F by setting $F(f) = 1$.

We start off by setting F_0 equal to the empty function. Let $F_{<\alpha} = \bigcup_{\gamma<\alpha} F_\gamma$. At stage α we proceed as follows:

If f is $F_{<\alpha}$-computable by a computation of length $\leqslant \alpha$ and $F_{<\alpha}(f)$ is undefined, let

(i) $F_\alpha(f) = 0$ if f is a pair $\langle f_1, f_2\rangle$ and f is (the characteristic function of) a code for a set x such that $\text{rank}(x) \leqslant \alpha$ and $x \in R$.

(ii) $F_\alpha(f) = 1$ otherwise.

We separate the proof in four steps.

8.3.14. R is w-$\Delta^*(F)$.

Let $x \in M_a(F)$. Then x will have a code A in $M_a(F)$, see Lemma 8.3.4. We also have that $\alpha = \text{rank}(x) \in M_a(F)$.

There will be a set $B \in M_a(F)$ which has not been computed before stage α. Let f_1, f_2 be the characteristic functions of B, A respectively, and put $f = \langle f_1, f_2 \rangle$. f will then be in $M_a(F)$ and $x \in R$ iff $F(f) = 0$. For any a such that $x \in M_a$ we then see that

$$\begin{aligned} x \in R \quad &\text{iff} \quad \exists f = \langle f_1, f_2 \rangle \in M_a(F)[f_2 \text{ is a code for } x \text{ and} \\ &\qquad\qquad f \notin \langle M_b^{\text{rank}(x)}(F)\rangle_{b \in I} \text{ and } F(f) = 0], \\ &\text{iff} \quad \forall f = \langle f_1, f_2 \rangle \in M_a(F)[f_2 \text{ is a code for } x \text{ and} \\ &\qquad\qquad f \notin \langle M_b^{\text{rank}(x)}(F)\rangle_{b \in I} \Rightarrow F(f) = 0]. \end{aligned}$$

Thus R is w-$\Delta^*(F)$ and by Lemma 8.3.12 Spec$(I; R) \subseteq$ Spec$(I; F)$.

The reader should note that this is a point where weak definability is needed. An arbitrary $x \in R$ need not be an element of I. We are not able to prove for arbitrary x that there is a code for x in $M_{\{x\}}$. Therefore, we started out with a set $M_a(F)$ where $a \in I$. Then, as we pointed out, every $x \in M_a$ will have a code in M_a, and codes are needed for the construction of F.

8.3.15. The relation

"$\{e_1\}(\sigma, {}^{k+2}E, F) \simeq n$ by a computation of length less than α"

is R-recursive.

The proof is by an analysis of the definition of F using the recursion theorem for set-recursion relative to R. From this and Σ^*-collection over I we further prove:

8.3.16. If $\{e_1\}(\sigma, {}^{k+2}E, F)\downarrow$, then the length of this computation will be in $M_a(R)$, where a codes the input sequence σ.

We are now ready for the final stage of the argument.

8.3.17. F is w-$\Delta^*(R)$.

Let $f \in M_a(R)$, then by the first part of the proof, $f \in M_a(F)$, hence for some α, $F_\alpha(f)$ is defined. By 8.3.16 we can find such an α in $M_a(R)$. We can then write for $f \in M_a(R)$

$$\begin{aligned} F(f) = 0 \quad &\text{iff} \quad \exists \alpha \in M_a(R) \cdot F_\alpha(f) = 0, \\ &\text{iff} \quad \forall \alpha \in M_a(R)(F_\alpha(f) \text{ is defined} \Rightarrow F_\alpha(f) = 0). \end{aligned}$$

To complete the proof we note that if $\alpha \in M_a(R)$, then F_α is R-recursive in α, I, hence by Lemma 8.3.11 F_α will be w-$\Delta^*(R)$.

8.3.18. Remark. If we are in the case $I = \mathrm{Tp}(k)$, $k > 0$, then well-foundedness is set-recursive relative to I. This fact simplifies the proof of Theorem 8.3.13 considerably, simply define F by

$$F(f) = \begin{cases} 0 & \text{if } f \text{ is a code for a set } x \in R, \\ 1 & \text{otherwise.} \end{cases}$$

8.4 Degrees of Functionals

We shall in this section use the theory of set-recursion to give a priority argument relative to ^3E. This is but an introduction to a vast topic which the reader is invited to investigate for him- or herself. In general the setting will be $I = \mathrm{Tp}(k)$ and we study set-recursion relative to some relation R.

8.4.1 Definition. Let $I = \mathrm{Tp}(k)$ and R be given.

$A \leqslant_R B$ iff A is set-recursive in I and some individual $a \in I$ relative to R and B.

This is a very liberal reducibility notion. Harrington in his thesis [53] used *subindividuals* instead of *individuals* (see Sacks [144] for a very informative exposition). We shall comment on this below.

To simplify the discussion we retreat at once to the case $I = \mathrm{Tp}(1)$. In this case the continuum is assumed given, i.e. we are concerned with the structure $\mathrm{Spec}(^3E; I) = \langle M_a \rangle_{a \in I}$, where $I = \mathrm{Tp}(1)$. In order to obtain any results at all we shall have to introduce the following

8.4.2 Assumption. We assume for the rest of this section that $V = L$, i.e. there is a wellordering $<$ of I which is recursive in ^3E and has length $\aleph_1$. We let $\| \ \|$ be the norm associated with $<$.

This assumption is needed if we insist on the liberality of Definition 8.4.1. If we had restricted ourselves to reducibility relative to *sub*individuals there would have been no need of introducing $V = L$. This is tied up with the fact that the set of subindividuals is *strongly* finite, hence we have an associated admissible structure, and it is possible to use the full arsenal of techniques of admissibility theory, such as e.g. the blocking technique (see Chapter 6). On the other hand, Theorem 8.4.6 fails in the presence of strong assumptions of determinacy. We would not be surprised if the same theorem proved independent of ZF; see the similar situation in 6.3.3.

We shall need some technical results derived from the well-ordering $<$ on I.

8.4.3 Definition. For $a \in I$ let a', the *a-jump*, be the $<$-least b such that $b \notin M_a$.

8.4.4 Lemma.

(a) *If* $a < b$, *then* $M_a \subseteq M_b$.
(b) $\|a'\|$ *is the least ordinal not in* M_a.
(c) $M_a \in M_{a'}$.
(d) $M_a <_{\Sigma_1} M_{a'}$, *i.e.* M_a *is a* Σ_1-*substructure of* $M_{a'}$.

For the proof of (a) we note that $\{c \in I : c < b\}$ is countable and thus can be enumerated by an element of I. Using $<$ we may find such an enumeration $\{c_i\}_{i\in\omega}$ in M_b. Then $a = c_i$ for some $i \in \omega$, therefore $a \in M_b$ and thus $M_a \subseteq M_b$.

The proof of (b) follows directly from the definition of the a-jump and the fact that $b \in M_a$ iff $\|b\| \in M_a$.

Part (c) then follows since M_a can be recursively enumerated up to the ordinal $\|a'\|$; thus $M_a \in M_{a'}$.

The last part of the theorem is less trivial, being an application of further reflection, see Theorem 7.1.7. Let c be the characteristic function of a complete $\Sigma_1(M_a)$ subset of ω (= the subindividuals). We see that $c \notin M_a$, so $a' \leqslant c$. On the other hand, $c \in M_{a'}$, since $M_a \in M_{a'}$, therefore $M_{a'} = M_c$. Further reflection is now essentially the statement $M_a <_{\Sigma_1} M_c$, see the Compactness property 7.1.7 and the proof.

We shall need one more technical result. In Section 8.3, see in particular Theorem 8.3.13, we showed how to translate from set-recursion back to Kleene-recursion in higher types. In the proof we had to worry about *weak*-definability, a technical nuisance was the fact that for an arbitrary $x \in M$ there is not necessarily a code for x in $M_{\{x\}}$ (see the remark after 8.3.14). We get around this complication by introducing the set

$$^1M = \{\langle a, x\rangle : x \in M_a\}.$$

^1M is a Σ^*-subset of M and for subsets of ^1M the notions $\mathbf{\Sigma}^*$ and weak-$\mathbf{\Sigma}^*$ coincide. The use of ^1M is unnecessary in the priority argument, but is required for restating the result in terms of Kleene-recursion in higher types. The facts we use are:

(i) If $x \in {}^1M$ then there is an $a \in I$ such that $M_a = M(\{x, I\})$.
(ii) If $x \in {}^1M$, then $M(\{x, I\})$ is locally of type $k + 1$.
(For definitions see in particular 8.3.3.)

We are in the case $I = \text{Tp}(k)$, where $k > 0$. We can now use Remark 8.3.18 to simplify the construction of a functional F_Q corresponding to a given relation Q. In detail: let $Q \subseteq I \times M$ and set

$$A_Q = \{\langle a, f\rangle : f \text{ is the characteristic function of a code for a set } x \text{ and } \langle a, x\rangle \in Q\}.$$

And let

$$F_Q = \text{the characteristic function of } A_Q.$$

By a suitable coding F_Q is a functional of type-2 over the domain I, in our case F_Q is of type 3.

8.4.5 Proposition. *Let* $I = \mathrm{Tp}(k)$, $k > 0$, *and* R *a relation. Let* $\mathrm{Spec}(R; I) = \langle M_a \rangle_{a \in I}$ *and let* $M = \bigcup_{a \in I} M_a$, ${}^1M = \{\langle a, x \rangle : x \in M_a\}$.

Assume that $Q \subseteq M$, *then*

(a) Q, $A_a \cap M$ *and* $F_Q \cap M$ *are* w-$\Delta^*(R)$ *in each other.*

(b) *If* $Q \in w$-$\Delta^*_a(R)$, *where* $a \in I$, *then* F_Q *is weakly Kleene-recursive in* F_R, ${}^{k+2}E$, a.

(c) *If* $Q \subseteq {}^1M$ *and* $Q \in \Sigma^*_a(R)$, *then* A_Q *is Kleene-semirecursive in* F_R, ${}^{k+2}E$, a.

For the notion of weakly Kleene-recursive see Definition 4.1.7.

We observe that part (a) is immediate since each M_b in the spectrum is locally of type $k+1$. To prove part (b) we must find an index e in Kleene-recursion such that

$$F_Q(\lambda b \cdot \{e'\}_K(F_R, {}^{k+2}E, a, b, \sigma)) \simeq \{e\}_K(F_R, {}^{k+2}E, a, \sigma, e').$$

The proof is a bit messy, but the ideas are rather straightforward and by now familiar. Q is w-$\Delta^*_a(R)$. We first observe that Δ_0-formulas can be handled by ${}^{k+2}E$. And the unbounded quantifiers over $M_{a,\sigma}$ needed in the w-Δ^*_a-definition of F_Q from R can be replaced by unbounded quantifiers over $k+1$-$\mathrm{sc}(F_R, {}^{k+2}E, a, \sigma)$. But objects in the $k+1$-section have numerical codes, and we may in a familiar way use Gandy-selection over ω to complete the proof. The proof of (c) is similar taking account of technical facts (i) and (ii) above.

Before stating the main theorem we note that Assumption 8.4.2 implies that there is a wellordering of ${}^1M$ recursive in ${}^3E$. Since knowledge of this wellordering is important for the proof of the main theorem, we shall give a fairly detailed description of it. The spectrum $\langle M_a \rangle_{a \in I}$ has an approximation $\langle M_a^\alpha \rangle_{a \in I}$, where

$$M_a^\alpha = \{\{e\}(a, I) : \|e, a, I\| \leqslant \alpha\}.$$

Let ${}^1M^\alpha = \{\langle a, x \rangle : x \in M_a^\alpha\}$.

We note that in a standard way we can introduce a wellordering $<^*$ on ${}^1M^\alpha - \bigcup_{\beta < \alpha} {}^1M^\beta$ of ordertype $\aleph_1$; this wellordering is induced from the given wellordering of I. Let $\alpha(x) =$ least ordinal α such that $x \in {}^1M^\alpha$. We then set

$x <^1 y$ iff $\alpha(x) < \alpha(y)$ or $\alpha(x) = \alpha(y) = \alpha$ and x is less than y in the wellordering $<^*$.

We use $\| \;\|^1$ for the associated norm. Note that $\| \;\|^1$ will be a set-recursive function on ${}^1M$ with a set-recursive inverse.

We can now state the theorem.

8.4.6 Theorem. ($V = L$). *There is a* Σ^**definable subset* Q *of* ${}^1M$ *such that* $M(Q; I) = M$, *but* Q *is not* $\mathbf{\Delta}^*$*-definable over* M.

8.4.7 Corollary. ($V = L$). *There is a set* $A \subseteq \mathrm{Tp}(2)$ *semirecursive in* ${}^3E$ *such that*

(i) *A is not recursive in ${}^3E$ and a function.*

(ii) *If $B \subseteq I$ is recursive in A, ${}^3E$ and a function, then B is recursive in ${}^3E$ and a function.*

(iii) *No complete semirecursive subset of I is recursive in A, ${}^3E$ and a function.*

The corollary follows immediately from the theorem using the appropriate parts of Proposition 8.4.5:

We have constructed a set $Q \subseteq {}^1M$, let $A = A_Q$. By (c) in 8.4.5, A is semirecursive in ${}^3E$, and since Q is not Δ^* in parameters from I, A is not recursive in ${}^3E$ and a function. Part (ii) of the corollary follows from the equality $M(Q; I) = M$. The same equality also gives (iii), but this needs a small supplementary argument. Let C be a complete semirecursive subset of I, and assume that C is recursive in A, ${}^3E$ and a function. Then there will be an $a \in I$ such that $C \in M_a(Q; I)$. But since C is complete, $C \notin M$, i.e. we have the sought for violation of equality $M(Q; I) = M$, which proves (iii).

We turn now to a proof of the theorem.

Let $W_{e,a}$ be an enumeration of the Σ_a^*-subsets of M and let $W_{e,a}^\sigma$ be recursive approximations, $\sigma \in \mathrm{On} \cap M$. We shall have to worry about two kinds of conditions:

$1 \cdot e \cdot a$	$M - Q \neq W_{e,a}$
$2 \cdot e \cdot a$	Protect the computation $\{e\}_Q(a, I)$.

The first set of conditions will secure that Q is not Δ^*-definable, the second will give us the necessary control over $\mathrm{Spec}(Q; I)$.

Each condition will be coded as a pair $\langle a, n\rangle = \langle a, \langle e, i\rangle\rangle$, $i = 1, 2$, and we order the pairs by the lexicographical ordering on $I \times \omega$. The order type will be $\aleph_1$. We shall let ν denote both a condition and its place in the ordering. This ordering determines our priorities.

We shall also use the following terminology. Let $\nu < \aleph_1$, we say that y is in row ν, or $y \in \mathrm{row}\ \nu$, if $y \in {}^1M$ and for some ordinal β, $\|y\|^1 = \aleph_1 \cdot \beta + \nu$.

We shall also find the following terminology useful:

a-conditions:	$\nu = \langle a, n\rangle$
1-conditions:	$\nu = 1 \cdot e \cdot a$
2-conditions:	$\nu = 2 \cdot e \cdot a.$

Recall our dual use of ν. Notice that if $a < b$ then all the a-conditions have higher priority than the b-conditions.

Q^σ will be constructed by an induction on the stage $\sigma \in \mathrm{On} \cap M$ such that Q^σ will be uniformly set-recursive in σ. At each state σ we shall worry about one "relevant" 1-condition or all "relevant" 2-conditions.

The 2-conditions will be met by creating requirements freezing certain computations relative to Q. A requirement x for ν is *active* at stage σ if $x \cap Q^\sigma = \varnothing$.

To meet the 1-conditions $\nu = 1 \cdot e \cdot a$ we will see to it that $M - Q$ and $W_{e,a}$ differ on row ν. This will be done by putting an element from row $\nu \cap W_{e,a}$ into Q if possible. In doing this we must be careful not to destroy other parts of the construction. This leads to the following stipulation:

8.4.8. y is a *candidate* for ν at stage σ if

(i) $y \in {}^1M^\sigma$

(ii) $y \in \text{row}\ \nu$

(iii) If x is an active requirement for $\nu_1 < \nu$ at stage σ, then $y \notin x$

(iv) $\sigma \in M_y^{\sigma+1}$

(v) Let $\sigma_0 \leqslant \sigma$ be minimal such that $y \in {}^1M^{\sigma_0}$, then for all σ_1 such that $\sigma_0 \leqslant \sigma_1 \leqslant \sigma$ we require that $y \in M_{\sigma_1}^\sigma$.

The first four conditions on y are quite reasonable: since we want $Q \subseteq {}^1M$ we must insist on (i); (iii) expresses that we do not want to injure active requirements of higher priority, and (iv) expresses our intention to make Q Σ^*, therefore we will not put y into $Q^{\sigma+1}$ unless σ is in M_y. Only (v) needs a comment, it is included purely for technical reasons in order to preserve computations $\{e\}_Q(b, I)$ inside M_b, or $M_{b'}$ at worst.

8.4.9. A condition ν *requires attention* at stage σ if

(i) $\nu = 1 \cdot e \cdot a$, $Q^\sigma \cap \text{row}\ \nu = \varnothing$, and there is a candidate for ν at stage σ in $W_{e,a}^\sigma$, or

(ii) $\nu = 2 \cdot e \cdot a$, there is no active requirement for ν at stage σ, and $\{e\}_{Q^\sigma}(a, I)\downarrow$ by a computation of length $\leqslant \sigma$ which uses elements from M^σ only in the subcomputations.

We can now proceed to the construction of Q. As usual, set $Q^0 = \varnothing$ and $Q^\lambda = \bigcup_{\sigma<\lambda} Q^\sigma$ for λ a limit stage.

8.4.10. Assume that Q^σ is constructed.

If no condition requires attention at stage σ let $Q^{\sigma+1} = Q^\sigma$. Otherwise, let ν be the least condition that requires attention at stage σ. There are two cases

(i) $\nu = 1 \cdot e \cdot a$. Let y be the $<^1$-least candidate for ν in $W_{e,a}^\sigma$ and set $Q^{\sigma+1} = Q^\sigma \cup \{y\}$.

(ii) $\nu = 2 \cdot e \cdot a$. Let $Q^{\sigma+1} = Q^\sigma$ and create the following requirements: For each $\nu_1 = 2 \cdot e_1 \cdot a_1$, if ν_1 requires attention at stage σ, let $M^\sigma - Q^\sigma$ be a requirement for ν_1.

Finally, set $Q = \bigcup_{\sigma \in M} Q^\sigma$.

If x is a requirement and we put a $y \in x$ into Q, then we say that x is *injured*. A rather trivial cardinality argument leads to

8.4.11. If ν is a condition, there will be created at most countably many requirements for conditions $\leqslant \nu$.

A requirement for ν is injured at most countably many times.

This simple observation gives one half of the theorem.

8.4.12. $M - Q \neq W_{e,a}$ for all e, a.

Let $\nu = 1 \cdot e \cdot a$ and let σ be a stage at which all requirements for $\nu_1 < \nu$ ever to be constructed are constructed and all 1-conditions $\nu_1 < \nu$ to be met are met. σ exists by 8.4.11 and we may assume that $\sigma = \kappa_0^b$ for some $b \geqslant a$. (Recall that $\kappa_0^b = \sup(\mathrm{On} \cap M_b)$.)

Assume that $M - Q$ and $W_{e,a}$ coincide on row ν. Let $y \in \text{row } \nu$ be the element with norm $\langle \sigma, \nu \rangle$ in $\| \ \|^1$. If $y \in Q$ then y must also be an element of $W_{e,a}$, otherwise it would never be put into Q. But this goes against our assumption, therefore $y \in M - Q$ and hence also $y \in W_{e,a}$. Since $W_{e,a}$ is Σ^*, there will be a $\sigma_1 \in M_{a,y}$ such that $y \in W_{e,a}^{\sigma_1}$. We show that we can choose $\sigma_1 \geqslant \sigma$ and $\sigma_1 \in M_y$ by proving the

Claim. $a \in M_y$ and $\sigma \in M_y$.

Let $y \in {}^1M$ be of the form $\langle b', y_1 \rangle$. Then σ is minimal such that $y_1 \in M_{b'}^{\sigma}$ so $\sigma \in M_{b'} = M_y$. Clearly $a \in M_\nu$ and since $y \in \text{row } \nu$, we may compute ν from σ and y.

So assume that $\sigma_1 \geqslant \sigma$, $\sigma_1 \in M_y$, and $y \in W^{\sigma_1}$. We shall argue that at stage σ_1 y is a candidate for ν. The troublesome part is (v) of 8.4.8. But by the choice of σ and y we first see that σ is minimal such that $y \in {}^1M^\sigma$. It remains to verify that if $\sigma' \geqslant \sigma$ then $y, \sigma \in M_{\sigma'}^{\sigma}$, but this follows immediately from the fact that σ is chosen to be of the form κ_0^b for suitable b. Thus at stage σ_1 y is a candidate for ν, and ν thus requires attention. By choice of σ we do not pay attention to any $\nu_1 < \nu$, so at stage σ_1 something from row ν is put into Q^{σ_1+1}. But then, after all, $M - Q$ and $W_{e,a}$ must differ.

The other half of the theorem requires more detailed information about the construction. (And it is in the proof of the following lemma that the usefulness of (v) in 8.4.8 will become clear.)

8.4.13. (a) Let x be a requirement for ν created at stage $\sigma \in M_a$, where $x, \nu \in M_a$. If x is injured, then x is injured before stage κ_0^a.

(b) If $a = b'$ and $\nu = 2 \cdot e \cdot a_0$ for some $a_0 \leqslant a$, then there is a stage $\sigma \in M_a$ such that if $\sigma \leqslant \sigma_1 < \kappa_0^a$ no 1-condition $\nu_1 < \nu$ is met at stage σ_1.

For the proof of (a) assume that x is injured by putting a y from row ν_1 into Q at stage σ_1. Since $x = M^\sigma - Q^\sigma$, y will be an element of ${}^1M^\sigma$, so, in particular,

$$y \in {}^1M^{\kappa_0^a}.$$

If $\sigma_1 \geqslant \kappa_0^{a'}$, then $y \notin M_{a'}$ by clause (iv) in 8.4.8. By clause (v) of the same definition $y \in M_{\kappa_0^a}^{\sigma_1} \subseteq M_{a'}$. This contradiction shows that $\sigma_1 < \kappa_0^{a'}$. Then

$$M_{a'} \models \exists\sigma_1 \exists y \in \text{row}\ \nu_1 \cap Q^{\sigma_1+1}.$$

By reflection ((d) of 8.4.4) we get

$$M_a \models \exists\sigma_1 \exists y \in \text{row}\ \nu_1 \cap Q^{\sigma_1+1}.$$

This shows that the injury took place before stage κ_0^a, which proves part (a) of the lemma.

To prove (b) let $\nu_1 < \nu$ be a 1-condition. If ν_1 is a c-condition for some $c < a$ and if we meet ν_1 inside M_a, we also meet ν_1 inside M_b since $M_b <_{\Sigma_1} M_a$ and $\nu_1 \in M_b$.

Thus there could be at most finitely many 1-conditions $\nu_1 < \nu$ which we meet between κ_0^b and κ_0^a. This could happen only when $a_0 = a$, i.e. $\nu = \langle a, n\rangle$, and ν_1 is of the form $\nu_1 = \langle a, m\rangle$, where $m < n$. If this were the case, choose $\sigma > \kappa_0^b$ such that all these conditions are met before stage σ.

8.4.14. Let $x \in M$, we say that

$$x \in M_a(Q),$$

is *finally protected* at stage σ if for some $e \in \omega$ the computation $\{e\}_Q(a, I) = x$ is protected by a requirement active at stage σ which is never injured.

8.4.15. Assume that $\{e\}_Q(x_1, \ldots, x_n)\downarrow$. Let $a, c \in I$ and δ be an ordinal in $M_{\langle a,c\rangle}$. Assume that the statements

$$x_1, \ldots, x_n \in M_a(Q),$$

are finally protected at stage δ. Then there is a $\sigma > \delta$ in $M_{\langle a,c\rangle}$ such that

$$\exists x \in M_a^\sigma(Q^\sigma; I)\cdot\{e\}_{Q^\sigma}(x_1, \ldots, x_n) = x.$$

In applications of this lemma $x_1, \ldots, x_n$ will come from the set $I \cup \{I\}$, in which case the assumption is trivially true. The assumption seems, however, necessary in order to make the inductive proof work.

The proof is by induction on the length of the computation $\{e\}_Q(x_1, \ldots, x_n)$. We give case iv and case v of 8.1.1, the other cases are either similar or simpler.

Case iv. Here

$$\{e\}_Q(x_1, \ldots, x_n) \simeq \bigcup_{y \in x_1} \{e_1\}_Q(y, x_2, \ldots, x_n),$$

where $x_1 \in M_a(Q), \ldots, x_n \in M_a(Q)$ all are finally protected at stage δ.

First note that when x_1 is computed from a, I, there will be a function f mapping I onto x_1, uniformly recursive in the computation of x_1. For each $y = f(b) \in x_1$, $y \in M_{a,b}(Q)$ will be finally protected at stage δ.

From the induction hypothesis and Σ^*-collection we can now claim

$$\forall c \forall_\gamma \in M_{a,c} \exists \sigma \in M_{a,c} \forall b \exists \sigma_b [\gamma < \sigma_b < \sigma \wedge \exists x_b \in M^{\sigma_b}_{a,b,c} \cdot \{e_1\}_{Q^{\sigma_b}}(f(b), x_2, \ldots, x_n) \simeq x_b].$$

We may now define a strictly increasing sequence of ordinals $\langle \delta_b \rangle_{b \in I} \in M_{a,c}$ such that if $\|b_2\| = \|b_1\| + 1$, then

$$\forall b \in I \exists \sigma_b [\delta_{b_1} < \sigma_b \leqslant \delta_{b_2} \wedge \exists x_b \in M^{\sigma_b}_{a,b,b_1} \cdot \{e_1\}_{Q^{\sigma_b}}(f(b), x_2, \ldots, x_n) \simeq x_b].$$

Let $\sigma = \sup\{\delta_b : b \in I\}$. Since the cofinality of σ is $\aleph_1$, we may apply the cardinality argument used in 8.4.11 to the construction up to stage σ. The δ_b's are chosen such that for each $b_i \in I$ we will cofinally often below σ try to protect the computation $\{e_1\}_Q(f(b_i), x_2, \ldots, x_n)$. Thus at stage σ they will all be protected by active requirements. But then

$$\exists x \in M^{\sigma}_{a,c}(Q^\sigma) \cdot x = \bigcup_{b \in I} \{e_1\}_{Q^\sigma}(f(b), x_2, \ldots, x_n),$$

which completes case iv. (At this point the reader will see the usefulness of having $\leqslant$ rather than $<$ in the definition of the approximations M^{α}_{a}, x is a computation of length σ.)

Case v. We have composition

$$\{e\}_Q(x_1, \ldots, x_n) = \{e_0\}_Q(\{e_1\}_Q(x_1, \ldots, x_n), \ldots, \{e_m\}_Q(x_1, \ldots, x_n), x_1, \ldots, x_n).$$

To be able to use the induction hypothesis we must find a stage where all statements

$$\{e_i\}_Q(x_1, \ldots, x_n) \in M_a(Q),$$

are finally protected. Since reflection is available we have some freedom to manoeuver.

Let $\delta_0 = \kappa^{a,c}_0$. By the induction hypothesis there will be stages $\delta_1, \ldots, \delta_m$ in $M_{\langle a,c \rangle'}$ such that for $1 \leqslant i \leqslant m$

$$\exists y_i \in M^{\delta_i}_{a}(Q^{\delta_i}) \cdot \{e_1\}_{Q^{\delta_i}}(x_1, \ldots, x_n) \simeq y_i.$$

The associated conditions will be a-conditions. By 8.4.13 these conditions will be met by requirements which are never injured at a stage $\delta_{m+1} \in M_{\langle a,c \rangle'}$. Thus at stage δ_{m+1} all statements

$$y_i \in M_a(Q),$$

are finally protected, where $y_i = \{e_i\}_Q(x_1, \ldots, x_n)$.

By the induction hypothesis, once more, there is a $\delta_{m+2} \geqslant \delta_{m+1}$ in $M_{\langle a,c\rangle'}$ such that

$$\exists x \in M_a^{\delta_{m+2}}(Q^{\delta_{m+2}}) \cdot \{e_0\}_{Q^{\delta_{m+2}}}) y_1, \ldots, y_m, x_1, \ldots, x_n) \simeq x.$$

And such that

$$\exists x \in M_a^{\delta_{m+2}}(Q^{\delta_{m+2}}) \cdot \{e\}_{Q^{\delta_{m+2}}}(x_1, \ldots, x_n) \simeq x.$$

Since $M_{\langle a,c\rangle} <_{\Sigma_1} M_{\langle a,c\rangle'}$ we find an σ in $M_{\langle a,c\rangle}$ with the same property as δ_{m+2} above.

This completes case v.

The following lemma will now complete the proof of the theorem.

8.4.16. If $a = b'$, then $M_a(Q; I) = M_a$.

If $x \in M_a(Q; I)$ there is an index e such that $x = \{e\}_Q(a, I)$. I and a are elements of M_a protected at all stages. Let $\nu = 2 \cdot e \cdot a$.

By Lemma 8.4.13 (b) there is a $\sigma_0 \in M_a$ such that between σ_0 and κ_0^a we do not meet any 1-condition $\nu_1 < \nu$. By Lemma 8.4.15 there is a $\sigma_1 > \sigma_0$ in M_a such that

$$\exists x \in M_a^{\sigma_1} \cdot \{e\}_{Q^{\sigma_1}}(a, I) \simeq x.$$

We would then at stage σ_1 create a requirement to protect this computation, and by Lemma 8.4.13 (a) this requirement is never injured.

Thus

$$\{e\}_Q(a, I) = \{e\}_{Q^{\sigma_1}}(a, I) \in M_a,$$

and the lemma is proved.

8.4.17 Remark. As noted in the introduction to this chapter we have concentrated on a fairly simple priority argument to demonstrate the "naturalness" of set-recursion as a computation theory for the study of degrees of functionals.

Theorem 8.4.6 is due to Dag Normann and is not published in this form elsewhere. More advanced results, such as e.g. the splitting theorem, can be found in Normann [125, 127].

Reducibility relative to subindividuals has been studied by Harrington [53], see the exposition in Sacks [144].

There are many open problems, e.g. we would like to have a set-recursive version of the *density theorem*, see 6.3.1.

8.4.18 Remark. G. E. Sacks has in a forthcoming paper *Post's problem, absoluteness and recursion in higher types* (**Kleene symposium**, North-Holland, to appear) shown that in the case $I = \mathrm{Tp}(1)$ the set $M = \bigcup_{a \in I} M_a$ is *not* Σ_b^*-definable for any $b \in I$. It is, however, weakly-Σ^* definable. Thus there was a need for introducing the notion of weak-Σ^* definability and the set ${}^1M$ in our arguments above.

8.5 Epilogue

It is time to take our farewell of the reader. And let us do so by casting a quick glance back on the territory covered, pointing out some open problems, some omissions, and some areas of future research.

8.5.1. *Computing in an algebraic context.* Except for the introductory Pons Asinorum we have concentrated on traditional "hard core" recursion theory. But recursion-theoretic ideas could have wider application, beyond the usual tie-up with definability theory and descriptive set theory. Algebra is one possibility, we gave an example and several references in 0.3.4. There is, too, an interesting study by W. Hodges [63] who uses the Jensen-Karp theory of primitive recursive set functions to study the effectivity of some field constructions.

8.5.2. Part A of the book gave a reasonably thorough *axiomatic analysis of the notion of computation.* Whilst we used many combinatorial tricks first developed in the context of the λ-calculus, we did not look at recursion theory from the point of view of the λ-calculus, so let us make good this omission giving a reference to G. Mitschke, *λ-Kalkül, δ-Konversion und Axiomatische Rekursionstheorie* [104], in which the relationship is explicitly studied; see also Barendregt [9].

We also remarked in Section 2.7 on the difficulties in studying computations relative to a *partial* higher type objects. We gave some examples, but did not have an abstract axiomatic analysis to offer. This deserves further study. The reader should consult the exposition of Platek's thesis in Moldestad [105], and also a recent study by S. C. Kleene [86]. See also part II of [86] and a paper by D. P. Kierstead, *A semantics for Kleene's j-expressions*, to appear in the KLEENE SYMPOSIUM, North-Holland.

8.5.3. In Chapter 3 we gave an outline of the connection between Spector theories and the theory of inductive definability. Spector theories is our general version of hyperarithmetic theory, i.e. of the *effective* theory of Borel sets, and has many applications to *descriptive set theory*, see Moschovakis [118]. Applications beyond the first levels of the projective hierarchy require extra set-theoretic assumptions (e.g. the axiom of constructibility, the existence of measurable cardinals, or the axiom of determinacy). We find these "applications" a bit problematic from a philosophic point of view, and are more fascinated by the recent "sharper" applications of the "absolute" theory, see e.g. Louveau [96, 97]. And see also Fenstad-Normann [32] and references therein, where the interest is how far one can go with certain problems of descriptive set theory within the accepted set-theoretic foundation.

In Chapter 4 we gave a brief introduction to second-order definability theory, and referred to Kechris [76] for a fuller exposition. Here one would like to see applications to "real-life" mathematics.

Our treatment of general hyperarithmetic theory has been both highly selective and rather brief. We can make good one omission by drawing the reader's attention to the "omitting types" paper by Grilliot [51].

8.5.4. In Part C on admissible prewellorderings and degree structure we were on several occasions led to *the borderline between recursion theory and set theory.* We shall not repeat this discussion here. Interesting problems remain in determining the "true" domain for degree theory. The reader is referred back to Section 6.3 and the examples, problems, and references there given.

8.5.5. Our discussion of recursion in higher types in Chapters 4, 7, and 8 was again selective and rather abstract, the reader should consult the book by Hinman [61] for many interesting examples and further developments.

One example we must mention is the *superjump* introduced by Gandy [38]. The study of this object has played an important role in shaping the general theory. The superjump S has a simple definition

$$S(F, \alpha, e) \simeq \begin{cases} 0 & \text{if } \{e\}(\alpha, F)\downarrow \\ 1 & \text{otherwise.} \end{cases}$$

The basic result is that 1-sc(S) is exactly $L_{\rho_0} \cap 2^{\omega}$, where ρ_0 is the first recursively Mahlo ordinal. There is an extensive literature, Gandy [38], Aczel-Hinman [8], Harrington [53, 54], Normann [126], and Lavori [95].

Another important concrete example is the type three object ${}^3CL$ which is essentially *the diagonalization operator for arbitrary inductive definitions on* ω. This was studied by Harrington (unpublished, but see a brief reference in Kechris [76]). A main result is that 1-sc$({}^2E, {}^3CL) = L_{\sigma(\pi_0)} \cap 2^{\omega}$, where π_0 is the least non-projectible ordinal and $\sigma(\pi_0)$ is the least ordinal stable in π_0. There is also an interesting connection to the Kolmogorov R-operator.

To continue our list of omissions: We have said almost nothing about *hierarchies.* Two basic references on the positive aspects of hierarchies in higher recursion theory are Shoenfield [149], and Wainer [170], where "good" hierarchies (in the sense of a genuine building up from below) are given for recursion in an arbitrary total type-2 object; the normal case is due to Shoenfield, the general case to Wainer.

Moving up in types the situation is more problematic. One can always "after the fact" extract a hierarchy, since we have ordinals associated with computations. But a genuine building up from below in the presence of ${}^3E$ does not exist, see Schwichtenberg-Wainer [146] for a good discussion. Notice that for certain type-3 objects, e.g. S and ${}^3CL$ mentioned above, in which ${}^3E$ is not recursive, we do have interesting hierarchies.

Let us conclude this list of omissions by drawing attention to the existence of *gap phenomena* in computations in higher types, see Moldestad [105] for an introduction. This is a fascinating area which invites further study. One should also not neglect the further study of *reflection phenomena* in higher types, see the brief introduction in Section 7.1 and the exposition in Kechris [74]. *Basis results* are discussed in Moldestad-Normann [107].

8.5.6. One growing and important area of general recursion theory has been entirely absent from our discussion, *the theory of countable or continuous functionals.*

By way of an introduction let us discuss the following example of Grilliot [50]. The setting is the total type-2 functionals over ω.

Theorem. *^2E is recursive in F iff F is effectively discontinuous.*

We indicate the proof of one half of the theorem, viz. that ^2E is recursive in F if F is effectively discontinuous. So suppose that $\langle g_i \rangle_{i\in\omega}$ and f are recursive in F, that $f = \lim_i g_i$, but $F(f) \neq \lim_i F(g_i)$.

By thinning out the sequence $\langle g_i \rangle_{i\in\omega}$ we may assume that $F(f) \neq F(g_i)$, for all $i \in \omega$ and, further, that $g_i(j) = f(j)$ for $j \leqslant i$.

Introduce an operator J by

$$J(h) = \begin{cases} f & \text{if } \forall x(h(x) \neq 0) \\ g_i & \text{if } h(i) = 0 \wedge \forall x < i(h(x) \neq 0). \end{cases}$$

J is recursive in F as can be seen from the following equation

$$J(h)(j) = \begin{cases} g_j(j) & \text{if } \forall x \leqslant j(h(x) \neq 0) \\ g_i(j) & \text{if } i \leqslant j \wedge h(i) = 0 \wedge \forall x < i(h(x) \neq 0). \end{cases}$$

But then $^2E(h) = 1$ iff $F(J(h)) = F(f)$, i.e. ^2E is recursive in F.

A corollary of this result is that ^2E is recursive in F and a function iff F is discontinuous.

The topology is here the usual product topology on Baire-space ω^ω. This topology is determined by "finite information", i.e. by a neighborhood basis consisting of sets $N_u = \{\alpha \in \omega^\omega : u \subseteq \alpha\}$, where u is a sequence number and $u \subseteq \alpha$ means that $\alpha(i) = u_i$, $i < \mathrm{lh}(u)$.

We have the following well-known observation

Proposition. *F is a continuous map from Baire-space into ω iff there is a function α_F, called an* associate *of F, which satisfies*

(i) $\forall\alpha\exists n\alpha_F(\bar{\beta}(n)) > 0$,
(ii) $\forall\beta\forall n(\alpha_F(\bar{\beta}(n)) > 0 \Rightarrow \alpha_F(\bar{\beta}(n)) = F(\beta) + 1)$.

We see how ^2E divides the higher recursion theory: either ^2E is recursive in F and then F is discontinuous and we are in the case of normal recursion in higher types, i.e. finite theories in the sense of Chapters 3 and 4 or ^2E is *not* recursive in F which leads to the non-normal case. And a major part of the non-normal case is concerned with the *countable* or *continuous* functionals, where computations are determined by finite information, i.e. by an associate α_F in the sense of the above proposition.

We have mostly been concerned with the full type structure in Part D. But in the countable/continuous case it makes good sense to go to a thinner hierarchy. This is the appropriate notion. A *type structure* is a collection of sets $\{A_\tau : \tau$ a type symbol} such that (i) $A_0 = \omega$, and (ii) if τ is of the form $\tau_1 \times \ldots \times \tau_n \to 0$, then A_τ is a set of maps (but not necessarily all maps) from $A_{\tau_1} \times \ldots \times A_{\tau_n}$ to ω.

One important example of a type hierarchy different from the full type structure is the hierarchy

$$\mathbf{C} = \{C_\tau : \tau \text{ is a type symbol}\},$$

of countable objects where $C_0 = \omega$, $C_1 = \mathrm{Tp}(1)$, but where, from C_2 on, we restrict ourselves by roughly requiring that a map $F: C_n \to \omega$ should be allowed in C_{n+1} only if its value at a g in C_n is determined by a finite amount of information about g. In the case of C_2 we can take this to mean that F should have an associate α_F, or, equivalently, that F is continuous.

At higher types the connection with topology is more problematic. Lifting the idea of having an associate presents, however, no difficulties. We let a map $\Phi: C_2 \to \omega$ belong to C_3 iff it has an associate α_Φ satisfying

(i) $\forall\beta\forall F \in C_2[\beta$ is an associate for $F \Rightarrow \exists n\alpha_\Phi(\bar{\beta}(n)) > 0]$,

(ii) $\forall\beta\forall n\forall F \in C_2[\beta$ is an associate for $F \wedge \alpha_\Phi(\bar{\beta}(n)) > 0 \Rightarrow \alpha_\Phi(\bar{\beta}(n)) = \Phi(F) + 1]$.

And so we continue.

Associated with the hierarchy $\mathbf{C}$ we have two natural notions of recursiveness. One can be quickly explained:

1. A countable functional F is *recursively countable* if it has a recursive associate α_F.

The other notion makes sense for arbitrary type structures $\{A_\tau : \tau$ a type symbol$\}$, provided the structure is sufficiently closed. It is simply the notion of computation obtained by relativizing the standard Kleene schemes S1–S9 (see Kleene [83] or Chapter 4) from the full hierarchy to the thin hierarchy $\{A_\tau\}$. It is easy to convince oneself that $\mathbf{C}$ is sufficiently closed, hence we have a notion of Kleene-computation relative to $\mathbf{C}$.

2. A countable functional F is *Kleene computable* if it has an index, i.e. there is an $e \in \omega$ such that $F(\sigma) = \{e\}(\sigma)$, where the right-hand side is determined by the relativized schemes S1–S9 over $\mathbf{C}$.

Both notions are natural. But they do not coincide. It is easy to show that Kleene computable functionals have recursive associates. The converse is false.

Let $[\gamma] = \{\beta \in C_1 : \forall x \in \omega(\beta(x) \leqslant \gamma(x))\}$. We see that $[\gamma]$ for any $\gamma \in C_1$ is a compact subset of C_1. The *fan functional* $\Phi(F, \gamma)$ computes a uniform *modulus of continuity* for F on $[\gamma]$, i.e.

$$\Phi(F, \gamma) = (\mu n)(\forall\beta, \beta' \in [\gamma])[\bar{\beta}(n) = \bar{\beta}'(n) \to F(\beta) = F(\beta')].$$

Φ is recursively countable (by König's lemma), but the fan functional is not Kleene computable. We indicate the proof.

Let $\gamma \equiv 1$ and $F \equiv 0$, then $\Phi(F, \gamma) = 0$. Assume that Φ is Kleene computable, i.e. there is an index e such that

$$(*) \qquad \{e\}(G, \gamma) = \Phi(G, \gamma)$$

for all G. Select a non-recursive δ in $[\gamma]$. Choose a "restricted associate" α for F, i.e. an α such that: (a) For all recursive β, $\exists n \alpha(\bar{\beta}(n)) > 0$, but (b) $\forall n \alpha(\bar{\delta}(n)) = 0$.

One can now prove that there exist sequence numbers $\sigma_1, \ldots, \sigma_k$ such that

(i) $\alpha(\sigma_i) > 0, \quad i = 1, \ldots, k,$
(ii) if $G = F$ on $N_{\sigma_1} \cup \ldots \cup N_{\sigma_k}$, then $\{e\}(G, \gamma) = \{e\}(F, \gamma)$.

This one proves by reflecting on the meaning of $\{e\}(F, \gamma)\downarrow$, taking into account that α is a restricted associate of F.

We can now choose an n such that $N_{\bar{\delta}(n)} \cap N_{\sigma_i} = \varnothing$ for $i = 1, \ldots, k$. Define a functional G by

$$G(f) = \begin{cases} 1 & \text{if } f \in N_{\bar{\delta}(n)} \\ 0 & \text{ow.} \end{cases}$$

It is immediate from the definition of the fan functional Φ that $\Phi(G, \gamma) = n$. But from (ii) above it equally follows that $\{e\}(G, \gamma) = \{e\}(F, \gamma) = \Phi(F, \gamma) = 0$. This contradicts (*) and shows that Φ is *not* Kleene computable.

A good epilogue is always an introduction to something beyond. And an introduction is an invitation not a complete story. In particular, in the case of countable functionals we urge the reader to go beyond these introductory remarks and here are a few basic references on this topic.

The study of countable functionals was opened up by the papers of Kleene [82] and Kreisel [88] in 1959 which introduced the countable hierarchy, the notion of an associate and the two ways of approaching the notion of recursiveness in **C**. The non-computability of the fan functional is due to Tait (unpublished), for an exposition see Gandy-Hyland [42] which is an excellent introduction to the field. Hinman [60] taking a lead from the theorem of Grilliot quoted above made a first contribution to the degree theory of continuous functionals.

The theory was further advanced by several contributions of Yu. L. Ershov [22, 23]. The "obvious" axiomatization problem, i.e. how to extend S1–S9 to a set of schemes giving all recursively countable functionals has been discussed by Feferman [25], see also Hyland [66]. Further results here may have an interesting feed-back on the general axiomatics of the notion of *computation*.

Of many more recent contributions we mention the theses of Bergstra [15] and Hyland [64], and the further contributions of Gandy-Hyland [42], Hyland [67], Normann [123, 128] and Normann-Wainer [131]. A systematic introduction to part of the *pure theory* can be found in the Lecture Notes of Normann [129]. There is also an *applied part* of the theory, viz. the application of countable functionals to constructivity and proof theory, see the original paper of G. Kreisel [88]. A survey is given in Troelstra [167]; a recent contribution is Hyland [65].

References

The list of references is not an attempt at a complete bibliography of general recursion theory but contains only items referred to in the main body of the text. For an extensive bibliography of recursion theory up to the mid 1960's see Rogers [136]. An "uncritical bibliography" of generalized recursion theory up to the early 1970's can be found in Fenstad-Hinman [31]. There is also useful bibliographic information in the **Handbook** [12], in particular, Shore [152] for α-recursion theory. For degree theory in ORT see the bibliography in Soare [157]. The books Barwise [11] and Hinman [61] contain much supplementary bibliographic information on admissible sets and structures and on recursion theoretic hierarchies. Moschovakis [118] has an extensive bibliography of descriptive set theory. One may also consult various proceedings' volumes such as Gandy-Yates [44] and Fenstad-Gandy-Sacks [30] for additional references.

Aanderaa, S.

1. Inductive definitions and their closure ordinals. In: Fenstad-Hinman [31], 207–220

Aczel, P.

2. An axiomatic approach to recursive function theory on the integers. (Circulated notes, 1969)
3. Representability in some systems of second-order arithmetic. Israel J. Math. *8*, 309–328 (1970)
4. Stage comparison theorems and game playing with inductive definitions. (Circulated notes, 1972)
5. An axiomatic approach to recursion on admissible ordinals and the Kreisel-Sacks construction of meta-recursion theory. Recursive Function Theory Newsletter, 1974
6. Quantifiers, games, and inductive definitions. In: Kanger [73], 1–14
7. An introduction to inductive definitions. In: Barwise [12], 739–782

Aczel, P., Hinman, P. G.

8. Recursion in the superjump. In: Fenstad-Hinman [31], 3–41

Barendregt, H.

9. Normed uniformly reflexive structures. In: Böhm [16], 272–286

Barwise, K. J.

10. Admissible sets over models of set theory. In: Fenstad-Hinman [31], 97–122
11. **Admissible Sets and Structures, Perspectives in Mathematical Logic.** Berlin, Heidelberg, New York: Springer, 1975, 394 pp.
12. **Handbook of Mathematical Logic.** Amsterdam: North-Holland, 1977, 1165 pp.
13. Monotone quantifiers and admissible sets. In: Fenstad-Gandy-Sacks [30], 1–38

Barwise, K. J., Gandy, R., Moschovakis, Y. N.

14. The next admissible set. J. symbolic Logic *36*, 108–120 (1971)

Bergstra, J.
15. Computability and continuity in finite types. Utrecht: Ph.D. thesis, 1976

Böhm, C.
16. **λ-Calculus and Computer Science Theory** (Proceedings of the Symposium held in Rome, 1975, editor). Berlin, Heidelberg, New York: Springer, 1975, 370 pp.

Butts, R. E., Hintikka, J.
17. **Logic, Foundations of Mathematics and Computability Theory** (Proceedings of Fifth International Congress of Logic, Methodology and Philosophy of Science, London, Ontario: 1975, editors). Dordrecht: D. Reidel Publishing Company 1977, 406 pp.

Cenzer, D.
18. Ordinal recursion and inductive definitions. In: Fenstad-Hinman [31], 221–264

Devlin, K. J.
19. **Aspects of Constructibility.** Berlin, Heidelberg, New York: Springer 1973, 240 pp.

Diller, J., Müller, G. H.
20. **Proof Theory Symposium, Kiel 1974** (Proceedings of the International Summer Institute and Logic Colloquium, editors). Berlin, Heidelberg, New York: Springer 1975, 383 pp.

Driscoll, G. C., Jr.
21. Metarecursively enumerable sets and their metadegrees. J. symbolic Logic *33*, 389–411 (1968)

Ershov, Y. L.
22. Maximal and everywhere defined functionals. Algebra and Logic *13*, 210–225 (1974)
23. Model C of partial continuous functionals. In: Gandy-Hyland [41], 455–467

Feferman, S.
24. Some applications of the notions of forcing and generic sets. Fund. Math. *56*, 325–345 (1964/65)
25. Inductive schemata and recursively continuous functionals. In: Gandy-Hyland [41], 373–392

Fenstad, J. E.
26. On axiomatizing recursion theory. In: Fenstad-Hinman [31], 385–404
27. Computation theories: an axiomatic approach to recursion on general structures. In: Müller et al. [119], 143–168
28. Between recursion theory and set theory. In: Gandy-Hyland [41], 393–406
29. On the foundation of general recursion theory: Computations versus inductive definability. In: Fenstad-Gandy-Sacks [30], 99–110

Fenstad, J. E., Gandy, R. O., Sacks, G. E.
30. **Generalized Recursion Theory II** (Proceedings of the 1977 Oslo Symposium, editors). Amsterdam: North-Holland 1978, 417 pp.

Fenstad, J. E., Hinman, P. G.
31. **Generalized Recursion Theory** (Proceedings of the 1972 Oslo Symposium, editors). Amsterdam: North-Holland 1974, 456 pp.

Fenstad, J. E., Normann, D.
32. On absolutely measurable sets. Fund. Math. *81*, 91–98 (1974)

Friedman, H.
33. Axiomatic recursive function theory. In: Gandy-Yates [44], 113–137

34. Algorithmic procedures, generalized Turing algorithms, and elementary recursion theory. In: Gandy-Yates [44], 316–389

Friedman, S. D.
35. An introduction to β-recursion theory. In: Fenstad-Gandy-Sacks [30], 111–126
36. Negative solutions to Post's problem, I. In: Fenstad-Gandy-Sacks [30], 127–133

Friedman, S. D., Sacks, G. E.
37. Inadmissible recursion theory. Bull. Amer. math. Soc. *82*, 255–256 (1977)

Gandy, R. O.
38. General recursive functionals of finite type and hierarchies of functions. Ann. Fac. Sci. Univ. Clermont-Ferrand *35*, 5–24 (1967)
39. Inductive definitions. In: Fenstad-Hinman [31], 265–299
40. Set-theoretic functions for elementary syntax. In: Jech [69], 103–126

Gandy, R. O., Hyland, J. M. E.
41. **Logic Colloquium '76** (Proceedings of the Summer School and Colloquium in Mathematical Logic, Oxford: 1976, editors). Amsterdam: North-Holland 1977, 612 pp.
42. Computable and recursively countable functions of higher type. In: Gandy-Hyland [41], 407–438

Gandy, R. O., Sacks, G. E.
43. A minimal hyperdegree. Fund. Math. *61*, 215–223 (1967)

Gandy, R. O., Yates, C. E. M.
44. **Logic Colloquium '69** (Proceedings of the Summer School and Colloquium in Mathematical Logic, Manchester: 1969, editors). Amsterdam: North-Holland 1971, 451 pp.

Gordon, C. E.
45. A comparison of abstract computability theories. Los Angeles: Ph.D. thesis 1968

Gregory, J.
46. On a finiteness condition for infinitary languages. Maryland: Ph.D. thesis 1969
47. On a finiteness condition for infinitary languages. In: Kueker [94], 143–206

Grilliot, T. J.
48. Selection functions for recursive functionals. Notre Dame J. formal Logic *10*, 333–346 (1969)
49. Inductive definitions and computability. Trans. Amer. math. Soc. *158*, 309–317 (1971)
50. On effectively discontinuous type-2 objects. J. symbolic Logic *36*, 245–248 (1971)
51. Omitting types: applications to recursion theory. J. symbolic Logic *37*, 81–89 (1972)

Gurrik, P. K.
52. Om E-rekursjonsteori [in Norwegian]. Oslo: Cand. Real. thesis 1978

Harrington, L. A.
53. Contributions to recursion theory in higher types. MIT: Ph.D. thesis 1973
54. The superjump and the first recursively Mahlo ordinal. In: Fenstad-Hinman [31], 43–52

Harrington, L. A., MacQueen, D. B.
55. Selection in abstract recursion theory. J. symbolic Logic *41*, 153–158 (1976)

Harrington, L. A., Kechris, A. S.
56. On characterizing Spector Classes. J. symbolic Logic *40*, 19–24 (1975)
57. On monotone vs. nonmonotone induction. Bull. Amer. math. Soc. *82*, 888–890 (1976)

Harrington, L. A., Kechris, A. S., Simpson, S. G.
58. 1-envelopes of type-2 objects. Amer. math. Soc. Notices *20*, A-587 (1973)

Hinman, P. G.
59. Hierarchies of effective descriptive set theory. Trans. Amer. math. Soc. *142*, 111–140 (1969)
60. Degrees of continuous functionals. J. symbolic Logic *38*, 393–395 (1973)
61. **Recursion-Theoretic Hierarchies, Perspectives in Mathematical Logic.** Berlin, Heidelberg, New York: Springer 1978, 480 pp.

Hinman, P. G., Moschovakis, Y. N.
62. Computability over the continuum. In: Gandy-Yates [44], 77–105

Hodges, W.
63. On the effectivity of some field constructions. Proc. London math. Soc. (3), *32*, 133–162 (1976)

Hyland, J. M. E.
64. Recursion theory on the countable functionals. Oxford: Ph.D. thesis 1975
65. Aspects of constructivity in mathematics. In: Gandy-Hyland [41], 439–454
66. The intrinsic recursion theory on the countable or continuous functionals. In: Fenstad-Gandy-Sacks [30], 135–145
67. Filter spaces and continuous functionals. Ann. math. Logic (to appear)

Heyting, A.
68. **Constructivity in Mathematics** (Proceedings of the Colloquium held at Amsterdam, 1957, editor). Amsterdam: North-Holland 1959, 297 pp.

Jech, T. J.
69. **Axiomatic Set Theory II** (Proceedings of the Thirteenth Symposium in Pure Mathematics of the Amer. math. Soc., Los Angeles: 1967, editor). Amer. math. Soc., Providence, R.I.: 1974, 222 pp.

Jensen, R. B.
70. Admissible sets. (Circulated notes, 1969)
71. The fine structure of the constructible hierarchy. Ann. math. Logic *4*, 229–308 (1972)

Jensen, R. B., Karp, C.
72. Primitive recursive set functions. In: Scott [147], 143–176

Kanger, S.
73. **Proceedings of the Third Scandinavian Logic Symposium** (Uppsala, Sweden: 1973, editor). Amsterdam: North-Holland 1975, 214 pp.

Kechris, A. S.
74. **The Structure of Envelopes: a Survey of Recursion Theory in Higher Types.** MIT Logic Seminar Notes, 1973
75. **On Spector Classes.** In: Kechris, A. S. and Moschovakis, Y. N. (editors), CABAL SEMINAR. Berlin, Heidelberg, New York: Springer 1978, 245–277
76. Spector second order classes and reflection. In: Fenstad-Gandy-Sacks [30], 147–183

Kechris, A. S., Moschovakis, Y. N.
77. Recursion in higher types. In: Barwise [12], 681–737

Kleene, S. C.
78. **Introduction to Metamathematics.** Amsterdam: North-Holland; Groningen: P. Noordhoff; New York: van Nostrand Co. 1952

79. Arithmetic predicates and function quantifiers. Trans. Amer. math. Soc. *79*, 312–340 (1955)
80. On the forms of the predicates in the theory of constructive ordinals (second paper). Amer. J. math. *77*, 405–428 (1955)
81. Hierarchies of number-theoretic predicates. Bull. Amer. math. Soc. *61*, 193–213 (1955)
82. Countable functionals. In: Heyting [68], 81–100
83. Recursive functionals and quantifiers of finite types, I. Trans. Amer. math. Soc. *91*, 1–52 (1959)
84. Quantification of number-theoretic functions. Compositio math. *14*, 23–41 (1959)
85. Recursive functionals and quantifiers of finite types, II. Trans. Amer. math. Soc. *108*, 106–142 (1963)
86. Recursive functionals and quantifiers of finite types, revisited, I. In: Fenstad-Gandy-Sacks [30], 185–222

Kolaitis, P. G.
87. Recursion in E on a structure versus positive elementary induction. J. symbolic Logic (to appear)

Kreisel, G.
88. Interpretation of analysis by means of functionals of finite type. In: Heyting [68], 101–128
89. Set theoretic problems suggested by the notion of potential totality. In: **Infinistic Methods** (Proceedings of the 1959 Warsaw Symposium). Oxford: Pergamon Press 1961, pp. 103–140
90. Some reasons for generalizing recursion theory. In: Gandy-Yates [44], 139–198

Kreisel, G., Sacks, G. E.
91. Metarecursive sets. J. symbolic Logic *30*, 318–338 (1965)

Kripke, S.
92. Transfinite recursions on admissible ordinals, I, II (abstracts). J. symbolic Logic *29*, 161–162 (1964)
93. Admissible ordinals and the analytic hierarchy (abstract). J. symbolic Logic *29*, 162 (1964)

Kueker, D. W.
94. **Infinitary Logic: In Memoriam Carol Karp** (A Collection of papers, editor). Berlin, Heidelberg, New York: Springer 1975, 206 pp.

Lavori, P.
95. Recursion in the extended superjump. Illinois J. Math. *21*, 752–758 (1977)

Louveau, A.
96. La hierarchie borelienne des ensemble Δ_1^1. C.R. Acad. Sc. Paris *285*, 601–604 (1977)
97. Recursivity and compactness. In: Müller-Scott [120], 303–337

MacQueen, D. B.
98. Post's problem for recursion in higher types. MIT: Ph.D. thesis 1972

Malcev, A. I.
99. **Algebraic Systems.** Berlin, Heidelberg, New York: Springer 1973, 317 pp.

Maass, W.
100. Inadmissibility, tame r.e. sets and the admissible collapse. Ann. math. Logic *13*, 149–170 (1978).
101. Fine structure theory for the constructible universe in α- and β-recursion theory. In: Müller-Scott [120], 339–359
102. Recursively invariant β-recursion theory. (In preparation)

Miettinen, S., Väänänen, J.
103. **Proceedings of the Symposiums on Mathematical Logic, Oulu 1974 and Helsinki 1975.** Helsinki: 1977, 103 pp.

Mitschke, G.
104. **λ-Kalkiil, δ-Konversion und axiomatische Rekursionstheorie.** Darmstadt: Habilitationschrift 1976

Moldestad, J.
105. **Computations in Higher Types.** Berlin, Heidelberg, New York: Springer 1977, 203 pp.
106. On the role of the successor function in recursion theory. In: Fenstad-Gandy-Sacks [30], 283–301

Moldestad, J., Normann, D.
107. Models for recursion theory. J. symbolic Logic *41*, 719–729 (1976)

Moldestad, J., Stoltenberg-Hansen, V., Tucker, J. V.
108. Finite algorithmic procedures and inductive definability. Math. Scand. (to appear)
109. Finite algorithmic procedures and computation theories. Math. Scand. (to appear)

Moldestad, J., Tucker, J. V.
110. On the classification of computable functions in an abstract setting. (In preparation)

Moschovakis, Y. N.
111. Hyperanalytic predicates. Trans. Amer. math. Soc. *129*, 249–282 (1967)
112. Abstract first-order computability I, II. Trans. Amer. math. Soc. *138*, 427–464, 465–504 (1969)
113. Axioms for computation theories—first draft. In: Gandy-Yates [44], 199–255
114. Structural characterizations of classes of relations. In: Fenstad-Hinman [31], 53–79
115. **Elementary Induction on Abstract Structures.** Amsterdam: North-Holland 1974, 218 pp.
116. On non-monotone inductive definability. Fundamenta Math. *82*, 39–83 (1974)
117. On the basic notions in the theory of induction. In: Butts-Hintikka [17], 207–236
118. **Descriptive Set Theory.** Amsterdam: North-Holland (to appear)

Müller, G. H., Oberschelp, A., Potthoff, K.
119. **Logic Conference Kiel, 1974** (Proceedings of the International Summer Institute and Logic Colloquium, editors). Berlin, Heidelberg, New York: Springer 1975, 651 pp.

Müller, G. H., Scott, D. S.
120. **Higher Set Theory** (Proceedings Oberwolfach 1977, editors). Berlin, Heidelberg, New York: Springer 1978, 476 pp.

Normann, D.
121. On abstract 1-sections. Synthese *27*, 259–263 (1974)
122. Imbedding of higher type theories. Oslo preprint 1974
123. A continuous functional with non-collapsing hierarchy. J. symbolic Logic *43*, 487–491 (1978)
124. Set recursion. In: Fenstad-Gandy-Sacks [30], 303–320
125. Recursion in ${}^3E$ and a splitting theorem. In: Hintikka, J., Niiniluoto, I, Saarinen, E. (editors), ESSAYS ON MATHEMATICAL AND PHILOSOPHICAL LOGIC. Dordrecht: D. Reidel Publishing Company 1979, 275–285
126. A jump operator in set recursion. Z. math. Logik und Grundl. Math. (to appear)
127. Degrees of functionals. Ann. math. Logic (to appear)
128. Countable functionals and the analytic hierarchy. J. symbolic Logic (to appear)
129. **Recursion in the Countable Functionals.** Springer Lecture Notes (to appear)

Normann, D., Stoltenberg-Hansen, V.
130. A non-adequate admissible set with a good degree-structure. In: Fenstad-Gandy-Yates [30], 321–329

Normann, D., Wainer, S.
131. The 1-section of a countable functional. J. symbolic Logic (to appear)

Nyberg, A.
132. Inductive operators on resolvable structures. In: Miettinen-Väänänen [103], 91–100

Platek, R. A.
133. Foundations of Recursion theory. Stanford: Ph.D. thesis and supplement 1966

Richter, W.
134. Recursively Mahlo ordinals and inductive definitions. In: Gandy-Yates [44], 273–288

Richter, W., Aczel, P.
135. Inductive definitions and reflecting properties of admissible ordinals. In: Fenstad-Hinman [31], 301–381

Rogers, H., Jr.
136. **Theory of Recursive Functions and Effective Computability.** New York: McGraw-Hill 1967, 482 pp.

Rose, H. E., Shepherdson, J. C.
137. **Logic Colloquium '73** (Proceedings of the Summer School and Colloquium in Mathematical Logic, Bristol: 1973, editors). Amsterdam: North-Holland 1975, 513 pp.

Sacks, G. E.
138. **Degrees of Unsolvability.** Princeton: Ann. math. Studies, No. *55*, 1963
139. The recursively enumerable degrees are dense. Ann. Math. *80*, 300–312 (1964)
140. Post's problem, admissible ordinals and regularity. Trans. Amer. math. Soc. *124*, 1–23 (1966)
141. Forcing with perfect closed sets. In: Scott [147], 331–355
142. The 1-section of a type *n* object. In: Fenstad-Hinman [31], 81–93
143. The *k*-section of a type-*n* object. Amer. J. Math. *99*, 901–917 (1977)
144. R.e. sets higher up. In: Butts-Hintikka [17], 173–194

Sasso, L. P.
145. Degrees of unsolvability of partial functions. Berkeley: Ph.D. thesis 1971

Schwichtenberg, H., Wainer, S. S.
146. Infinite terms and recursion in higher types. In: Diller-Müller [20], 314–364

Scott, D. S.
147. **Axiomatic Set Theory I** (Proceedings of the Thirteenth Symposium in Pure Mathematics of the Amer. math. Soc., Los Angeles: 1967, editor). Providence, R.I.: Amer. math. Soc. 1971, 474 pp.

Shepherdson, J. C.
148. Computations over abstract structures: serial and parallel procedures and Friedman's effective definitional schemes. In: Rose-Shepherdson [137], 445–513

Shoenfield, J. R.
149. A hierarchy based on type two objects. Trans. Amer. math. Soc. *134*, 103–108 (1968)

Shore, R. A.
150. Splitting an α-recursively enumerable set. Trans. Amer. math. Soc. *204*, 65–78 (1975)

151. The recursively enumerable α-degrees are dense. Ann. math. Logic *9*, 123–155 (1976)
152. α-recursion theory. In: Barwise [12], 653–680

Simpson, S. G.
153. Admissible ordinals and recursion theory. MIT: Ph.D. thesis 1971
154. Degree theory on admissible ordinals. In: Fenstad-Hinman [31], 165–193
155. Post's problem for admissible sets. In: Fenstad-Hinman [31], 437–441
156. Short course on admissible recursion theory. In: Fenstad-Gandy-Sacks [30], 355–390

Soare, R. I.
157. Recursively enumerable sets and degrees. Bull. Amer. math. Soc. *84*, 1149–1181 (1978)

Spector, C.
158. Recursive well-orderings. J. symbolic Logic *20*, 151–163 (1955)
159. Recursive ordinals and predicative set theory. In: Summaries of talk presented at the Summer Institute for Symbolic Logic, Cornell: 1957
160. Hyperarithmetic quantifiers. Fundamenta Math. *48*, 313–320 (1959)
161. Inductively defined sets of natural numbers. In: **Infinistic Methods** (Proceedings of the 1959 Warsaw Symposium). Oxford: Pergamon Press 1961, 97–102

Stoltenberg-Hansen, V.
162. On priority arguments in Friedberg theories. Toronto: Ph.D. thesis 1973
163. Finite injury arguments in infinite computation theories. Ann. math. Logic *16*, 57–80 (1979)
164. A regular set theorem for infinite computation theories. Oslo preprint 1977
165. Weakly inadmissible recursion theory. In: Fenstad-Gandy-Sacks [30], 391–405

Strong, H. R.
166. Algebraically generalized recursive function theory. IBM J. Res. Devel. *12*, 465–475 (1968)

Troelstra, A.S.
167. **Metamathematical Investigations of Intuitionistic Analysis, Arithmetic, and Analysis** (A collection of papers, editor). Berlin, Heidelberg, New York: Springer 1973, 485 pp.

Tucker, J. V.
168. Computing in algebraic systems. Oslo preprint 1978

Wagner, E. G.
169. Uniform reflexive structures: on the nature of Gödelizations and relative computability. Trans. Amer. math. Soc. *144*, 1–41 (1969)

Wainer, S. S.
170. A hierarchy for the 1-section of any type two object. J. symbolic Logic *39*, 88–94 (1974)

Wang, H.
171. Remarks on constructive ordinals and set theory. In: Summaries of talks presented at the Summer Institute for Symbolic Logic. Cornell: 1957

Notation

Chapter 1

$\mathfrak{A} = \langle A, C; 0, 1\rangle$	computation domain 19
C	code set 19
σ, τ	sequences of elements from A 20
$\mathrm{lh}(\sigma)$	length of σ 20
Θ	computation set 20
$\{a\}_\theta^n$	Θ-computable function with code a 20, 21
$f(\sigma) \simeq z$	the computation tuple $(\hat{f}, \sigma, z) \in \Theta$ 21
φ	functional 21
DC	definition by cases 21
$\mathbf{C}^n$	composition 22
$\mathbf{P}_{n,i}^m$	permutation 22
S_m^n	the s-n-m mapping 22
$f(\sigma)\downarrow$	$f(\sigma)$ is defined 26
$\mathfrak{A} = \langle A, C, N; s, M, K, L\rangle$	computation domain 30
$\langle M, K, L\rangle$	pairing structure 30
$\langle x_1, \ldots, x_n\rangle$	ordered n-tuple 31
$\Gamma_{\mathbf{f}}$	inductive operator 31
$\mathrm{PR}[\mathbf{f}]$	prime computation set in $\mathbf{f}$ 32
$\lvert a, \sigma, z\rvert_{\mathrm{PR}[\mathbf{f}]}$	length of computation 32
$\Theta \leqslant H$	extension of precomputation theories 34
$\Theta \sim H$	equivalence of theories 34

Chapter 2

$\langle\Theta, <_\Theta\rangle$	computation structure 44
$\mathbf{S}_{(a,\sigma,z)}$	set of subcomputations 44
$\lvert a, \sigma, z\rvert_\Theta$	length of computation 44
$\Gamma_{f,\varphi}$	inductive operator 45
$\mathrm{PR}[f, \varphi]$	prime computation set in f, φ 46
$\lvert a, \sigma, z\rvert_{\mathrm{PR}[f,\varphi]}$	length of computation 46
$<_{\mathrm{PR}[f,\varphi]}$	subcomputation relation 47
$H[\mathbf{f}, \boldsymbol{\varphi}]$	theory generated from H and $\mathbf{f}$, $\boldsymbol{\varphi}$ 47
$\langle\Theta, <_\Theta\rangle \leqslant \langle H, <_H\rangle$	extension of computation theories 48
$\langle\Theta, <_\Theta\rangle \sim \langle H, <_H\rangle$	equivalence of theories 48
$\mathbf{E}_B$	functional defining Θ-finiteness 52
WDC	weak definition by cases 59

Chapter 3

$\lvert x\rvert_\Theta$	length of computation $x \in \Theta$ 65
$\mathbf{E}'_S$	functional defining weak Θ-finiteness 66
$\mathbf{Q}$	monotone quantifier 67
$\mathbf{F}_{\mathbf{Q}}^{\#}$	functional associated to quantifier 67
$\mathrm{sc}^*(\Theta)$	extended section 73
$\mathrm{sc}(\Theta)$	section 73
$\mathrm{en}^*(\Theta)$	extended envelope 73
$\mathrm{en}(\Theta)$	envelope 73
$\Gamma: 2^A \to 2^A$	inductive operator 79
Γ_α	stage of inductive operator 79
Γ_∞	fixed-point of inductive operator 79
$\lvert\Gamma\rvert$	ordinal of inductive operator 79
$\mathrm{IND}(\mathbf{C})$	the $\mathbf{C}$-inductive relations 80
$\lvert\mathbf{C}\rvert$	the ordinal of the class $\mathbf{C}$ 80
$\mathrm{HYP}(\mathbf{C})$	the $\mathbf{C}$-hyperdefinable relations 80

Chapter 4

$\mathfrak{A} = \langle A, S, \mathbf{S}\rangle$	computation domain on two types 90

$\mathbf{S} = \langle N, s, M, K, L\rangle$	coding scheme	90
$\mathrm{Tp}(S)$	the set ω^S	90
PRF	primitive recursive functions	91
$\mathrm{PR}(\mathbf{L})$	partial recursive in $\mathbf{L}$	93
$\lvert e, \sigma, x\rvert_{\mathbf{L}}$	length of computation	94
$\kappa_{\mathbf{L}}$	ordinal of computations in $\mathbf{L}$	94
$\langle e, \sigma\rangle \downarrow$	the computation $\{e\}_{\mathbf{L}}(\sigma)$ is defined	94
$\langle e, \sigma\rangle \uparrow$	the computation $\{e\}_{\mathbf{L}}(\sigma)$ is undefined	94
$\mathbf{C}_{\mathbf{L}}$	coded set of $\mathbf{L}$-computations	97

Chapter 5

$(\mathfrak{A}, \preccurlyeq)$	computation domain with prewellordering	111
$\lvert\preccurlyeq\rvert$	ordinal of pwo $\preccurlyeq$	111
$\mathbf{E}^{\preccurlyeq}$	functional computing the pwo $\preccurlyeq$	111
$\lVert\Theta\rVert$	ordinal of the theory Θ	111
$\Theta^{\lvert w\rvert}$	Θ-computations of length $\lvert w\rvert$	111
$\mathbf{E}^{1/2}$	functional associated with a selection operator	112
$\mathbf{L}(\mathfrak{A}, \preccurlyeq, \mathbf{R})$	first-order language associated with $\mathfrak{A}, \preccurlyeq, \mathbf{R}$	113
$\Delta_0(\preccurlyeq, \mathbf{R})$, $\Sigma_1(\preccurlyeq, \mathbf{R})$, $\Pi_1(\preccurlyeq, \mathbf{R})$, $\Delta_1(\preccurlyeq, \mathbf{R})$	classification of relations with respect to the language $\mathbf{L}(\mathfrak{A}, \preccurlyeq, \mathbf{R})$	113
$\Gamma_\theta(X)$	inductive operator induced by the formula $\theta(\sigma, X)$	113
Γ_∞	fixed-point of Γ_θ	113
$\lvert\Gamma_\theta\rvert$	ordinal of Γ_θ	113
$\mathrm{PR}[[\preccurlyeq, \mathbf{R}]]$	computation theory generated by $\mathbf{R}$ over $(\mathfrak{A}, \preccurlyeq)$	119
$R(x, \alpha)$, $R_\Theta(x, \alpha)$	"universal" relation in Θ	125, 129
α_Θ, α_F	ordinal of Θ, $\mathrm{PR}[F]$	127
$\Psi <_1 \Theta$	extension of Spector theories	127
Θ_τ	Θ-computations of length $<\tau$	129
$m(\mathrm{sc}(\Theta))$	abstract 1-section associated to Θ	135
$\Vdash$	forcing relation	137

Chapter 6

$\lambda x \cdot W^x$	$\preccurlyeq$-enumeration	140
$\lambda ax. W^x_a$	$\preccurlyeq$-parametrization	141
L^β	elements of ordinal $<\beta$ in the pwo $\preccurlyeq$	143
$\lvert\preccurlyeq\rvert^*$	projectum of Θ	143
$\lvert\preccurlyeq\rvert^+$	r.e.-projectum of Θ	144
$\lambda z \cdot K_z$	enumeration of Θ-finite sets	144
$\leqslant_w$	weakly Θ-computable in	144
$\leqslant$	Θ-computable in	144
$\equiv$	equivalence relation derived from $\leqslant$	145
$\leqslant_d$	Θ-definable in	145
$B^w \prec B^z$	every element of B^w is $\prec$-less than every element of B^z	148
$\leqslant_m$	many-one reducible	149
B'	jump of B	149
$\deg(A)$	degree of A	150
$\mathbf{a}$	degree	150
$\Sigma_0, \Pi_0, \Sigma_n, \Pi_n$	definability classification	151
$\lim_\alpha f$	limit of f	151
$\Sigma_2\text{-cf}(\alpha)$	Σ_2-cofinality of α	152
$\Sigma_1\text{-cf}(\beta)$	Σ_1-cofinality of β	160
$\mathfrak{A}_\beta$	the admissible collapse of L_β	163

Chapter 7

$\mathrm{Ord}(X)$	ordinals of Θ-computations in X	168
κ^X	the ordinal of Θ-computations in X	168
λ^X	ordertype of $\mathrm{Ord}(X)$	168
$\mathrm{sc}(\Theta)$	Θ-computable subsets of A	171
$\mathrm{sc}(\Theta, \sigma)$	subsets of A Θ-computable in σ	171
$\mathrm{en}(\Theta)$	Θ-semicomputable subsets of A	171
$S\text{-en}(\Theta)$	Θ-semicomputable subsets of S	171
$\Psi <_1 \Theta$	extension of theories on two types	180

Chapter 8

$\{e\}_R(\sigma)$	set-recursive computation relative to R	183
Θ_R	the set-recursive computation theory	183
$\lVert a, \sigma, z\rVert$	length function in Θ_R	183
$M(A; R)$	set-recursive closure of A relative to R	185
${}^fA$	finite subsets of A	185

$\langle M(B; R)\rangle_{B \in \mathscr{I}_A}$ the splitting of $M(A; R)$ 185
$\Sigma_B^*(R)$, $\Delta_B^*(R)$ definability classification 185
$\mathrm{Spec}(R; I)$ the spectrum of R over $I = \mathrm{Tp}(k)$ 187
$M_a(I; R)$ notation for $M(\{a, I\}; R)$ 188
$M(I; R)$ notation for $\bigcup M_a(I; R)$ 188
$M^\alpha(A; R)$ the α-approximation to $M(A; R)$ 191
$w\text{-}\Sigma_a^*(R)$, $w\text{-}\Delta_a(R)$ weak definability classification 191
$\leqslant_R$ set-recursive reducibility notion 194
a' the a-jump 195
${}^1M$ notation for $\{\langle a, x\rangle : x \in M_a\}$ 195
κ_0^a the ordinal $\mathrm{On} \cap M_a$ 199

Author Index

Subject Index

D. van Dalen

Logic and Structure

Universitext

1980. Approx. 180 pages
ISBN 3-540-09893-3

This book provides an efficient introduction to logic students of mathematics. The central theme is that part of first order logic which can be handled directly on the basis of **derivation** and **validity.** Emphasis is placed on notions that play a role in every-day mathematics, such as models, truth, relativized quantifiers, consistency, Skolem functions, and extension by definition. Following a self-contained presentation of propositional logic (including completeness), predicate logic – with applications to elementary algebra – is treated systematically, leading to an exposition of the first principles of model theory. A unique feature of this book is the systematic use of Gentzen's system of **Natural Deduction**. Closer to natural informal reasoning than an axiomatic approach, it enables the student to devise derivations as a simple exercise. Inductive definitions have been employed wherever appropriate. Model-theoretic topics include the main facts of compactness, non-standard models of arithmetic and the reals, and, in a special section, some of the properties of second-order logic. The material is illustrated by many exercises and demands a minimum background in mathematics of the reader.

Springer-Verlag
Berlin
Heidelberg
New York